PYTHON PROGRAMMING USING PROBLEM SOLVING

Python Programming Using Problem Solving

Harsh Bhasin

Mercury Learning and Information
Dulles, Virginia
Boston, Massachusetts
New Delhi

Publisher: David Pallai
MERCURY LEARNING AND INFORMATION
22841 Quicksilver Drive
Dulles, VA 20166
info@merclearning.com
www.merclearning.com
800-232-0223

H. Bhasin. *Python Programming Using Problem Solving.*
ISBN: 978-1-68392-862-1

Library of Congress Control Number: 2023934762

232425321   This book is printed on acid-free paper in the United States of America.

To
My Mother

CONTENTS

PREFACE

Python is a robust, procedural, object-oriented, and functional language. The features of the language make it tremendously valuable for web development, gaming, and scientific programming. Lately, the language has become incredibly popular. The popularity of the language can be gauged from the fact that it is currently being used by *Google*, *YouTube*, *Bit Torrent* and many other companies.

This book deals with problem-solving and programming in Python. Programming is the soul of computer science, and designing a program requires involved expertise of the paradigms, along with the ability to use the standard procedures. To become a virtuous programmer one must, therefore, not only learn the syntax of the language but also develop an ability to apply the mastered concepts to solve problems. It may be stated here that a programming language and its syntax learned by a professional is of no use until the algorithms that are to be implemented have been well designed. Thus, a basic knowledge of data structures and algorithms is also essential. This is the reason that the first section of this book has been dedicated to problem-solving. One of the most important goals of the book is to make the readers understand Python's discriminative features. The ability of Python to deal with multi-dimensional arrays via `NumPy` has been included. Python also helps in visualization via `matplotlib`. The topic has also been presented in section IV.

Harsh Bhasin
June 2023

SECTION I

ALGORITHMIC PROBLEM-SOLVING AND PYTHON FUNDAMENTALS

This section deals with algorithms and introduces Python. It contains three chapters namely "Introduction to Algorithms," "Introduction to Python," and "Fundamentals." The first chapter presents the definition of an algorithm, the features of a good algorithm, and the ways of writing an algorithm. The asymptotic complexity has also been discussed in the chapter. The chapter also discusses the differences between recursive and iterative algorithms. Algorithms of some common problems have also been included in this chapter. The second chapter introduces Python. The features of the language, its chronology, and its applications have been discussed in the third chapter. The chapter also presents a brief overview of the control structures used in Python. It also describes the installation of Anaconda, which is an immensely popular Data Science platform. These chapters are the building blocks of the chapters that follow.

1

ALGORITHMIC PROBLEM-SOLVING

Objectives

After reading the chapter, the reader should be able to
- Understand the importance of algorithms.
- Understand the features of a good algorithm.
- Understand the ways of writing an algorithm.
- Understand asymptotic notations.
- Differentiate between recursive and iterative algorithms.

1.1 INTRODUCTION

This chapter introduces problem-solving and algorithms. Let us begin our discussion by understanding the term algorithm. The word algorithm comes from "algorithmi," from the title "Algoritmi De Numero Indorum," a book written by Muhammad Ibn Musa Al-Khwarizmi, who was a Persian Mathematician. The word was corrupted and became "Algorism." Finally, in the nineteenth century, it became algorithm. Interestingly, the book stated above was on Indian numerals. Lately, the word algorithm is identified with any procedure applied to accomplish a given computing task.

An algorithm directs how to solve a problem and there can be many algorithms to solve the same problem. However, not all of them are effective and efficient. Also, it is desirable that in the sequence of steps for accomplishing a task, each step should be as basic as possible. The task should be completed in a finite number of steps. So, a good algorithm should be finite, and each instruction should be unambiguous.

Algorithms are implemented using programming languages. However, designing an algorithm cannot be automated as it is, an art [1]. Art cannot be automated, but you can at least learn approaches like Divide and Conquer, Backtracking, Branch and Bound, Dynamic programming, Greedy approaches, etc. Learning these approaches would not only help you in Computer Science but also help in other disciplines like Computational Biology, Finance, etc.

Algorithms are used everywhere, right from your set-top box to the machine that gathers biometric data. The advancements in the field of algorithms have changed the lives of millions. The page rank algorithm of Larry Page has helped in the creation of Google, which is a part of our life. The routing algorithms allowed packets to be transferred from one computer to another via the shortest paths and helped in the advancement of communication. Likewise, the pre-processing algorithms for magnetic resonance imaging have helped scientists to develop computer-aided techniques for the diagnosis of diseases. The conventional techniques clubbed together with the latest advancements like Deep Learning have been able to solve many problems in the society.

1.2 DEFINITION AND CHARACTERISTICS

Having seen the importance of algorithms, let us now move to their formal definition and understand their features. An algorithm is a sequence of steps used to accomplish a given task. It processes the input and generates some output. The most essential elements of an algorithm are input, output, correctness, efficiency, and definiteness.

The number of input arguments can even be zero. For example, some of the pseudo-random number generators, do not take any argument to generate a random number. However, there must be at least one output. The first thing that should be taken care of while designing an algorithm is its correctness. An algorithm that is not correct, is of no use. No amount of fancy controls or sophisticated techniques can replace correctness. Also, there can be numerous ways to solve a given problem but not all of them are equally efficient.

The efficiency of an algorithm is also important. The algorithm should be efficient both in terms of time and space. That is, it should take the minimum possible time and space. For example, linear search and binary search are the two most important techniques of searching. The first takes $O(n)$ time, while the second takes $O(\log n)$ time. That is, a list having 1024 elements would take time of order of 1024 units, in the case of linear search, and would take time proportional to 10 units, in the case of binary search. Hence, the choice of an efficient algorithm is immensely important to make an effective model. The meaning of O has been explained in Section 1.5.

The above discussion can be summarized as follows:

- an algorithm takes zero or more input,
- it produces at least one output,
- it must be correct,
- it should be efficient, both in terms of memory and space, and
- it should not be ambiguous.

1.3   NOTATIONS: PSEUDOCODE AND FLOW CHART

In order to appreciate the discussion, let us come back to the example of linear search. In linear search, a list is searched for an item by looking for the item iteratively, that is at each position of the list. The algorithm for linear search can be stated as follows (Algorithm 1). Table 1.1 shows the conventions used for writing algorithms, in this book. The algorithm can also be represented as a flow chart as shown in Figure 1.1. Pseudocode and flow chart are the two most commonly used ways of representing an algorithm. The third way of writing an algorithm can be English-like. However, this is not preferred, as it can be ambiguous.

Algorithm 1: Linear Search

Input:

```
List: L
Length of the list: n
Item to be searched: item
```
Algorithm: Linear Search
```
Set i=0;
Set Flag=0;
While (i<n)
    {
    if (L[i]==item)
        {
        Print("Item found at ",i);
        Flag=1;
```

```
        }
    i++;
    }
if(Flag==0)
    {
    Print("Not found");
    }
}
```

Linear Search: Flow Chart

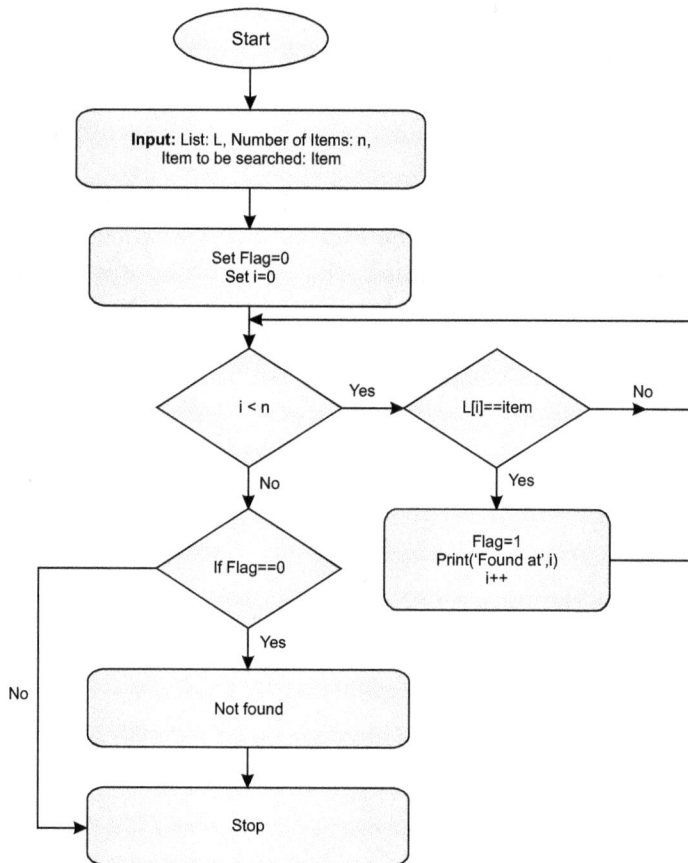

FIGURE 1.1 Flowchart for linear search.

Linear Search: English-Like

1. Set Flag=0

2. Set the value of i to 0 and start scanning the items of the list. If the item to be searched is found, set Flag to 1.

3. If Flag==0, then print "Not Found".

TABLE 1.1 Conventions used in algorithms.

Convention	Description
//	Single line comment
/*…*/	Multiple line comment
{ }	Block
<variable name> = <value>	Assignment
a < b	Less than operator
a > b	Greater than operator
a <= b	Less than or equal to operator
a >= b	Greater than or equal to operator
a == b	Checking equality
a != b	Checking the values of the two variables are not equal
&&	AND operator
||	OR operator

1.4 STRATEGIES FOR PROBLEM-SOLVING: RECURSION VERSUS ITERATION

Algorithms can be recursive or iterative. Recursion is the invocation of a function inside that function. In order to develop a code using recursion, one must express a function in terms of itself (with reduced values of the parameters) and should specify the base condition(s) to serve as the stopping criteria. For example, the nth Fibonacci term can be expressed as the sum of the $(n-1)$th and $(n-2)$th Fibonacci terms. Since, the evaluation of nth Fibonacci term requires two previous results of evaluation, two base conditions must be specified. The first and the second terms of this sequence are 1 and 1. Therefore the function can be written as follows.

$$fib(n) = fib(n - 1) + fib(n - 2)$$
$$fib(1) = 1$$
$$fib(2) = 1$$

That is, to find the fifth Fibonacci term, we need to find the sum of the fourth and the third term. The fourth Fibonacci term can be found by adding the third and the second term. The third Fibonacci term can be found by adding the second and the first term, both of which are 1. The process has been depicted in Figure 1.2.

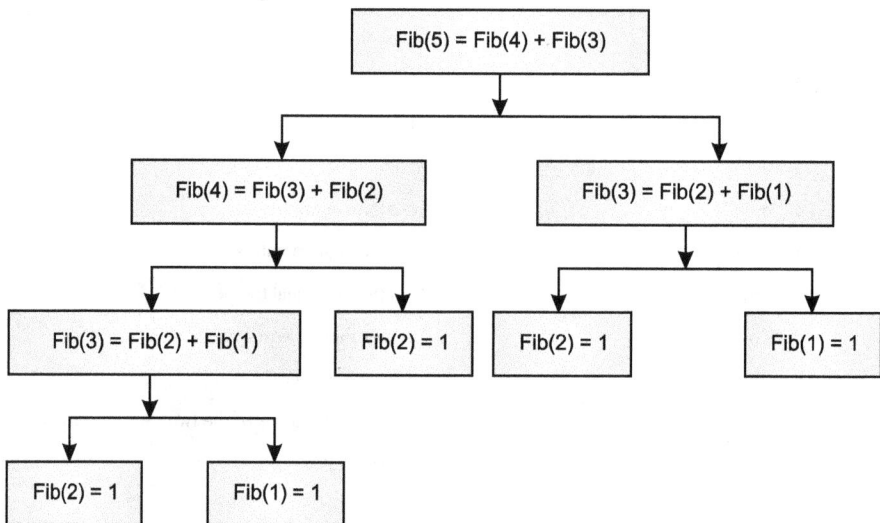

FIGURE 1.2 The evaluation of the fifth Fibonacci term.

Though recursion provides graceful solutions and a way to define a function mathematically, it is hard to get hold of the way problem is solved using recursion. As a matter of fact, debugging an intricate program, that uses recursion, is difficult. Moreover, it is inefficient as it uses function calls.

However, some problems like the in-order, the pre-order, and the post-order traversals of trees; the Depth First Search and Breadth First Search of graphs; binary search, Quicksort and Merge sort, etc., can be easily solved using recursion. It may be stated here that a problem that can be solved using recursion can also be solved iteratively.

Recursion uses run-time-stack, which provides a Last-In-First-Out access. In order to understand this, let us consider the example of calculating factorial using recursion. The factorial of a number can be calculated using the following formula.

$$fac(n) = n * fac(n - 1)$$
$$fac(1) = 1$$

Evaluating $fac(3)$ would require multiplying $fac(2)$ with 3. The evaluation of $fac(2)$ would require multiplying $fac(1)$ with 2. The value of $fac(1)$ is 1. So, in the above example, $fac(3)$ calls $fac(2)$ which calls $fac(1)$. When $fac(1)$ ends, the run-time-stack stores the location of $fac(2)$ to return to $fac(3)$. Likewise, when $fac(2)$ ends the run-time-stack stores the location of $fac(3)$ to return to $fac(4)$. The run-time-stack makes backtracking easy. Chapter 7 deals with recursion in detail.

Another example of recursion is the calculation of the power of a given number. The power of a number can be calculated using iterative algorithms as shown in program 1. In the program, the variable p is initialized to 1. The loop runs b times and each time a is multiplied by p. It may be stated here that the syntax and the nitty-gritty of programming have been introduced in the following chapters. However, the reader can revisit this section after completing the next two units.

Program:

```
a=int(input('Enter the first number\t'))
b=int(input('Enter the second number\t:'))
p=1
i=1
while (i<=b):
  p=p*a
  i+=1
print(a, ' to the power of ', b, 'is ',p)
```

Output:

```
Enter the first number     : 2
Enter the second number    :10
2 to the power of 10 is 1024
```

The above task can also be accomplished using recursion. The following formula can be used to find a to the power of b.

$$a^b = (a^{b/2})^2, \text{ if } b \text{ is even and}$$

$$a^b = (a^{(b-1)/2})^2 \times b, \text{ if } b \text{ is odd}$$

The logic has been implemented in the following program. The output follows.

Program:

```
def pow (a, b):
 if b==1:
    return a
 elif b%2==0:
    return (pow(a, b/2)**2)
 else:
    return ((pow(a, int(b/2))**2)*a)
pow(5,1)
pow(5,2)
pow(5,3)
pow(5,4)
```

Output:

```
5
25
125
625
3125
```

1.5   ASYMPTOTIC NOTATION

An algorithm can be analyzed by considering its best and worst case. For example, in linear search, the best case would be finding the element at the very first position of the list. The worst case of this algorithm would be the case where the element is found at the last position or is not found.

The asymptotic growth of a function can be defined in terms of its input size, for a sufficiently large value of the input size, n. The asymptotic notations can be used to compare the running time or the space requirement of algorithms. The best-case running time of an algorithm can be depicted by its lower bound and the worst-case running time can be depicted by its upper bound. The lower bound would be henceforth represented by big omega, that is, $\Omega()$. The upper bound would be henceforth represented by big Oh, that is, $O()$. The formal definition of the two follows.

Big O: $O()$

The worst-case behavior of an algorithm is depicted by the asymptotic upper bound notation. For any two functions $f(n)$ and $g(n)$

$$O(g(n)) = f(n), \text{ for all } n > 0 \text{ and}$$

$$f(n) \le c \times g(n)$$

Omega: $\Omega()$

The best-case behavior of an algorithm is depicted by the asymptotic lower-bound notation. For any two functions $f(n)$ and $g(n)$

$$\Omega(g(n)) = f(n), \text{ for all } n > 0 \text{ and}$$

$$f(n) \ge c \times g(n)$$

Theta: $\Theta()$

The *asymptotically tight bound* for a function $f(n)$ can be defined as follows. For any two functions $f(n)$ and $g(n)$.

$$\theta(g(n)) = f(n), \text{ for all } n > 0 \text{ and}$$

$$c_1 \times g(n) \le f(n) \le c_2 \times g(n)$$

It may also be noted that,

$$f(n) = O(g(n)) \text{ and}$$

$$f(n) = \Omega(g(n))$$

$$then,$$

$$f(n) = \theta(g(n))$$

1.6 COMPLEXITY

The algorithm should be efficient both in terms of memory and time. That is, an algorithm should take the least amount of space and time. In order to understand the concept, let us consider five different algorithms to solve the same problem. Assume that the number of elements given as input to the algorithm is **n**. The first algorithm takes time proportional to **n** to accomplish the given task ($O(n)$), the second algorithm takes time proportional to $\mathbf{n}^2$ to do the same task ($O(n^2)$), the third takes time proportional to $\mathbf{n}^3$($O(\mathbf{n}^3)$), the fourth takes time proportional to $\log(n)$ ($O(\log n)$)and the fifth takes time proportional to $n \log(n)$, that is $O(n \log n)$. This implies that, if the number of elements doubles, the time taken to accomplish the given task by the first algorithm would double, by the second algorithm would be four times, the third algorithm eight times, and the increase in time of the fourth would be less than the increase in the first and the increase in the time by the fifth would be less than the increase in the second. Therefore, the order of the time complexity would be as follows:

$$O(\log n) < O(n) < O(n \log n) < O(n^2) < O(n^3)$$

For example, linear search described in the following section takes $O(n)$ time whereas binary search takes $O(\log n)$ time. Therefore, binary search takes lesser time as compared to linear search. Merge sort and bubble sort are the two most popular algorithms for sorting. The merge sort takes $O(n \log n)$ time and bubble sort takes $O(n^2)$ time. Therefore, Merge sort has lesser time complexity vis-à-vis bubble sort and is hence better.

1.7 ILLUSTRATIONS

Having seen the definition, characteristics, and notations used for writing an algorithm, let us now move to some basic examples. This section presents four problems and their solutions.

1.7.1 Minimum in a List

Illustration 1.1:

Given a List, L. Write an algorithm to find the minimum valued element in the list.

Solution:

Let the first element of the list be the minimum valued element ("min" = $L[0]$). The list is scanned from left to right. At any point, if we are able to find an element having a value less than the value stored in **"min,"** the value of that element is stored in the variable **"min,"** The **min1** function performs the requisite task.

Algorithm:

```
def min1(L):
     {
     min=L[0];
     i=0;
     while(i<len(L))
          {
          if(L[i]<min)
                {
                min=L[i];
                }
          i+=1;
          }
     return min;
     }
```

Test:
```
min([51,12,71,91,13,19])
```

Output:

```
12
```

1.7.2 **Insert a Card in a Pack of Cards** (Or Insert an element in a sorted list). There are ten cards in the pack, numbered from 1 to 10.

Illustration 1.2:

It is required to insert a card in an ordered pack of cards. The above problem can also be stated as follows. Given a sorted list, insert an item at its appropriate position.

Solution:

The given list is sorted and the given item is to be inserted at its appropriate position. We begin with the last element and shift each element one position to the right, till an item, smaller than the given item is found. This is followed by the insertion of the given item at the position.

Algorithm:

```
def insert(L, item)
    {
    //L is a sorted list and item is the number to be inserted
    n=len(L); //The len function finds the length of the given
    list
    i=n;//set i to the last position
    while(L[i]>item)
        {
        L[i+1]=L[i];
        i=i-1;
        }
    L[i+1]=item;
    print(L)
    }
```

Test:
```
insert([1,3,4,6,8,9],7)
```

Expected Output (Implementation in Python):

```
[1, 3, 4, 6, 7, 8, 9]
```

1.7.3   Guess a Number in a Given Range

Illustration 1.3:

The computer generates a number, in a given range and you are required to guess it within 10 trials.

Solution:

The algorithm requires a pseudo-random number generator. The user enters a range and the program generates a random number in that range. The computer guides the user, telling the user if the correct number is lesser or greater than the number guessed by the user. The user is allowed only ten trials.

Algorithm:

```
GuessNumber()
{
 import random;
n = 0;
//ask the user to enter two numbers
```

```
print('Hi there! I will guess a number between the range
entered by you\t:');
a=int(input('Enter the first number\t:'));
b=int(input('Enter the second number\t:'));
print('Generating a number between ',a, ' and ',b);
number = random.randint(a, b);
while (n < 10)
{
        guess = int(input('Take a guess\t:'));
        if (guess < number)
                {
                print('Think of a higher number\t:');
                }
        if (guess > number)
                {
                print('Think of a lower number\t:');
                }
        if (guess == number)
                {
                print('Congratulations you win!')
                break;
                }
            n = n + 1;
        if (guess == number)
                {
                print('Won in ', (n+1), 'chance(s)');
                }
        if (guess != number)
            {
        print('The number was ', number);
                }
}
```

Expected Output

Hi there! I will guess a number between the range entered by you:

Enter the first number   : 3

Enter the second number   : 10

Generating a number between 3 and 10 :

Take a guess	: 9
Think of a lower number	:
Take a guess	: 8
Think of a lower number	:
Take a guess	: 7
Think of a lower number	:
Take a guess	: 6
Think of a lower number	:
Take a guess	: 5
Think of a lower number	:
Take a guess	: 4
Think of a lower number	:

The number was 3

1.7.4 Tower of Hanoi

Tower of Hanoi requires the transfer of **n** disks of increasing sizes kept in the source peg to the destination peg, moving one disk at a time, in a way so that a larger disk should not be placed on a smaller disk at any point in time. The following example illustrates the process. The value of **n** in this example is three. Note that in none of the steps, a larger disk is placed above a smaller disk (Figures 1.3–1.9).

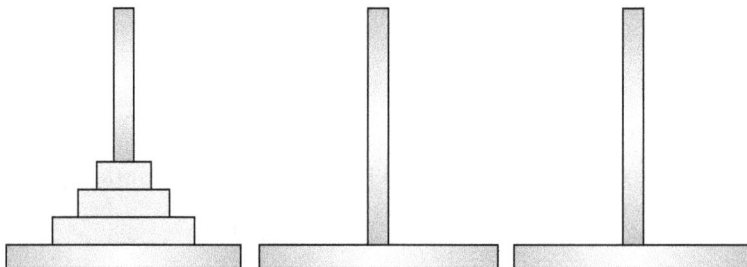

FIGURE 1.3 Initially the first peg (source) has all three disks which are to be transferred to the second peg (destination).

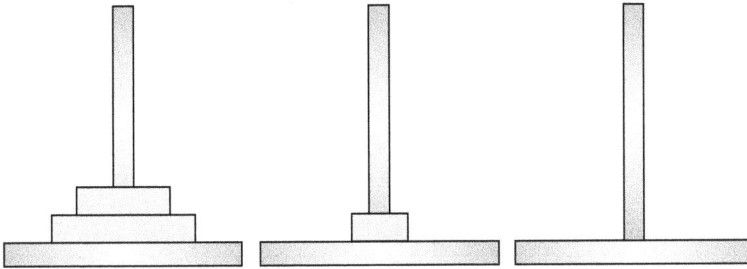

FIGURE 1.4 Move the smallest disk to the second peg.

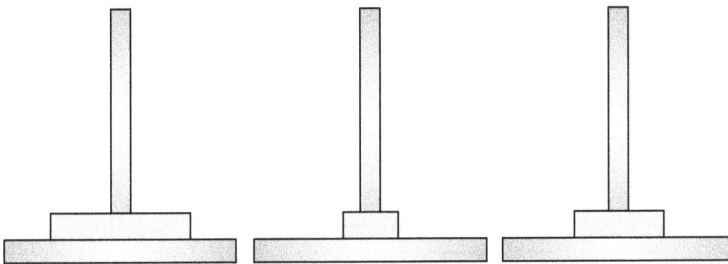

FIGURE 1.5 Move the second largest disk to the third peg.

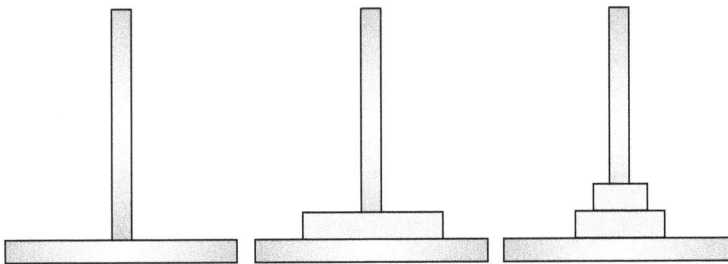

FIGURE 1.6 Move the smallest disk to the third peg and the largest disk to the second peg.

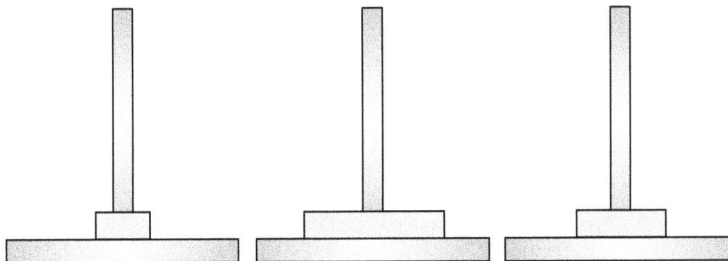

FIGURE 1.7 Now move the smallest disk to the first peg.

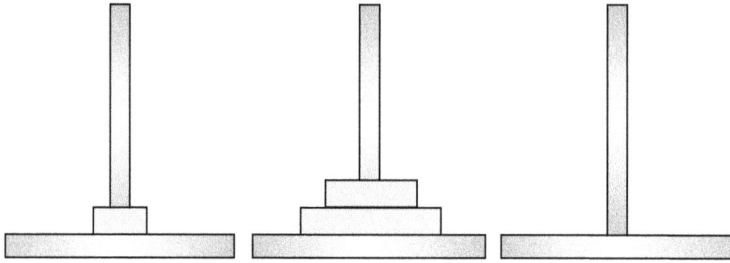

FIGURE 1.8 Move the second largest disk to the second peg and place it above the largest disk.

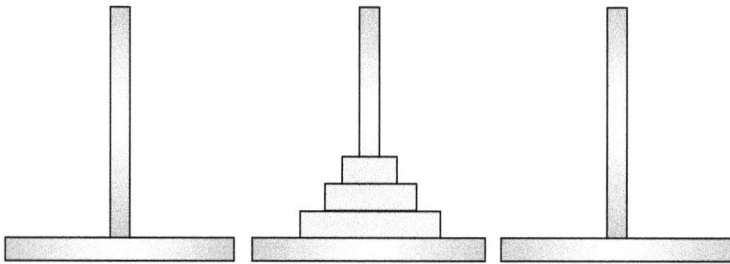

FIGURE 1.9 Now place the smallest disk above the second largest disk.

Illustration 1.4:

Write an algorithm for the solution to the Tower of Hanoi problem.

Algorithm:

```
def towerOfHanoi(n, source, destination, intermediate)
{
    if(n==1)
    {
        print("Move ",n," from ", source, " to ",destination);
    }
        else
    {
        towerOfHanoi(n-1, source, intermediate, destination);
        print("Move ",n," from ", source, " to ",destination);
        towerOfHanoi(n-1, intermediate, destination, source);
    }
}
```

Test:

```
towerOfHanoi(1,'A','B', 'C')
```

Expected Output

Move 1 from A to B

Test:

```
towerOfHanoi(2,'A','B', 'C')
```

Expected Output

```
Move 1 from A to C
Move 2 from A to B
Move 1 from C to B
```

Test:

```
towerOfHanoi(3,'A','B', 'C')
```

Expected Output

```
Move 1 from A to B
Move 2 from A to C
Move 1 from B to C
Move 3 from A to B
Move 1 from C to A
Move 2 from C to B
Move 1 from A to B
```

Test:

```
towerOfHanoi(4,'A','B', 'C')
```

Expected Output

```
Move 1 from A to C
Move 2 from A to B
Move 1 from C to B
Move 3 from A to C
Move 1 from B to A
Move 2 from B to C
Move 1 from A to C
Move 4 from A to B
Move 1 from C to B
```

```
Move 2 from C to A
Move 1 from B to A
Move 3 from C to B
Move 1 from A to C
Move 2 from A to B
Move 1 from C to B
```

1.8   CONCLUSION

Algorithms are the set of steps used to accomplish a given task, efficiently and effectively. An algorithm must be definite, and each instruction should be unambiguous. There are many ways of representing the steps to accomplish the given task: by simply writing the instruction in English, by drawing the flowchart, or by writing the pseudocode. An algorithm must generate at least one output. Moreover, a good algorithm should be efficient both in terms of memory and computation time. The asymptotic complexity helps us to ascertain the efficiency of an algorithm. Problem-solving using algorithms is an involved process and requires due deliberation and intricate analysis. Some algorithms like linear search, binary search, etc., have been presented in this chapter. Readers, new to programming, may find it difficult to get hold of some of the procedures presented in this chapter. However, they should cover the procedural programming presented in Section II of this book and revisit this chapter. This chapter is a door to the fascinating world of problem-solving and Python would be your friend in this long journey. So, let us meet our new friend "Python" in the next chapter.

GLOSSARY

Algorithm: Set of steps written to accomplish a given computing task.

Features of a good algorithm: Correctness, Definiteness, Unambiguity, Input, and Output.

POINTS TO REMEMBER

- Correctness should be the priority in designing an algorithm.
- An algorithm should be
    - definite,
    - unambiguous, and
    - efficient.

- Binary search is better for a sorted list.
- Big Oh notation represents the upper bound.
- Theta is the tight bound.

EXERCISES

Multiple Choice Questions

1. An algorithm should be
   - **(a)** Definite
   - **(b)** Unambiguous
   - **(c)** Both (*a*) and (*b*)
   - **(d)** None of the above

2. In which of the following algorithms, an input argument may not be required?
   - **(a)** Linear search
   - **(b)** Binary search
   - **(c)** Pseudo-random number generator
   - **(d)** None of the above

3. Which of the following can be used, if the input list is sorted?
   - **(a)** Linear search
   - **(b)** Binary search
   - **(c)** Both are equally efficient
   - **(d)** Depends on the input constraints

4. Which of the following would not work if the input array is not sorted?
   - **(a)** Linear search
   - **(b)** Binary search
   - **(c)** Both (*a*) and (*b*)
   - **(d)** Depends on the problem

5. In a recursive algorithm
   - **(a)** The function needs to be expressed in terms of itself
   - **(b)** The base case must be stated
   - **(c)** Both of the above
   - **(d)** None of the above

6. Generally, which of the following has more time complexity?

   (a) Recursive algorithm   (b) Iterative algorithm

   (c) Both (a) and (b)   (d) None of the above

7. Which of the following represents the upper bound?

   (a) Big Oh notation   (b) Omega notation

   (c) Theta notation   (d) None of the above

8. Which of the following represents the lower bound?

   (a) Big Oh notation   (b) Omega notation

   (c) Theta notation   (d) None of the above

9. Which of the following represents tight bound?

   (a) Big Oh notation   (b) Omega notation

   (c) Theta notation   (d) None of the above

10. If an algorithm is $O(n)$ then it is

    (a) $O(n^2)$ (b) $O(n^3)$

    (c) $O(2^n)$ (d) All of the above

11. If an algorithm is $O(n^2)$ then it is not

    (a) $O(n)$ (b) $O(n^3)$

    (c) $O(2^n)$ (d) All of the above

12. If an algorithm is $\Omega(n^3)$ then it is

    (a) $\Omega(n^2)$ (b) $\Omega(n)$

    (c) $\Omega(1)$ (d) All of the above

13. If an algorithm is $\Omega(n^2)$ then it is not

    (a) $\Omega(n)$ (b) $\Omega(n^3)$

    (c) $\Omega(2^n)$ (d) All of the above

14. Who was the brain behind the famous page rank algorithm?

    (a) Larry Page   (b) Alan Turing

    (c) Trump   (d) None of the above

15. Who discovered Turing Machine?

(**a**) Einstein   (**b**) Newton

(**c**) George Boole   (**d**) None of the above

Theory

1. Define the term algorithm. Also, state the features of a good algorithm.

2. What is space and time complexity? Also, discuss the importance of the two.

3. Define the following asymptotic notations.

(**a**) big Oh,

(**b**) omega,

(**c**) theta.

4. What are the different ways of writing an algorithm?

5. Differentiate between an algorithm and a program.

6. Why should design precede implementation?

7. What are the various design approaches?

8. Discuss the time-memory trade-off.

Application

1. Write an algorithm for linear search.

2. If the given list is sorted, an element can be searched as follows. We start by looking at the first, last, and middle position of the list. If the element is not found at these positions, the list is divided into two halves. If the element to be searched is less than the element present in the middle, then the process is repeated on the left part of the list; otherwise, it is repeated on the right part. At the end, a single element is left, which can be easily checked. Write a formal algorithm for this process.

3. In the above case, if the list is divided into four parts, instead of two, write the algorithm.

4. Write an algorithm to sort a given list.

5. Write an algorithm to find the minimum element from a given list.

6. Write an algorithm to find the maximum element from a given list.

7. Write an algorithm to find the second maximum element from a given list.

8. Write an algorithm to find the sum of all the elements of a given list. Also find the average, standard deviation, and quartile deviation of the elements in that list.

9. Write an algorithm to find if a list contains repeated elements.

10. Write an algorithm to reverse a given list.

2

INTRODUCTION TO PYTHON

Objectives

After reading this chapter, the reader should be able to

- Understand the principles of Python
- Appreciate the importance and features of Python
- Enlist the areas in which Python is used
- Install Anaconda
- Understand the control flow in Python

2.1 INTRODUCTION

Art is an expression of human creative skills, and hence programming is an art. Therefore, the choice of programming language acts as a tool in the hands of an artist. This book introduces Python, which would help you to become a great artist. A. J. Perlis, who was a professor at Purdue University, USA, was also the recipient of the first Turing award, stated

A language that doesn't affect the way you think about programming is not worth knowing.

Python is worth knowing. Learning Python would not only motivate you to do highly complex tasks in the simplest manner but would also demolish the myths of conventional programming paradigms. Moreover, it is a language that would change how you look at a problem.

This book aims to explore the elements of Python programming. Though we would use Python for programming, most of the concepts presented in this book are general. One must appreciate the fact that Computer Science is now also being used in solving the problems of society, and the language you learn should take you forward toward your goal of contributing to society.

As stated in the previous chapter, a program is a set of instructions, and we cannot use English-like instructions as they are ambiguous. On the other hand, programming languages like Python are unambiguous. The Python interpreter would interpret the instructions fed to it.

Python is a strong, procedural, object-oriented, and functional language crafted in the late 1980s by Guido Van Rossum. The language is named after Monty Python, a comedy group. The language is currently being used in diverse application domains. These include software development, web development, desktop Graphical User Interface (GUI) development, and education and scientific applications. So, it spans almost all the facets of development. Its popularity is primarily owing to its simplicity and robustness, though many other factors are discussed in the following chapters.

There are many third-party modules for accomplishing the above tasks. For example, Django, an immensely popular web framework dedicated to clean and fast development, is developed on Python. This, along with the support for HTML, e-mails, FTP, etc., makes it apt for web development.

Third-party libraries are also available for software development. One of the most common examples is Scions, which is used for building controls. Clubbed with the in-built features and support, Python also works miracles for GUI development and for developing mobile applications, for example, Kivy is used for developing multi-touch applications.

Python also finds its applications in Scientific Analysis. SciPy is used for Engineering and Mathematics, and IPython is used for parallel computing. Those of you working in statistics and Machine Learning would find some libraries extremely useful and easy to use. For example, SciPy provides MATLAB like features and can be used for processing multidimensional arrays. Figure 2.1 summarizes the above discussion.

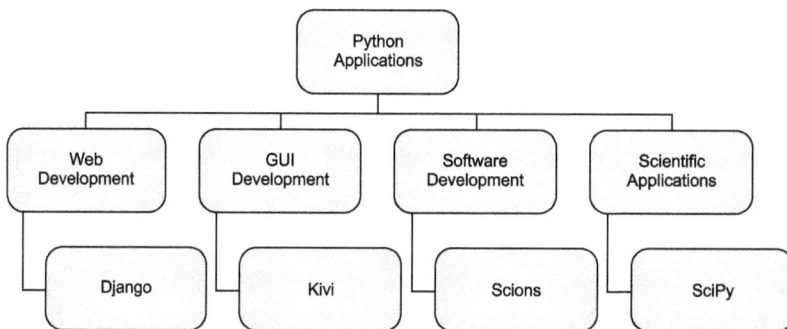

FIGURE 2.1 Some of the applications of Python.

The chapter has been organized as follows. Section 2.2 discusses the features of Python. Section 2.3 discusses the programming language paradigms supported by Python, and Section 2.4 discusses the chronology and the uses of Python. The installation of the Anaconda has been presented in Section 2.5. Section 2.6 presents a brief discussion of variables, statements, etc. The last section concludes.

2.2 FEATURES OF PYTHON

As stated earlier, Python is a simple but powerful language. It is portable and free. It has built-in object types and many libraries. This section briefly discusses the features of Python.

2.2.1 Easy

Python is easy to learn and understand. As a matter of fact, if you are from a programming background, you would find it elegant and uncluttered. The removal of braces and parenthesis makes the code short and sweet. Also, some of the tasks in Python are quite easy. For example, swapping numbers in Python is as easy as writing $(a, b) = (b, a)$.

It may also be stated here that learning something new is an involved and intricate task. However, the simplicity of Python makes this learning almost a cakewalk. Moreover, learning its advanced features, though a bit intricate, but worth the effort. It is also easy to understand a project written in this language. The code is concise and effective, which makes it understandable.

2.2.2 Type and Run

In most of the projects, testing something new requires scores of changes and hence recompilations and re-runs. This makes testing code a difficult and time-consuming task. In Python, code can be executed easily. As a matter of fact, we run scripts in Python. As we will see later in this chapter, it also provides users with an interactive environment to run independent commands.

2.2.3 Syntax

The Syntax of Python is easy; this makes the learning and understanding process easy. The three main features that make this language attractive are that it is simple, small, and flexible.

2.2.4 Mixing

If one is working on a big project, with a big team, it might be the case that some team members are good in other languages. This may lead to some of the modules, in some other language, wanting to be embedded with the Python code. Python, in fact, allows and even supports this.

2.2.5 Dynamic Typing

Python has its own way of managing memory associated with objects. When an object is created in Python, memory is dynamically allocated to it. When the object's life cycle ends, the memory is taken back from it. This memory management of Python makes the programs more efficient.

2.2.6 Built-in Object Types

As we will see in the next chapter Python has built-in object types. This makes the task, to be accomplished, easy, and manageable. Moreover, the issues related to these objects are beautifully handled by the language.

2.2.7 Numerous Libraries and Tools

In Python, the task to be accomplished becomes easy, really easy. This is because most of the common tasks (as a matter of fact, not so common tasks also), have already been handled in Python. For example, it has libraries that help users to develop GUIs, write mobile applications, incorporate security features, and even read MRIs. As we will see in the following chapters, the libraries and supporting tools make even intricate tasks like pattern recognition easy.

2.2.8 Portable

A program written in Python can run on almost every known platform. Be it Windows, Linux, or Mac. It may also be stated here that Python is written in C. However, some versions of this language are written in JAVA as well.

2.2.9 Free

Python is not propriety software. One can download Python compilers; from among the various available choices. Moreover, no known legal issues are involved in the distribution of the code developed in Python.

2.3   THE PARADIGMS

This section briefly introduces the three major paradigms. Note that Python fully supports the first two: procedural and object-oriented programming. However, it also supports some other features like tail optimization, etc.

2.3.1   Procedural

In procedural language, a program is a set of statements that execute sequentially. The only option a program has in terms of manageability is dividing the program into small modules. "C," for example, is a procedural language. Python supports procedural programming. The fifth section of this book deals with procedural programming.

2.3.2   Object-Oriented

This type of language primarily focuses on instances of a class. The instance of a class is called object. A class, here, is a real or a virtual entity, having importance to the problem at hand, and has sharp physical boundaries. For example, in a program that deals with student management, "student" can be a class. Its instances are made, and the task at hand can be accomplished by communicating via methods. Python is object-oriented language. Section III of this book deals with object-oriented programming.

2.3.3   Functional

In functional programming, each computation is treated as the result of the evaluation of a mathematical function. Python also supports functional programming. Moreover, Python supports immutable data, tail optimization, etc. This must be music to the ears of those from a functional programming background. Here, it may be stated that functional programming is beyond the scope of this book. However, some of the above features will be discussed in the chapters that follow.

2.4   CHRONOLOGY AND USES

This section briefly discusses Python's chronology and motivates the reader to bind with the language.

2.4.1   Chronology

Python, which is a multiparadigm language, was conceived in the late 1980s. The implementation of Python began in December 1989, when Guido Van

Rossum, who was working in Centrum Wiskunde & Informatica, decided to do something useful during his Christmas holidays. He actually wanted to work on the successor of the ABC programming language.

The next version, Python 2, was released on October 16, 2000, followed by Python 3, released on December 3, 2008. However, it was not backward compatible. In 2017, Google announced work on Python 2.7.

Some languages, like Perl, believe in providing options as more than one way of doing a task. On the other hand, Python believes in the fact that there is one obvious way of doing a task and therefore has a small core. Python supports object-oriented programming and procedural programming. It partly supports functional programming. Even though it has a small core, the standard library support is vast. It is simple, less cluttered, and has an extendible interpreter.

The principles on which Python is based can be seen by typing `import this` in the interpreter. This presents the "Zen of Python."

```
>>import this
```

Output:

```
The Zen of Python, by Tim Peters
Beautiful is better than ugly.
Explicit is better than implicit.
Simple is better than complex.
Complex is better than complicated.
Flat is better than nested.
Sparse is better than dense.
Readability counts.
Special cases aren't special enough to break the rules.
Although practicality beats purity.
Errors should never pass silently.
Unless explicitly silenced.
In the face of ambiguity, refuse the temptation to guess.
There should be one-- and preferably only one --obvious way to
    do it.
Although that way may not be obvious at first unless you're
    Dutch.
Now is better than never.
Although never is often better than *right* now.
If the implementation is hard to explain, it's a bad idea.
```

```
If the implementation is easy to explain, it may be a good idea.
Namespaces are one honking great idea -- let's do more of those!
```

It may also be stated that Python rejects the patches that provide marginal speed increases instead of a less understandable code. The features of Python have been discussed in the next section of this chapter. It may be stated that those following the above rules are said to use a Pytonic way of programming.

The continuous betterment of this language has been possible because of a dedicated group of people committed to supporting the cause of providing the world with an easy yet powerful language. The growth of this language has given rise to the creation of many interest groups and forums for Python. A change in the language can be bought about by what is generally referred to as PEP (Python Enhancement Project). The PSF (Python Software Foundation) takes care of this.

2.4.2 Uses

Python is being used to accomplish many tasks, the most important of which are as follows.

- GUI development
- Scripting Web Pages
- Database Programming
- Prototyping
- Gaming
- Component-based programming

If you are working in Unix or Linux, you need not install Python because it is generally pre-installed. If you work on Windows or Mac, you need to download Python. Once you have decided to download Python, look for its latest version. The reader is requested to ensure that the version they intend to download is not an alpha or a beta version. The next section briefly discusses the steps for downloading Anaconda, an open-source distribution.

Many development environments are available for Python. Some of them are as follows.

1. PyDev with Eclipse

2. Emacs

3. Vim

4. TextMate

5. Gedit

6. Idle

7. PIDA (Linux) (VIM Based)

8. NotePad++ (Windows)

9. BlueFish (Linux)

There are some more options available. However, this book uses IDLE and Anaconda. The next section presents the steps involved in the installation of Anaconda.

2.5 INSTALLATION OF ANACONDA

In order to install Anaconda, go to *https://docs.continuum.io/anaconda/install* and select the installer (Windows or Mac OS or Linux). This section presents the steps involved in installing Anaconda on the Windows Operating System.

First, one must choose the installer based on their processor (32-bit or 64-bit). After this, click on the selected installer and download the executable file. The installer would ask you to install the software on the default location. It may happen that during installation, you might have to disable the anti-virus software. Figure 2.2(*a*)–(*g*) takes the reader through the installation.

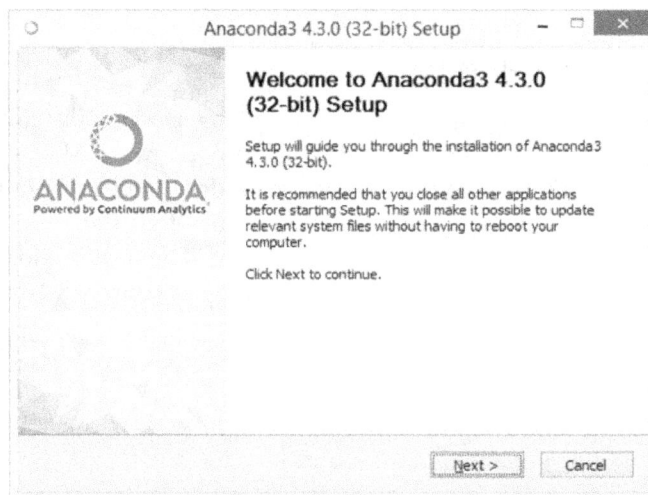

FIGURE 2.2(a) The welcome screen of the installer asks the user to close all running applications and then click Next.

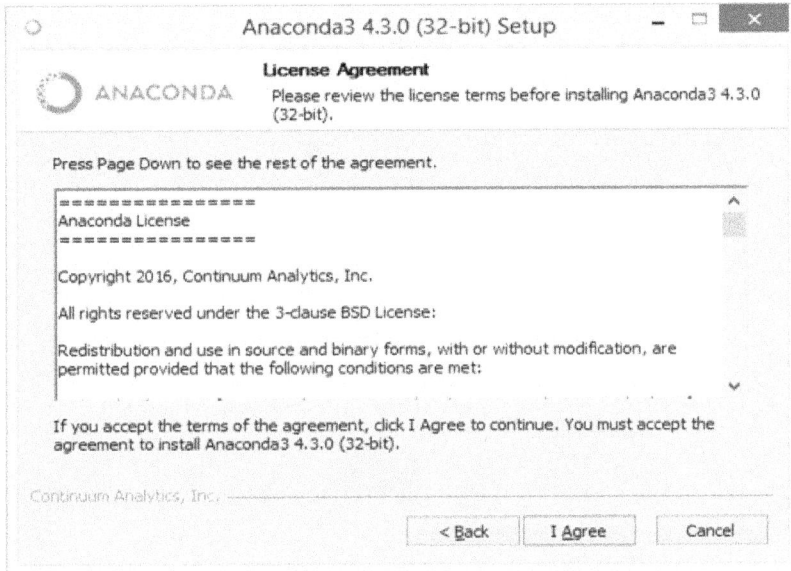

FIGURE 2.2(b) The Licence agreement to install
Anaconda3 4.3.0 (32-bit).

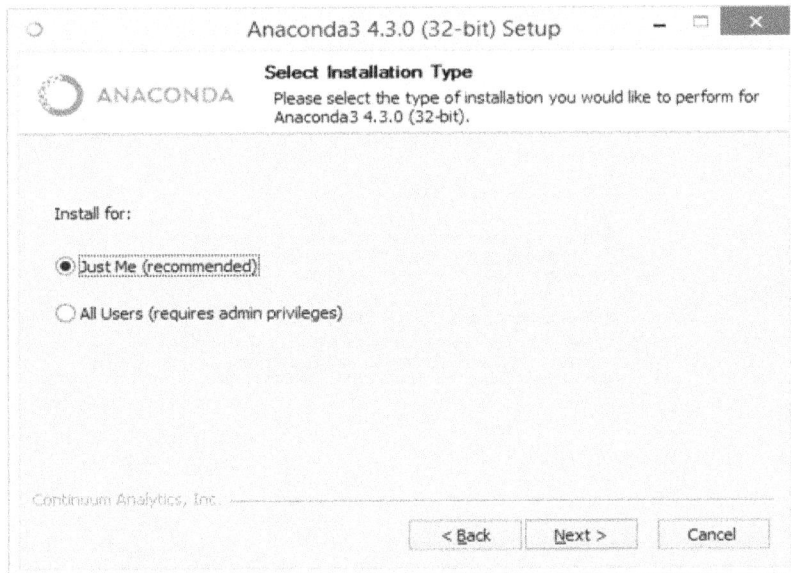

FIGURE 2.2(c) In the third step, the user is required to choose whether
he wants to install Anaconda for a single user or for all users.

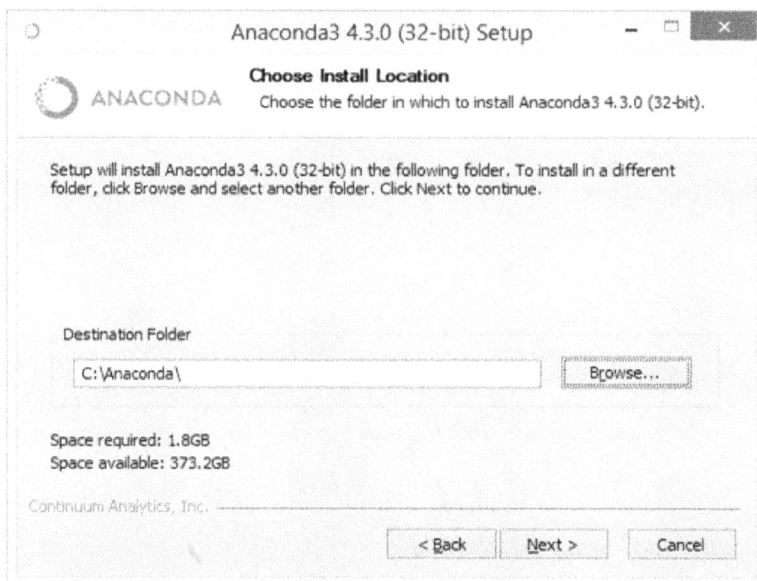

FIGURE 2.2(d) The user then needs to select the folder in which it will install.

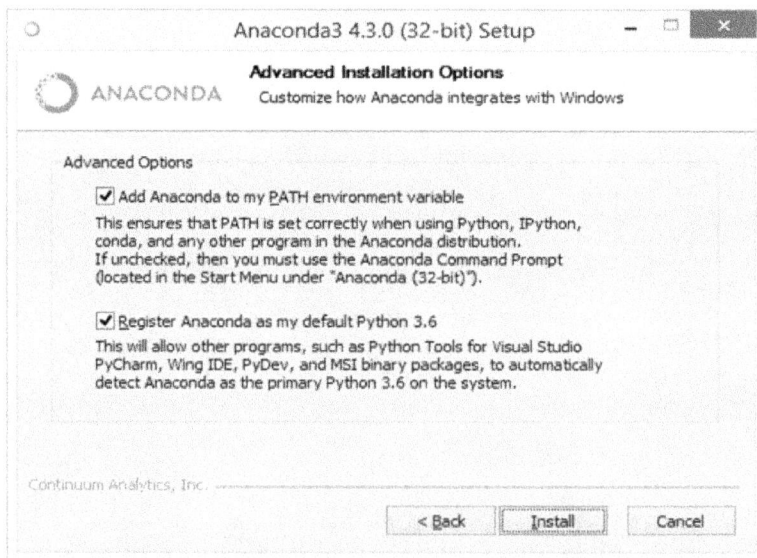

FIGURE 2.2(e) The user then must decide whether he wants to add Anaconda to Path environment variable and whether to register Anaconda as the default Python 3.6.

The installation then starts. After installation, the following screen will appear.

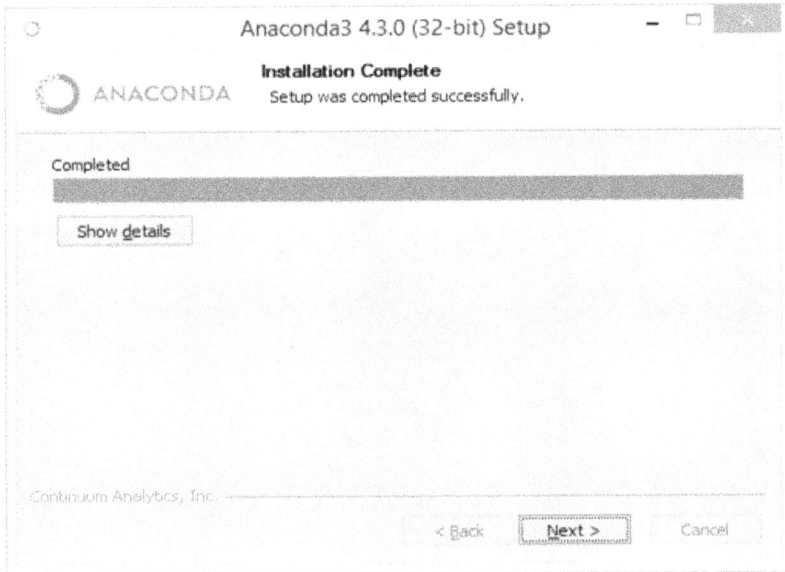

FIGURE 2.2(f) When the installation is complete, this screen appears.

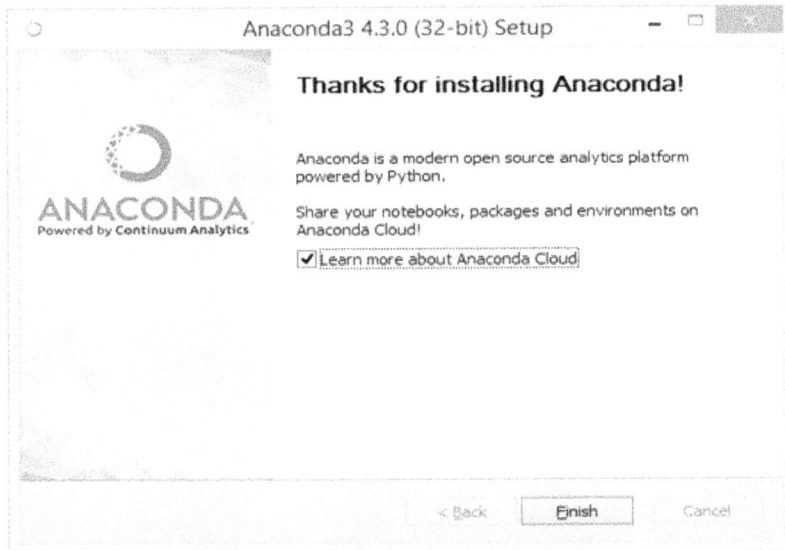

FIGURE 2.2(g) You can also share your notebooks on cloud.

Once Anaconda is installed, you can open Anaconda Navigator and run your scripts. Figure 2.3 shows the Anaconda Navigator. From the various options available, you can choose the appropriate option. For example, you can open the QTConsole and run the commands/scripts. Figure 2.4 shows the snapshot of QTConsole. The commands written may appear gibberish at this point but will become clear in the following chapters. The reader is advised to use the Jupyter Notebook to execute the code in this book.

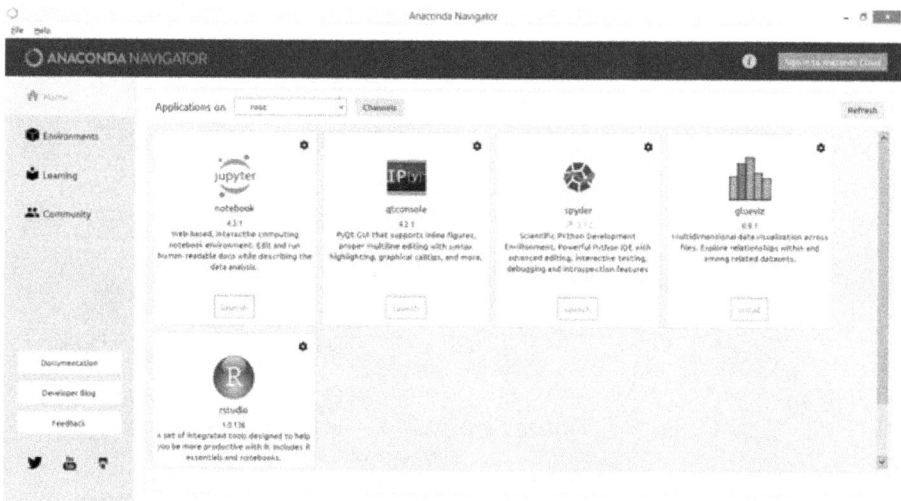

FIGURE 2.3 The Anaconda Navigator.

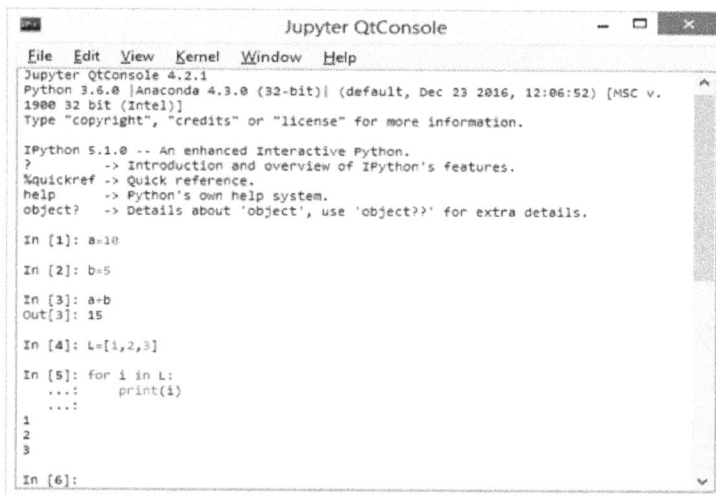

FIGURE 2.4 The QtConsole.

2.6 IMPLEMENTATION OF AN ALGORITHM: STATEMENT, STATE, CONTROL BLOCKS, AND FUNCTIONS

The definition, characteristics, and types of algorithms were discussed in the last chapter. This chapter introduces Python, which would help us to implement algorithms. This section briefly describes the building blocks of a program.

2.6.1 Statement

A program, as stated earlier, is a set of instructions. Each instruction directs the computer to what needs to be done. A statement in a program is the instructions that a Python interpreter can execute. A statement can be of many types, like assignment statements, conditional statements, etc. An assignment statement, for example, is of the form

<variable name> = <expression>

where <variable name> is any legal variable name and expression is any legal expression in Python, which when evaluated assigns value to the variable on the LHS. A legal variable name in Python begins with a letter or an underscore and cannot contain a space or a special character. The following chapters will explain each type of statement, in detail.

2.6.2 State

A program is called stateful if it remembers earlier interactions. The state of a program at one point in time is therefore the values of the variables and related details, at that time. A system can have any state from the set of possible states called the state space.

2.6.3 Control Flow

In a Python program the statements are executed one by one. If the flow of the program is to be altered, one of the legal control-flow statements can be used. Python provides the following control flow statements:

- if, if-else, if-elif ladder
- while loop, and
- for loop

Conditional Statement

Python also provides the **break** and **continue** keywords. The `if` statement helps us to execute a particular set of statements, if the given condition is true. The syntax of `if` is as follows.

Syntax

```
if <condition>:
    <block 1>
else:
    <block 2>
```

The first block `<block 1>` executes if the condition is true. If the condition is not true, then `<block 2>` is executed. The following example checks the value of the number entered by the user. If the number entered by the user is greater than 10, then "Hi" is printed else "Bye" is printed.

Code:

```
num=int(input('Enter a number Wt:'))
if num>10:
  print('Hi')
else:
  print('Bye')
```

Output:

```
Enter a number     :12
Hi
```

If there are multiple conditions, the if-elif ladder is used. Chapter 4 describes the conditional statements, in detail executes.

Loops

The repetition of a block is done using the `while` and `for` loops. The `while` loop repeats a block as long as a condition is true. In Python `while` can have an optional `else`. The syntax of the loop is as follows.

Syntax

```
while <condition>:
    <block 1>
else:
    <block 2>
```

The first block `<block 1>` is repeated till `<condition>` is true. If the condition is false, the second block `<block 2>` is executed. The example that follows prints the first ten multiples of the number entered by the user. The variable **i** is initialized to 1. The loop runs till the value of i becomes n. In each iteration i*n is printed. The loop is discussed in detail in Chapter 6 of this book. The `for` loop in Python is a bit different from C language as it also helps us iterate through a list, tuple, or string.

Code:

```
n = int(input('Enter number 'enter number it:'\t:'))
i=1
while(i<=10):
    print(i, ' * ',n,' = ',(i*n))
    i+=1
```

Output:

```
Enter number  :7
        1    *    7    =     7
        2    *    7    =    14
        3    *    7    =    21
        4    *    7    =    28
        5    *    7    =    35
        6    *    7    =    42
        7    *    7    =    49
        8    *    7    =    56
        9    *    7    =    63
       10    *    7    =    70
```

Function

A function is a named block that performs a specific task and may or may not return a value explicitly. Note that each function returns some value in Python at least NONE. A function, in Python, is defined using the `def` keyword. A function may have any number of parameters and can be called any number of times. Chapter 7 describes the topic in detail. The following code shows a function called `fun`. This prints the string "Turn Turn Turn." Note that `fun` has been called twice.

Code:

```
def fun():
    print('Turn Turn Turn')

fun()
fun()
```

Output:

```
Turn Turn Turn
Turn Turn Turn
```

2.7 CONCLUSION

Before proceeding any further, the reader must note that some of the features of Python are different as compared to any other language. Before proceeding any further, the following points must be mentioned to avoid any confusion.

- In Python statements do not end with any special character. Python considers the newline character as an indication of the fact that the statement has ended. If a statement is to span more than a single line, the next line must be preceded by a (\).
- In Python, indentation is used to detect the blocks. The loops, in Python, do not begin or end with delimiters or keywords.
- A file in Python is generally saved with a .py extension.
- Python shell can also be used as a handy calculator.
- The data type need not be mentioned in a program.

Choice at every step is good, but it can also be intimidating. As stated earlier, Python's core is small, and therefore it is easy to learn. Moreover, there are some of the things like if else, loops and exception handling, which is used in almost all the programs.

The chapter introduces Python and discusses its features of Python. One must appreciate the fact that Python supports all three paradigms: procedural, object-oriented, and functional. Also, this chapter paves the way for the topics presented in the following chapters. It may also be stated that the codes presented in the book would run on version 3.x.

GLOSSARY

PEP   Python Enhancement Project

PSF   Python Software Foundation

POINTS TO REMEMBER

- Python is a strong, procedural, object-oriented, and functional language crafted in the late 1980s by Guido Van Rossum
- Python is open source
- The Applications of Python include Software Development, Web Development, Desktop GUI development, Education, and Scientific Applications
- Python is popular due to its simplicity and robustness
- Python is easy to interface with C++ and JAVA
- SciPy is used for Engineering and Mathematics, IPython for parallel computing, etc., and Scions is used for build control.
- The various development environments for Python are PyDev with Eclipse, Emacs, Vim, TextMate, Gedit, Idle, PIDA (Linux) (VIM Based), NotePad++ (Windows), and BlueFish (Linux).

RESOURCES

- To download Python, visit *www.python.org*
- The documentation is available at *www.python.org/doc/*

EXERCISES

Multiple Choice Questions

1. Python can subclass a class made in

   (a) Python Only

   (b) Python, C++

   (c) Python, C++, C#, JAVA

   (d) None of the above

2. Who created Python?

   (a) Monty Python

   (b) Guido Van Rossum

   (c) Dennis Richie

   (d) None of the above

3. Monty Python was

   (a) Creator of Python Programming Language

   (b) British Comedy Group

   (c) American Band

   (d) Brother of Dosey Howser

4. In Python, libraries and tools are

   (a) Not supported   **(b)** Supported but not encouraged

   (c) Supported (only that of PSFs)   **(d)** None of the above

5. Python has

   (a) built-in object types   **(b)** Data types

   (c) Both (*a*) and (*b*)   **(d)** None of the above

6. Python is a

   (a) Procedural language   **(b)** Object-Oriented Language

   (c) Functional   **(d)** All of the above

7. In Python data type for a variable is not specified; therefore, it is applicable to the whole range of objects. This is called

   (a) Dynamic Binding   **(b)** Dynamic Typing

   (c) Dynamic Leadership   **(d)** None of the above

8. Which of the following is automatic memory management?

   (a) Automatically assigning memory to objects

   (b) Taking back the memory at the end of the life cycle

   (c) Both (*a*) and (*b*)

   (d) None of the above

9. PEP stands for

   (a) Python Ending Procedure   **(b)** Python Enhancement proposal

   (c) Python Endearment Project   **(d)** None of the above

10. PSF stands for

    (a) Python Software Foundation   (b) Python Selection Function

    (c) Python segregation function   (d) None of the above

11. Python can be used to create

    (a) GUI   (b) Internet Scripting

    (c) Games   (d) All of the above

12. What can be done using Python?

    (a) System programming

    (b) Component-based programming

    (c) Scientific programming

    (d) All of the above

13. Python is used by

    (a) Google   (b) Raspberry Pi

    (c) Bit Torrent   (d) All of the above

14. Python is used in

    (a) App Engine   (b) YouTube sharing

    (c) Real-time programming   (d) All of the above

15. Which is faster?

    (a) PyPy   (b) IDLE

    (c) Both are equally good   (d) Depends on the task

Theory

1. Write the names of three companies that are using Python.

2. Explain a few applications of Python.

3. What type of language is Python? (Procedural, Object-Oriented, or Functional)

4. What is PEP?

5. What is PSF?

6. Who manages Python?

7. Is Python open-source or proprietary?

8. What all languages can be integrated with Python?

9. Explain the chronology of the development of Python.

10. Name a few editors for Python.

11. What are the features of Python?

12. What is the advantage of using Python over other languages?

13. What is Dynamic Typing?

14. Does Python have data types?

15. How is Python different from JAVA?

16. What is meant by the state of a system?

17. Briley explains the importance of control structures in Python.

18. Which control statements can be used to repeat a given task?

19. What is the principle difference between Pearl and Python?

20. State the need and importance of functions.

3

FUNDAMENTALS

Objectives

After reading this chapter, the reader should be able to

- Understand how a program is executed.
- Learn various methods of running a program in **Python**.
- Understand the **print** and **input** functions in **Python**.
- Understand the elements of a **Jupyter Notebook**.
- Understand the importance of tokens, keywords, identifiers, and statements.

3.1   INTRODUCTION

Our journey toward becoming a programmer would be greatly facilitated by **Python**. Let us start the journey by exploring the modes in which we can write and execute a program in **Python**. This chapter discusses various ways of executing a **Python** program. Before starting the discussion, let us first have a look at how a program is executed in languages like **C**.

In C, the following steps are required to run a program (Figure 3.1):

- Compilation
- Linking
- Loading

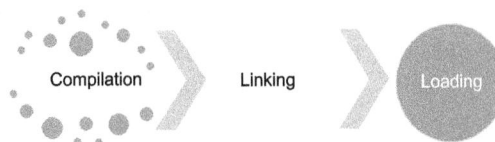

Compilation   Linking   Loading

FIGURE 3.1 In C, the compiler converts the source code to the object code. This is followed by links, and finally, memory is allocated to the process.

A compiler converts the source code into the object code. The linker converts the object code to the executable code and gives it to the loader. The loader primarily allocates memory to the executable file. The conversion of source code into object code requires the following steps (Figure 3.2):

■ Lexical analysis
■ Syntactic analysis
■ Semantic analysis
■ Intermediate code generation
■ Optimization
■ Final code generation

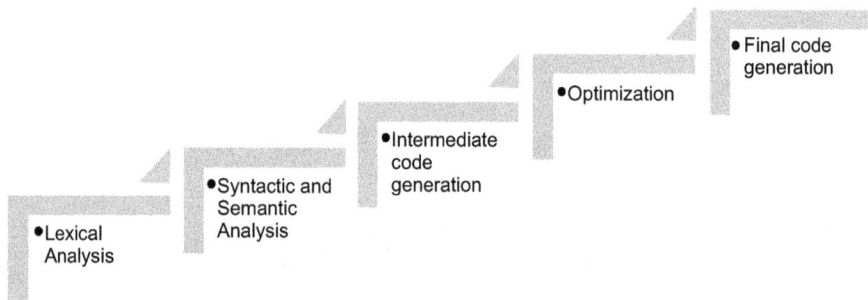

FIGURE 3.2 In C, the compiler converts the source code into object code. This is done in five steps. In the first step, the compiler converts the source code into tokens. This is followed by syntactic and semantic analyses. The intermediate code is then generated. Finally, the object code is generated after optimization.

The Lexical Analyzer converts the program into recognizable tokens. The second step checks the syntax. The analysis of semantics is also done in this step. The next step generates the intermediate code. This intermediate code is optimized, and the last step generates the final code. In some compilers, however, some steps are clubbed together. For example, the syntactic and semantic analysis is done in the same phase in most of the C compilers.

The above process produces optimized code, but the code generated may not be compatible. Therefore, the compiler produces intermediate code in languages like Java and C# to achieve portability. The rest of the process is done by JVM in JAVA and CLR in C#.

In **Python**, all of the above steps are not needed. Moreover, in **Python** a **Python** bytecode is generated. In **Python**, one can run a single comment, a procedure, or even a big program. There are many ways to do so, discussed in this chapter.

The chapter has been organized as follows. Section 3.2 presents a brief discussion on basic input output. Section 3.3 presents various ways of running a **Python** code in Windows. Section 3.4 also presents a brief overview of the **Jupyter notebook**. Section 3.5 discusses the **value** type and reference types. Section 3.6 discusses tokens, keywords, and identifiers. Section 3.7 briefly explains the types of statements. Section 3.8 introduces comments. Sections 3.9 and 3.10 discuss operators and the types of operators, respectively. Section 3.11 discusses the basic data types and Section 3.12 concludes.

3.2   BASIC INPUT OUTPUT

This section briefly discusses the basic input/output functions in **Python**, primarily the **input** and the **print** function. The **input** function prompts the user to enter a value and store the value in some variable. To print a string, the **print** function can be used.

3.2.1   Print Function

The **print** function prints the string given as the argument to the function. For example, the argument "Hi there" is printed in the following statements. Likewise, in the next statement, **"Turn Turn Turn! to everything, there is a season"** is printed.

Code:

```
print('Hi there')
```

Output:

```
'Hi there'
```

Code:

```
print('Turn Turn Turn! to everything, there is a season')
```

Output:

```
Turn Turn Turn! to everything, there is a season
```

The **print** function also accepts more than one argument, separated by commas. The function, in this case, prints the input string or the values of the variables (given as arguments). Note that all the arguments in the **print** function may belong to different data types. For example, in the following code, the values of the variables passed as the arguments, are printed.

Code:

```
a=10
b=3.678
x='harsh'
print(a, b, x)
```

Output:

```
10 3.678 harsh
```

3.2.2 Input

The **input** function in **Python** prompts the user to enter a value and store it in some variable. For example, in the following code, the string entered by the user is referred to by the variable called **name**. The **input** function takes a string as an argument, which prompts the user to enter his/her name.

Code:

```
name=input('Enter name\t:')
print('Hi ', name)
```

Output:

```
Enter name    : Harsh
Hi Harsh
```

The **int** function converts the input to an integer. That is, to take an integer input from the user, the input string, entered by the user, is converted into an integer using the **int** function. For example, the following code takes an integer input from the user and prints it.

Code:

```
num = int(input('Enter a number\t:'))
print(num)
```

Output:

```
Enter a number    : 45
45
```

Likewise, the **float** function converts the input to a float. To take a float input from the user, the string input, entered by the user, is converted into a float by the **float** function.

Code:

```
num = float(input('Enter a number\t:'))
print(num)
```

Output:

```
Enter a number    : 45.45
45.45
```

3.3   RUNNING A PROGRAM

Once you have installed **Python**, you can write and run **Python** commands in many ways. This section discusses some of the most common ways of executing a program written in **Python**.

3.3.1   Using the Command Prompt

To run your program or script in the command prompt, the following steps should be followed.

Step 1: Opening Command Prompt/ Changing directory: In the start menu type "Command Prompt." Once the **Command Prompt** appears, change the directory to the location where **Python** is saved.

Step 2: Changing directory: In case you installed **Anaconda**, as discussed in Chapter 2, you can change the directory to "Python" and start writing the commands.

Step 3: Writing Commands: Write the commands, and the interpreter will interpret the commands and show the results.

3.3.2 Executing Programs Written in .py Files

You can also execute a ".py" file in the **Command Prompt**. To do so, perform the following steps.

- Create a new file called helloworld.py (in the **C:\Users\<your account>\ Anaconda3** folder) and write the following code in it.

  print("Hello World")

- Write the following command in the **Command Prompt**
  >>C:\Users\Harsh\Anaconda3> python helloworld.py

- The following output will be displayed.
  Hello World

3.3.3   Using Anaconda Navigator

This book primarily uses **Jupyter**. It provides the following interfaces:

- **The Jupyter notebook:** This helps in writing and executing the code, combines code with text and equations. It also helps in visualizations. For a detailed explanation, please refer to the following link.
  https://jupyter-notebook.readthedocs.io/en/latest/
- **Jupyter Console:** As per the official site, "Jupyter Console is a terminal-based console for interactive computing."
- **Jupyter QT console**

The following steps may be performed to execute a **Python** code using **Jupyter**.

- Open the **Anaconda Navigator** from the start menu.
- Open the **Jupyter** notebook (Figure 3.3).
- Open a new **Python3** notebook.
- Write command as shown in the following screenshot (Figure 3.4).
- Run the scripts, written in a cell.
- The output is shown just after the cell.

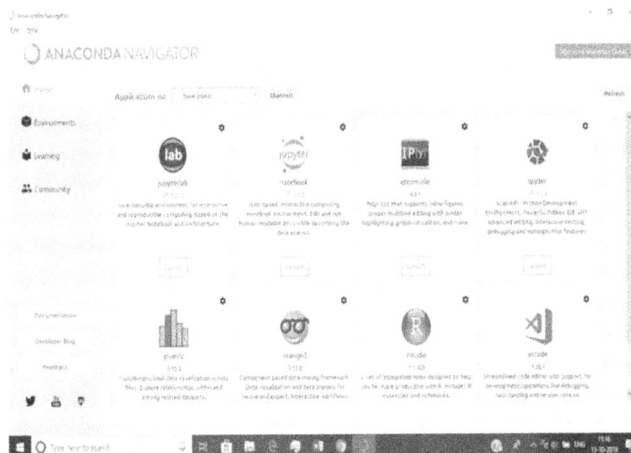

FIGURE 3.3 The **Anaconda** navigator provides you with many options, including the **Jupyter notebook**.

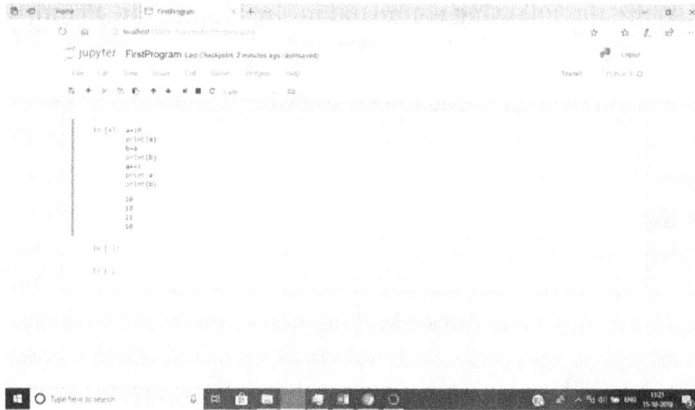

FIGURE 3.4 Running code in the **Jupyter** notebook.

For detailed documentation of **Anaconda** please visit docs.anaconda. com/anaconda/navigator/. In the documentation, you will also find a link explaining the details of the **Jupyter** Application.

3.4 THE JUPYTER NOTEBOOK

The installation of **Anaconda** has already been discussed in the last chapter. The steps to run a **Python** program in **Jupyter** are as follows:

Step 1: Click on the **Jupyter** icon on the desktop. In case you don't see the icon go to Start-> **Anaconda,** and when **Anaconda** opens click on **Jupyter**. (Figure 3.5)

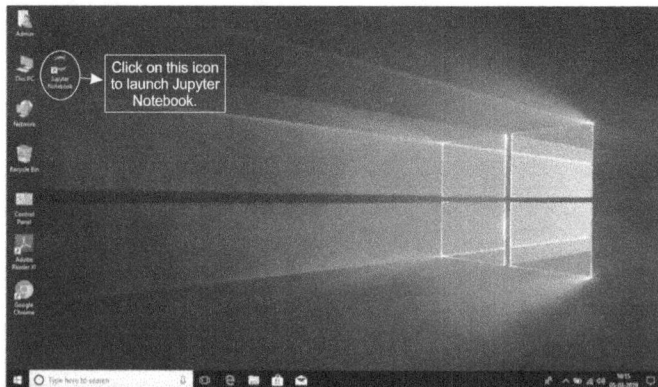

FIGURE 3.5 Click on the **Jupyter** icon on the desktop or go to Start -> **Anaconda** and click on **Jupyter**.

Step 2: The **Jupyter** Notebook opens, after which a new "Notebook" is created by clicking on "New" in the menu (Figure 3.6).

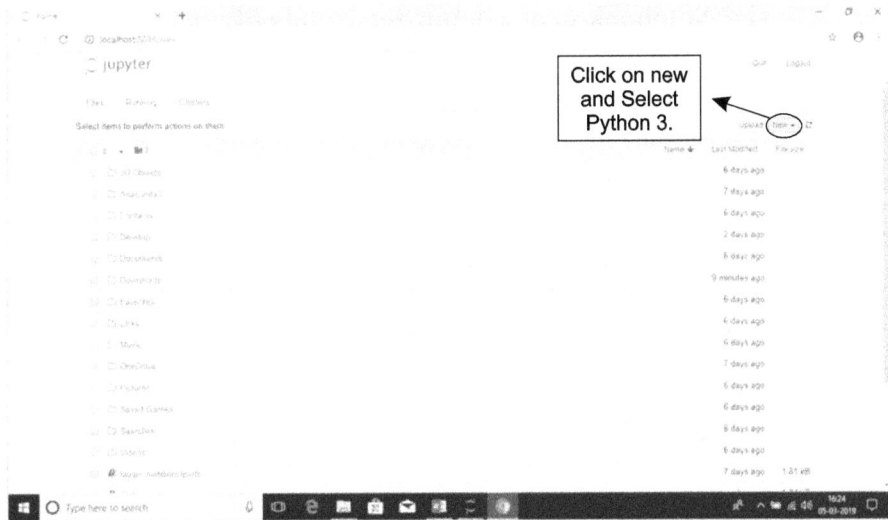

FIGURE 3.6 To open a new Notebook in **Jupyter**, click on **New** as shown in the figure.

Step 3: A new **Notebook** opens, as shown in Figure 3.7.

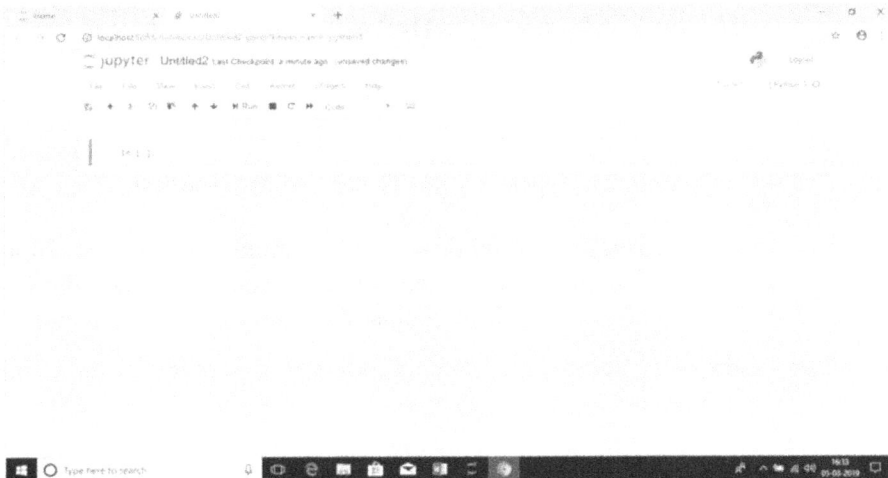

FIGURE 3.7 A new **Notebook** in **Jupyter**.

At this point in time, you need to understand the elements of the **Jupyter notebook**. The elements in this user interface are as follows:

1. **The Notebook name:** A **Jupyter Notebook** can be assigned any legal name. The name of the new notebook can be changed by clicking on the default name and replacing the text.

2. **The Menu bar:** It helps us to manage the **Notebook.** The elements in the menu bar are as follows:

    ▪ **File:** This option allows to create a new **Notebook**, open an existing **Notebook**, save a **Notebook**, and rename a **Notebook**.
    ▪ **Edit:** This option allows us to cut, delete, move, split, and merge **cells**.
    ▪ **Insert:** This option allows us to insert **cells** above or below the existing **cell**.
    ▪ **Cell:** This option allows us to write commands or scripts.
    ▪ **Kernel:** This option allows us to interrupt **cells**, restart **cells**, clear outputs, and shutdown.
    ▪ **Widget:** This option allows us to save, clear, download **Notebook** widget state, and also to embed widget.
    ▪ **Help:** This option presents **help**.

3. It contains icons for common actions such as **Save & checkpoint, Insert cell below, cut & copy a selected cell, paste cell below, move selected cell up & down, run cell, interrupt & restart the kernel.** In particular, the dropdown menu showing code lets you change the type of a cell.

   Once a **Notebook** is created, you can write a piece of code in a **cell. Cells** are mainly of two types, namely **Code** and **Markdown**.

1. **Markdown cell:** A **Markdown cell** contains the rich text. In addition to classic formatting options like bold or italics, we can add links, images, HTML elements, LaTeX mathematical equations, etc., in a **cell**.

2. **Code cell:** A code **cell** contains code to be executed by the kernel. The programming language corresponds to the kernel's language.

3.5 VALUE TYPE AND REFERENCE TYPE

This course aims to make the student capable of solving problems by writing programs. We need input from the user to write a program and generally store it somewhere. To store an input, we need variables. A variable in **Python** refers to a location where a value is placed; hence, they are **reference-type variables** against languages like **C**.

For example, in **C** language, to make an **integer-type variable** and store 7 in it, we write

$$int \; i = 7;$$

In the **C** language, this makes a variable called *i* and stores 7 in binary format. Whereas, in **Python** writing

$$i = 7$$

make a variable called *i* which refers to a location where 7 is stored. (Figure 3.8).

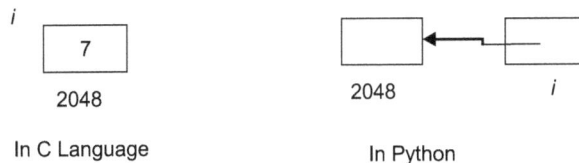

FIGURE 3.8 In C language, the variable *i* will contain value 7; in **Python** *i* points to a location that contains 7.

These variables are then processed using various statements, and the programs which contain these statements produce some output (Figure 3.9).

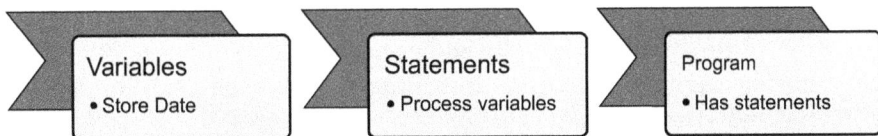

FIGURE 3.9 Variables store data, statements process variables, and a program contains statements.

The next section explains the concept of tokens, keywords, identifiers, operators, various data types, comments, assignments, and input/output in **Python**.

3.6 TOKENS, KEYWORDS, AND IDENTIFIERS

The **character set** of a language is the set of all legal characters in that language. The character set of **Python** has characters, digits, special symbols, and punctuation. These characters form **tokens**. The logical units in a program are called **tokens**. They may be **keywords**, **identifiers**, **operators**, or even **punctuation marks**. This section discusses various tokens in **Python**.

3.6.1 Python Keywords

Keywords are considered as special words, which the compiler reserves and convey some special meaning. Therefore, it should not be used as the name of any variable. Table 3.1 shows the **keywords** used in **Python**.

TABLE 3.1 Keywords in **Python**.

and	as	assert
break	class	continue
def	del	else
elif	except	for
from	finally	Global
if	lambda	none
not	pass	raise
return	sum	while

3.6.2 Python Identifiers

Identifiers are names given to different variables or objects in **Python**. They can be as long as possible and can contain an underscore. Furthermore, the name of an identifier must not begin with a digit but can contain a digit. Interestingly, in **Python**, the names have their scope, as discussed in the following chapters. Moreover, **identifiers** can also contain some special characters in **Python**. Also, **identifiers** are case-sensitive.

3.6.3 Python Escape Sequence

Escape sequences are used to print special symbols in strings. Here, it may be stated that one cannot directly print some characters in a string. For example, to print **"Hi How\s Mary?"**, the following string is used.

print("Hi How\\s Mary ?")

Note that to print "\" we need to precede "\" with another "\". Similarly, to print a string in two lines we need "\n" at the appropriate place. For example,

print("HI there!\nHow are you")

prints "HI there!" and "How are you in two different lines. The "\t" is used for inserting a tab in a given string. Table 3.2 shows the various **escape sequences** in **Python**.

TABLE 3.2 Escape sequences in **Python**.

Escape Sequence	Description
\\	Prints Backslash
\'	Prints Single Quote
\"	Prints Double Quotes
\n	Next Line
\t	Inserts Tab
\b	Removes one previous character
\a	Produces alert sound

3.7 STATEMENTS

Statements, in **Python**, are the instructions that can be interpreted and executed by the **Python interpreter**. Statements could be both single-line and multiline. We use "\" at the end of each statement to achieve a multiline statement.

3.7.1 Expression Statement

Expression statements are used to compute and write a value, or to call a function. For example, the following statement adds the values contained in variables **num1** and **num2** and stores the result in **num3**.

$$num3 = num1 + num2$$

3.7.2 Assignment Statements

Assignment statements bind names to values and modify the items of objects in which mutation is allowed. The first statement, for example, assigns the value stored in **num2** to **num1,** and the second statement assigns **4** to the second element of a list **L**:

$$num1 = num2$$
$$L[1] = 4$$

At this point, the above statements may not make much sense, but after studying variables and operators, the above statements would become clear.

3.7.3 The Assert Statements

The **assert statements** help us to insert debugging assertions into a program. These are used in debugging.

```
<something> = "assert" expression [" " ,expression]
```

3.7.4 The Pass Statements

The *pass* is a null operation. Actually, it does not perform any task, when executed. It is beneficial as a placeholder when a statement is obligatory but no code needs to be executed.

3.7.5 The Control Statements

The **control statements** are used to incorporate branching, looping, etc. in a program. The next two chapters focus on the control statements.

3.8 COMMENTS

At times, a programmer-readable explanation of a statement or a part of a code or a module is required. The programmer may want to write some description about that part, which is visible to him but not to the compiler or interpreter. These explanations are called **comments**. At times the document generators may also generate comments. The **comments** make the source code easy to understand. It may be noted that how a comment is written varies in different languages. In fact, the way of writing a **comment** is a part of the programming style. It may also be stated that unnecessary comments are undesirable.

Comments may be single-line comments or may span multiple lines. In **Python**, a single-line **comment** is preceded by a "#". The following code (Code 1) shows the usage of a single-line **comment**. Note that all the single-line **comments** are preceded by a "#". The rest of the code will become clear as we proceed with the course. The **multiple-line comment** may be written as a docstring. Code 2 displays the usage of multiple-line comments. The multiple-line comments are enclosed in """ """.

Code 1:

```
#Defining Student class
class Student():
    #Bound Method
    def display(self,something):
        print('\n',something)
#Instantiating Student
Hari=Student()
#Calling method
Hari.display('Hi I am Hari')
Student().display('Calling display again')
```

Code 2:

```
#Defining Student class
class Student():
    #Bound method
    def display(self,something):
    print('\n',something)
    #Another bound method
    def getdata(name,age):
    name=name
    age=age
    print('Name',name,'Age',age)
```

""" The following code instantiates the class Hari
Asks the user to enter the name and age of the Student.
This is followed by calling the getdata """

```
Hari=Student()
name=input('Enter the name of the student\t:')
age=int(input('Enter the age of the student\t:'))
Student.getdata(name,age)
```

3.9   OPERATORS

Python provides many operators, like arithmetic, comparison, assignment, binary logical, membership, and identity operators. Table 3.3 presents various operators, their usage, and their meaning. The usage of these operators has been demonstrated in the codes that follow. A briefing regarding the priority of operators follows this. Note that the code that follows **num1** and **num2** contains some values.

TABLE 3.3 Operators in **Python**.

Operator	Usage	Meaning
Comparison operators		
==	num1==num2	**True** if **num1** is equal to **num2**, else **False**
!=	num1!=num2	**True** if **num1** is not equal to **num2**, else **False**
<	num1<num2	**True** of **num1** is less than **num2**
>	num1>num2	**True** of **num1** is greater than **num2**
<=	num1<=num2	**True** of **num1** is less than or equal to **num2**
>=	num1>=num2	**True** of **num1** is greater than or equal to **num2**
Assignment operators		
+=	num1+=num2	num1=num1+num2
-=	num1-=num2	num1=num1-num2
*=	num1*=num2	num1=num1*num2
**=	num1**=num2	num1=num1**num2
//=	num1//=num2	num1=num1//num2
Binary operators		
&	num1&num2	Binary and
|	num1|num2	Binary or
^	num1^num2	Binary xor
~	~num1	Binary not
Membership operators		
in	x in L	**True** if **x** is present in **L** (L can be a list, tuple, string, etc.)
not in	x not in L	**True** if **x** is not present in **L**
Identity operators		
Is	a is b	**True** if **id(a)** is same as **id(b)**
is not	a is not b	**True** if **id(a)** is not same **as id(b)**

3.10 TYPES AND EXAMPLES OF OPERATORS

3.10.1 Arithmetic Operators

Python provides standard arithmetic operators for addition, subtraction, multiplication, division, modulo, and power. Table 3.4 shows the operators and their functions. The following code demonstrates the use of these operators with two integers entered by the user.

TABLE 3.4 Arithmetic operators in **Python**.

Operator	Function
+	Addition
–	Subtraction
*	Multiplication
/	Division
%	Modulo
**	Power

Code:

```
a=int(input('Enter the first number\t:'))
b=int(input('Enter the second number\t:'))
sum=a+b
prod=a*b
diff=a-b
mod=a%b
q=a/b
print(sum,' ',prod,' ',diff,' ',mod,' ',q)
r=a**b
print(r)
```

Output:

```
Enter the first number :45
Enter the second number :7
52 315 38 3 6.428571428571429
#Basic Operations in Python
373669453125
```

Code:

```
f1=float(input('Enter the first number\t:'))
f2=float(input('Enter the second number\t:'))
sum=f1+f2
prod=f1*f2
diff=f1-f2
```

```
mod=f1%f2
q=f1/f2
print(sum,' ',prod,' ',diff,' ',mod,' ',q)
```

Output:

```
Enter the first number :45.2
Enter the second number :5.32
50.52 240.46400000000003 39.88 2.6400000000000006
    8.496240601503759
```

3.10.2 String Operators

Python provides two operators for string manipulation: + and *. For strings, the + operator concatenates two strings and the * operator repeats the string n times. The code that follows demonstrates the use of these operators for strings.

Code:

```
str1=input('Enter the first string')
str2=input('Enter the second string')
str3=str1+str2
print(str3)
```

Output:

```
Enter the first string nikhil
Enter the second string miglani
Nikhilmiglani
Code:
str1*3
```

Output:

```
'nikhilnikhilnikhil'
```

3.10.3 Comparison Operators

The **comparison operator** is used to compare the values of two objects. It returns **True** if the values of the objects are the same and **False** if they are different. The following code illustrates the usage of the comparison operators. The following code creates two variables **num1** and **num2**. The value stored in **num1** is 5, and that stored in **num2** is **3**. The statement **num1==num2** results in a **False**, as the numbers are not equal. The statement **num1!=num2** results in a **True**, as the numbers are not equal. The statement **num1<num2** results in a **False**, as the first number is not less than the second number. The statement **num1>num2** results in a **True** as the first number exceeds the second. Likewise, the statement **num1<=num2** results in a **False**, as the first number is greater than the second number, and the statement **num1>=num2** results in a **True** as the first number is greater than the second number. The output follows.

Code:

```
num1=5
num2=3
print(num1==num2)
print(num1!=num2)
print(num1<num2)
print(num1>num2)
print(num1<=num2)
print(num1>=num2)
```

Output:

```
False
True
False
True
False
True
```

3.10.4 Assignment Operators

The following code illustrates the usage of the assignment operators. The following code creates two variables **num1** and **num2**. The value stored in **num1** is 5 and that stored in **num2** is 3. The statement **num1+=num2** assigns 8 (i.e., 5+3) to **num1** and **num2** remains unchanged. The statement **num1*=num2** assigns 24 (i.e., 8*3) to **num1** and **num2** remains unchanged. The statement **num1//=num2** (i.e., 24//3) assigns 8 to **num1** and **num2** remains unchanged. The statement **num1**=num2** (i.e., 8**3) assigns 512 to **num1** and **num2** remains unchanged. Finally, the statement **num1-=num2** (i.e., 512-3) assigns 509 to **num1,** and **num2** remains unchanged. The output follows.

Code:

```
num1=5
num2=3
num1+=num2 #num1= num1+num2=8
print(num1)
num1*=num2 #num1 = num1*num2=8*3=24
print(num1)
num1//=num2 #num1=num1//num2=24//3=8
print(num1)
num1**=num2 #num1=num1**num2=512
print(num1)
num1-=num2 #num1=num1-num2=512-3=509
print(num1)
```

Output:

```
8
24
8
512
509
```

3.10.5 Logical Operators

The logical operators like **or** and **and** have been discussed in detail in the chapter on Boolean Algebra. The **or** operator returns **TRUE** if any of the inputs are **TRUE**. The **and** operator returns a **TRUE** if both the inputs are **TRUE**. The **not** operator negates the value. The **xor** operator returns a **TRUE** if the inputs are alternate. The truth tables of the various operators follow (Tables 3.5 to 3.8). The following code illustrates the usage of binary logical operators. The following code creates two variables **num1** and **num2**. The value stored in **num1** is 0b101101 (note that the preceding value by 0b results in a binary number), and that stored in **num2** is 0b110110. The statement **num3=num1 & num2** assigns 0b100100 to **num3**. The statement **num1 ^ num2** takes the bitwise **xor** of **num1** and **num2** and stores the resultant number in **num5**. The statement **~num1** takes the bitwise **not** of **num1** and stores the resultant number in **num6**.

TABLE 3.5 Truth table of **AND**.

A	B	A and B
False	False	False
False	True	False
True	False	False
True	True	True

TABLE 3.6 Truth table of **OR**.

A	B	A or B
False	False	False
False	True	True
True	False	True
True	True	True

TABLE 3.7 Truth table of **NOT**.

A	not A
False	True
True	False

TABLE 3.8 Truth table of **XOR**.

A	B	A xor B
False	False	False
False	True	True
True	False	True
True	True	False

Code:

```
num1= 0b101101
num2= 0b110110
num3= num1&num2
num4= num1|num2
num5= num1^num2
num6= ~num1
print(bin(num3))
print(bin(num4))
print(bin(num5))
print(bin(num6))
```

Output:

```
0b100100
0b111111
0b11011
-0b101110
```

3.10.6 Priority of Operators

As stated in the above table, the operators have the following priorities. The ** operator has the highest priority, followed by the unary ~, +, −. The operators *, /, %, ‖ follow. The priorities of binary +, - are higher than the >>, <<, &, ^, |, <=, <, >, >=, <>, ==,!=, =, !=, is, not is, in, not in, not, or, and. Figure 3.10 presents the priority of the operators.

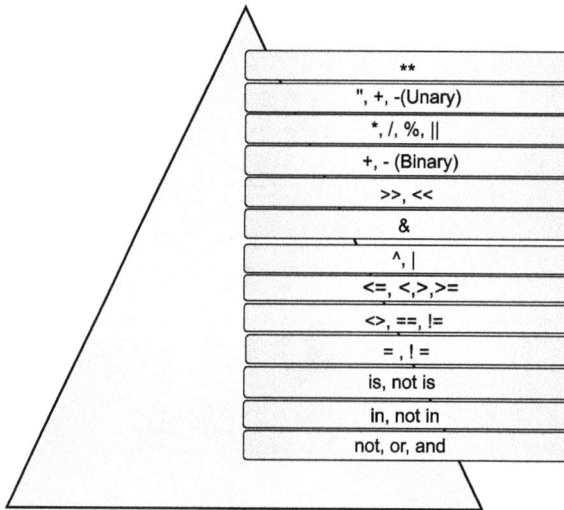

| ** |
| ", +, -(Unary) |
| *, /, %, ‖ |
| +, - (Binary) |
| >>, << |
| & |
| ^, | |
| <=, <,>,>= |
| <>, ==, != |
| = , ! = |
| is, not is |
| in, not in |
| not, or, and |

FIGURE 3.10 Priority of operators.

3.11 BASIC DATA TYPES

This section discusses the basic data types in **Python**. The following codes also illustrate the use of operators with these data types.

3.11.1 Integer

As stated earlier, the type of variable need not be specified in **Python**. So, to create an integer-type variable, one can simply assign an integer value to an object. For example, in the following snippet, **i** has been assigned the value **10**. Note that the type of a variable can be determined by using the **type()** function. Here, the type of **i** has been stored in a variable called **t**. The value of 't' <class, 'int'> is displayed n printing.

Code:

```
i=10
print('The value of i is ',i)
t=type(i)
print('The type of i is ',t)
```

Output:

```
The value of i is 10
The type of i is <class 'int'>
```

#Python automatically finds the data type of variable. We do not need to mention the type of variable.

One can use the **input()** function to take input from the user. The **input()** function takes a string as an argument displayed on the screen. The input of the user is taken as a string. To convert it into an integer, the **int()** function is used. In the snippet that follows, the number entered by the user is converted into an integer, and then a variable called **num** points to the location that stores the number.

Code:

```
num=int (input ('Enter a number\t:'))
print ('You have entered\t:', num)
```

Output:

```
Enter a number :34
You have entered: 34
```

3.11.2 Float

One can simply assign a float value to an object to create a float type variable. For example, in the following snippet, *i* has been assigned the value 2.35768.

Note that the type of a variable can be determined by using the **type()** function. Here, the type of *i* has been stored in a variable called **t**. The value of **t**<class, 'float'> is displayed on printing.

Code:

```
i=2.35768
print ('The value of i is ', i)
t=type(i)
print ('The type of i is ', t)
```

Output:

```
The value of i is 2.35768
The type of i is <class 'float'>
```

To convert a string into a **float,** the **float()** function is used. In the snippet that follows, the number entered by the user is converted into a **float,** and then a variable called **num** points to the location that stores the number.

Code:

```
num=float(input('Enter a number\t:'))
print('You have entered\t:',num)
```

Output:

```
Enter a number :34.567
You have entered : 34.567
```

3.11.3 String

To create a **string**-type variable, one can simply assign a string value to an object. For example, in the following snippet, **i** has been assigned the value "Nikhil." Note that the type of a variable can be determined by using the **type()** function. Here, the type of **i** has been stored in a variable called **t**. The value of **t** <class, 'str'> is displayed on printing.

Code:

```
i='Nikhil'
print('The value of i is ',i)
t=type(i)
print('The type of i is ',t)
```

Output:

```
The value of i is Nikhil
The type of i is <class 'str'>
```

As stated earlier, the **input()** function takes a string as input; hence, no further conversion is required, like in the case of an integer or a float. In the snippet that follows, a string entered by the user and a variable called **num** points to the location that stores the string.

Code:

```
name=input('Enter your name\t:')
print('Hi ',name)
```

Output:

```
Enter your name: Harsh
Hi Harsh
```

3.12 CONCLUSION

The C compiler converts the code into an object file. The steps in this conversion include lexical analysis, semantic and syntactic analysis, intermediate code generation, optimization, and final code generation. The machine-specific optimization renders the code un-portable. In the case of **Python**, the code is converted into **Python bytecode**. One can run a **Python** code using the **Command Prompt, Jupyter notebook, IDLE,** and using many other Integrated Development Environments (IDE's). This chapter presents some of the ways to run the code in **Python**.

This chapter also discusses the **input** and **print** functions. A program needs input. For taking a string-type input from the user, the **input()** function is used. For taking an integer-type input from the user, **int(input())** function is used, and for taking a float type input from the user, **float(input())** function is used. These inputs are then operated as per the requirement of the problem. For this, we need operators, and finally, the program's output is printed.

This chapter explained the above components in detail, forming the basis of the following chapters. The reader is advised to complete the exercise to understand better the topics discussed in the chapter.

EXERCISES

Multiple Choice Questions

1. In **Python,** what can be executed?
   (a) Single instruction   (b) Script
   (c) Full source code   (d) All of the above

2. Which of the following is used to display output on the screen?
   (a) print   (b) printf
   (c) WriteLine   (d) All of the above

3. In **Python3, print** is a
   (a) Function   (b) Command
   (c) None of the above

4. Which of the following supports value-type variables?
   (a) C#   (b) Python
   (c) Both   (d) None of the above

5. In which of the following variables need not be declared first before use?
   (a) C   (b) C++
   (c) C#   (d) Python

6. Which one of the following is not an operator in Python?
   (a) ++   (b) +=
   (c) -=   (d) None of the above

7. Which operator is used for integer division in Python?
   (a) /   (b) //
   (c) %   (d) None of the above

8. Which operator is used for finding the remainder?
   (a) /   (b) //
   (c) %   (d) None of the above

9. Which operator is used for comparing two numbers?

   (a) == **(b)** =

   (c) += **(d)** None of the above

10. Which operator is used for calculating power?

    (a) * **(b)** **

    (c) // **(d)** None of the above

11. With respect to the strings, what is the meaning of + operator?

    (a) Addition **(b)** Concatenation

    (c) Exception is raised **(d)** None of the above

12. Concerning strings, what is the meaning of * operator?

    (a) Multiplication **(b)** To print a string multiple times

    (c) Exception is raised **(d)** None of the above

13. In Python, which operator is used for performing logical AND?

    (a) && **(b)** &

    (c) ^ **(d)** None of the above

14. In Python, which operator is used for performing logical OR?

    (a) || **(b)** |

    (c) or **(d)** Both (*a*) and (*b*)

15. Which operator is used for finding if an element is in the list?

    (a) in **(b)** is

    (c) not in **(d)** not is

Theory

1. Is **Python** an interpreted language? Write arguments in support of your answer.

2. Write various ways of running a **Python** program in Windows.

3. Compare **Anaconda** and **IDLE**. (*Explore).

4. Explain how to run a **Python** script using a **Command Window**.

5. State the features of **Jupyter**.

6. Can we execute a single command in **Python**?

7. Explain the difference between a variable in **C** Language and **Python**.

8. Define the following:

   (i) Keyword

   (ii) Identifier

   (iii) Escape Sequence

   (iv) Expression Statement

   (v) Assignment Statement

9. What are the comments in **Python**? What are the various types of comments?

10. What are arithmetic operators in **Python**? Explain all the different types of arithmetic operators available in **Python**.

11. What are string operators in **Python**? Explain all the different types of string operators available in **Python**.

12. What are assignment operators in **Python**? Explain all the different types of assignment operators available in **Python**.

13. What are the logical operators in **Python**? Explain all the different types of logical operators available in **Python**.

14. How will you input an integer in **Python**?

15. How will you input a float in **Python**?

16. How will you input a string in **Python**?

Explore

- Steps to install Python in **MAC**.
- Features of **Jupyter Lab**.
- Features of **Python** Interpreter.
- How to restart the kernel in **Jupyter** and clear all outputs.

SECTION II
PROCEDURAL PROGRAMMING

This section deals with Python Objects, basic data types, and procedural programming elements. The section has eight chapters. The next two chapters introduce the reader to conditional statements and looping. The next chapter discusses functions and recursion. Chapters 7 and 8 discuss the most important topics namely list, tuple, dictionaries, iterators, and comprehensions. Chapter 9 deals with strings and Chapter 10 discuss File handling.

4

CONDITIONAL STATEMENTS

Objectives

After reading this chapter, the reader should be able to

- Use conditional statements in programs
- Appreciate the importance of the **if-else** construct
- Use the **if-elif-else** ladder
- Use the ternary operator
- Understand the importance of **&** and |

4.1 INTRODUCTION

The preceding chapters discussed the basic data types and simple statements in Python. The concepts studied so far are good for the execution of a program that has no branches. However, a programmer seldom finds a problem-solving approach devoid of branches.

Before proceeding any further, let us spend some time contemplating life. Can you move forward in life without making decisions? The answer is "NO." In the same way, the problem-solving approach would not yield results until the power of decision-making is incorporated. This is the reason why one must understand how to implement the process of decision-making and looping. This chapter describes the first concept. This is needed to craft a program that has branches. "Decision-making" empowers us to change the control flow of the program. In C, C++, JAVA, C#, etc., there are two major ways to accomplish the above task. The first is the **"if"** construct and the other is **"switch."** The **"if"** block in a program is executed if the **"test"** condition is true; otherwise, it is not executed. **Switch** is used to implement a scenario in which there are many **"test"** conditions, and the corresponding block executes in case a particular test condition is true.

This chapter introduces the concept of conditional statement, **if-elif** ladder, and finally, the **get** statement. The chapter assumes importance as conditional statements are used in every aspect of programming, be it client-side development, web development, or mobile application development.

The chapter has been organized as follows. The second section introduces the **"if"** construct. Section 4.3 introduces **"if-elif"** ladder. Section 4.4 discusses the use of logical operators. Section 4.5 introduces the ternary operator. Section 4.6 presents the **get** statement, and the last section concludes.

4.2 "IF," IF-ELSE, AND IF-ELIF-ELSE CONSTRUCTS

Implementing decision-making gives the power to incorporate branching in a program. As stated earlier, program is a set of instructions given to a computer. To accomplish the given task and most of them will require making decisions. So, conditional statements form an integral part of programming. The syntax of the construct is as follows.

General format

1. if

```
if <test condition>:
    <block if the test coniton is true>
```

2. if-else

```
if <test condition>:
    <block if the test condition is true>
else:
    <block if the test condition is not true>
...
```

3. If else ladder (discussed in the next section)

```
if      <test condition>:
        <block if the test condition is true>
elif    <test 2>:
        <second block>
```

```
elif     <test 3>:
         <third block>
else:
         <block if the test condition is true>
```

Note that, indentation is important, as Python recognizes a block through indentation. So, make sure that the **"if (<condition>)":** is followed by a block, each statement of which is at the same alignment. In order to understand the concept, let us consider a simple example. A student generally clears a university exam in India if he scores more than 40 percent. In order to implement the logic, the user is asked to enter the value of percentage. If the percentage entered is more than 40, then "Exam cleared" is printed; otherwise, "Failed" is printed. The following flowchart depicts the procedure for declaring the result when a user enters percentage (Figure 4.1).

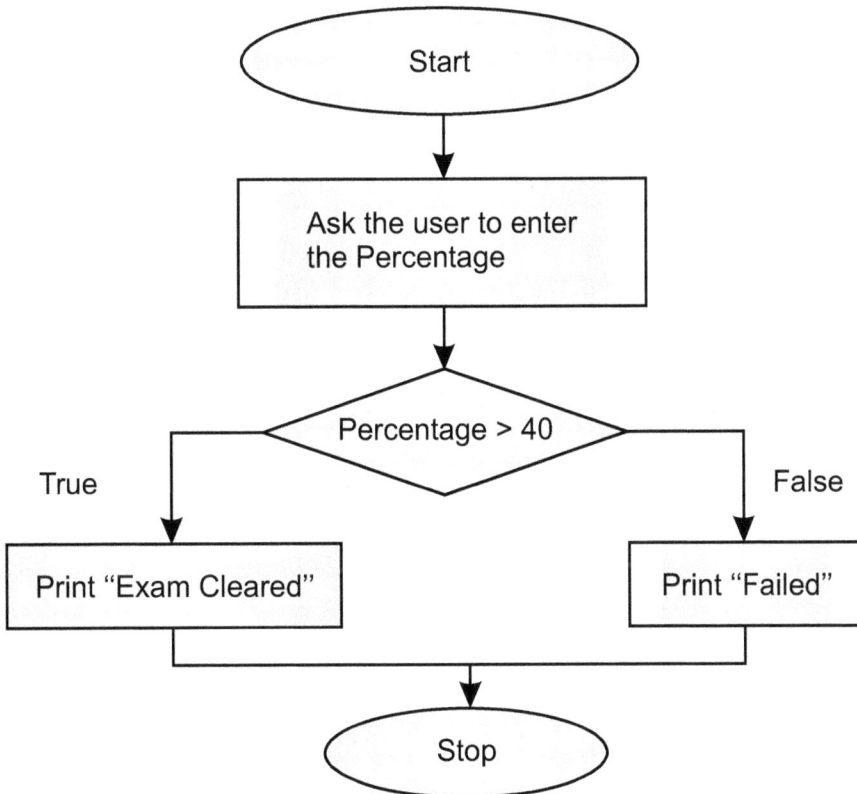

FIGURE 4.1 Flowchart for Example 1.

Illustration 4.1:

Ask the user to enter the marks, of a student, in a subject. If the marks entered are greater than 40, then print "pass", else print "fail".

Program:

```
a = input ("Enter marks : ")
if int(a)> 40:
    print('Pass')
else:
    print('Fail')
...
```

Output 1: Enter Marks : 50

```
Pass
```

Output 2: Enter Marks : 30

```
Fail
```

Let us have a look at another example. In the problem, the user is asked to enter a three-digit number, to find the number obtained by reversing the order of the digits of the number; then find the sum of the number and that obtained by reversing the order of the digits and finally, find whether this sum contains any digit in the original number. In order to accomplish the task, the following steps (presented in Illustration 4.2) must be carried out.

Illustration 4.2:

Ask the user to enter a three-digit number. Call it "num." Find the number obtained by reversing the order of the digits. Find the sum of the given number and that obtained by reversing the order of the digits. Finally, find if any digit in the sum obtained is same as that in the original number.

Solution:

The problem can be solved as follows.

- When the user enters a number, check whether it is between 100 and 999, both inclusive.

▨ Find the digits at units, tens, and hundreds place. Call them "*u*," "*t*," and "*h*," respectively.

▨ Find the number obtained by reversing the order of the digits (say, "rev") using the following formula.

- Number obtained by reversing the order of the digits, rev = h + t × 10 + *u* × 100

▨ Find the sum of the two numbers.

$$Sum = rev + num$$

▨ The sum may be a three-digit or a four-digit number. In any case, find the digits of this sum. Call them "u1," "t1," "h1," and "th1" (if required).

▨ Set "flag = 0."

▨ Check the following condition. If anyone is true make flag = 1. If "sum" is a three-digit number

$$u == u1$$

$$u == t1$$

$$u == h1$$

$$t == u1$$

$$t == h1$$

$$t == h1$$

$$h == u1$$

$$h == t1$$

$$h == h1$$

▨ If "sum" is a four-digit number, the above conditions need to be checked along with the following conditions.

$$u = = th1$$

$$h = = th1$$

$$t = = th1$$

▨ The above conditions would henceforth be referred to as "set 1." If the value of "flag" is 1, then print "true" else print "false."

▨ The process has been depicted in Figure 4.2.

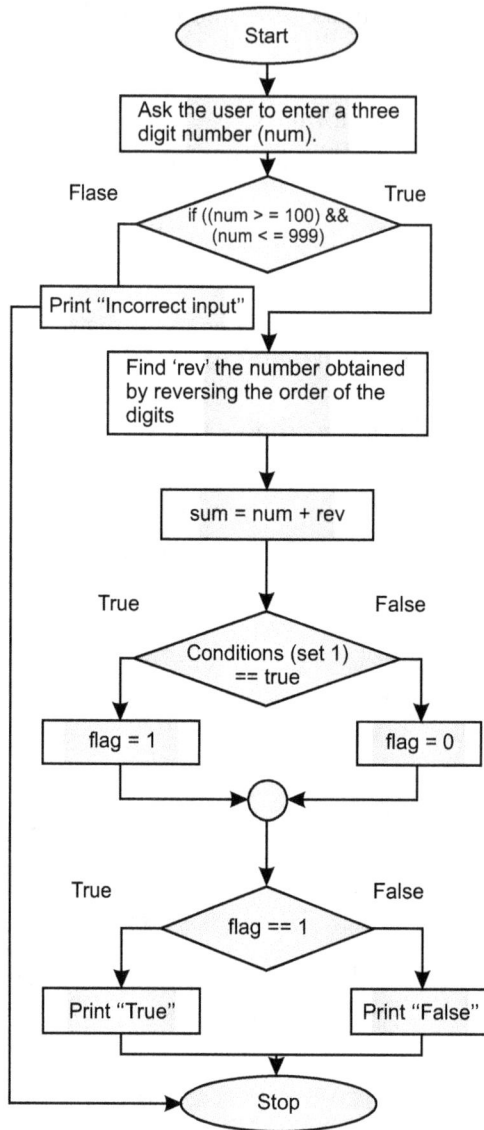

FIGURE 4.2 Flowchart for Illustration 4.2.

Program:

```
num=int(input('Enter a three digit number\t:'))
if ((num<100) | (num>999)):
    print('You have not entered a number between 100 and 999')
```

```
else:
    flag=0
    o=num%10
    t=int(num/10)%10
    h=int(num/100)%10
    print('o\t:',str(o),'t\t:',str(t),'h\t:',str(h))
    rev=h+t*10+o*100
    print('Number obtained by reversing the order of the
    digits\t:',str(rev))
    sum1=num+rev
    print('Sum of the number and that obtained by reversing the
    order of digits\t:',str(sum1))
    if sum1<1000:
        o1=sum1%10
        t1=int(sum1/10)%10
        h1=int(sum1/100)%10
        print('o1\t:',str(o1),'t1\t:',str(t1),'h1\t:',str(h1))
        if ((o==o1)|(o==t1)|(o==h1)|(t==o1)|(t==t1)|(t==h1)|(h==o
1)|(h==t1)|(h==h1)):
            print('Condition true')
            flag==1
    else:
        o1=sum1%10
        t1=int(sum1/10)%10
        h1=int(sum1/100)%10
        th1=int(sum1/1000)%10
        print('o1\t:',str(o1),'t1\t:',str(t1),'h1\
t:',str(h1),'t1\t:',str(t1))
        if
        ((o==o1)|(o==t1)|(o==h1)|(o==th1)|(t==o1)|(t==t1)|(t==h1)
|(t==th1)|(h==o1)|(h==t1)|(h==h1)|(h==th1)):
            print('Condition true')
            flag==1
```

Output: First run

```
Enter a three digit number :4
You have not entered a number between 100 and 999
>>>
```

Output: Second run

```
Enter a three digit number :343
o  : 3 t  : 4 h   : 3
Number obtained by reversing the order of the digits  : 343
No digit of the sum is same as the original number
>>>
```

Output: Third run

```
Enter a three digit number :435
o  : 5 t  : 3 h   : 4
Number obtained by reversing the order of the digits  : 534
No digit of the sum is same as the original number
>>>
```

Output: Fourth run

```
Enter a three digit number :121
o  : 1 t  : 2 h   : 1
Number obtained by reversing the order of the digits  : 121
Sum of the number and that obtained by reversing the order of
    digits: 242
o1 : 2 t1  : 4 h1 : 2
Condition true
>>>
```

.

TIP!   *One must be careful as regards the indentation, failing which the program would not compile. The indentation decides the beginning and end of a particular block in Python. It is advisable not to use a combination of spaces and tabs in indentation. Many versions, of Python, may treat this as a syntax error.*

The **if-elif** ladder can also be implemented using the get statement in case of dictionaries, explained later in the chapter. The important points as regards the conditional statements in Python are as follows.

▪ The **if <test>** is followed by a colon.
▪ There is no need of parenthesis for the test condition. Though, enclosing test in parenthesis would not result in an error.

- The nested blocks in Python are determined by indentation. Therefore, proper indentation in Python is essential. As a matter of fact, an inconsistent indentation or no indentation would result in errors.
- An **if** can have any number of **if**'s nested within.
- The test condition in **if** must result in a **True** or a **False**.

Illustration 4.3:

Write a program to find the greatest of the three numbers entered by the user.

Solution:

First of all, ask the user to enter three numbers (say **num1**, **num2**, and **num3**). This is followed by the condition checking (as depicted in the following program). Finally, the greatest number is displayed.

Program:

```
num1 = input('Enter the first number\t:')
num2 = input('Enter the second number\t:')
num3 = input('Enter the third number\t:')
if int(num1)> int(num2):
    if int(num1) > int(num3):
        big= int(num1)
    else:
        big = int(num2)
 else:
    if int(num2)> int(num3):
        big= num2
    else:
        big = num3

print(big)
```

4.3 THE IF-ELIF-ELSE LADDER

If there are multiple conditions and the outcomes decide the action, then **if-elif-else** ladder can be used. This section discusses the construct and presents the concept using appropriate examples. The syntax of this construct is as follows.

Syntax

```
if <test condition 1>:
    # The task to be performed if the condition 1 is true
elif <test condition 2>:
    # The task to be performed if the condition 2 is true
elif <test condition 3>:
    # The task to be performed if the condition 1 is true
else:
    # The task to be performed if none of the above condition is
    true
```

The flow of the program can be managed using the above construct. Figure 4.3 shows the diagram depicting the program's flow using the above constructs. In the figure, the left edge depicts the scenario where condition C1 is true, and the right edge depict the scenario where the condition is false. In the second graph, conditions C1, C2, C3, and C4 lead to different paths [*Programming in C#*, Harsh Bhasin, 2014].

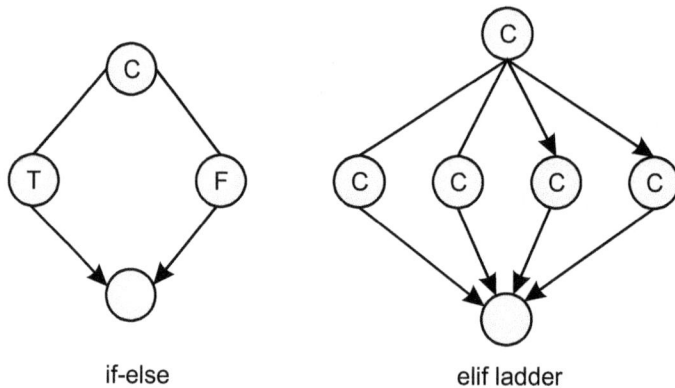

if-else   elif ladder

FIGURE 4.3 The flow graph of if and elif ladder.

The following section has programs that depict the use of **elif** ladder. It may be noted that if there are multiple **if** statements, then the **else** is taken along with the nearest **if**.

4.4 LOGICAL OPERATORS

In many cases, the execution of a block depends on the truth value of more than one statement. In such cases, the operators "**and**" ("**&**") and "**or**" ("**|**") come to our rescue. The first ("**and**") is used when the output is "**true**" when both conditions are "**true**." The second ("**or**") is used if the output is "**true**," if any of the condition is "**true**."

The truth table of "**and**" and "**or**" is given as follows. In the following tables, "**T**" stands for "**true**," and "**F**" stands for "**false**."

TABLE 4.1 **Truth table of a&b.**

a	b	a & b
T	T	T
T	F	F
F	T	F
F	F	F

TABLE 4.2 **Truth table of a|b.**

a	b	a|b
T	T	T
T	F	T
F	T	T
F	F	F

The above statements help the programmer to easily handle compound statement. As an example, consider a program to find the greatest of the three numbers entered by the user. The numbers entered by the user are (say) "**a**," "**b**," and "**c**," then "**a**" is greatest if (**a>b**) and (**a>c**). This can be written as follows:

```
if((a>b)&(a>c))
        print('The value of a greatest')
```

In the same way, the condition of "**b**" being greatest can be crafted. Another example can be that of a triangle. If all three sides of a triangle are equal, then it is an equilateral triangle. This condition can be stated as follows.

```
if((a==b)||(b==c))
        //The triangle is equilateral;
```

4.5 THE TERNARY OPERATOR

The conditional statements explained in the above section are immensely important for writing any program containing conditions. However, the code can still be reduced by using the ternary statements provided by Python. The ternary operator performs the same task as the if-else construct. However, it has the same disadvantage as in the case of C or C++. The problem is that each part caters to a single statement. The syntax of the statement is given as follows.

Syntax

<Output variable> = < The result when the condition is true > if < condition> else <The result when the condition is not true>

For example, the conditional operator can be used to check which of the two numbers entered by the user is greater.

```
great = a if (a>b) else b
```

Finding the greatest of the three given numbers is a bit intricate. The following statement puts the greatest of the three numbers in "**great**."

```
great = a if (a if (a > b) else c)) else(b if (b > c) else c))
```

The program that finds the greatest of the three numbers, entered by the user, using the ternary operator, is as follows.

Illustration 4.4:

Find the greatest of three numbers entered by the user, using ternary operator.

Program

```
a = int(input('Enter the first number\t:'))
b = int(input('Enter the second number\t:'))
c = int(input('Enter the third number\t:'))
big = (a if (a>c) else c) if (a>b) else (b if (b>c) else c)
print('The greatest of the three numbers is '+str(big))
>>>
```

Output:

```
Enter the first number  :2
Enter the second number    :3
```

```
Enter the third number : 4
The greatest of the three numbers is 4
>>>
```

4.6   THE GET CONSTRUCT

In C or C++ (even in C# and JAVA), switch is used in the case where different conditions lead to different actions. This can also be done using the **'if-elif'** ladder, as explained in the previous sections. The **get** construct greatly eases the task in the case of dictionaries.

In the example that follows, there are three conditions. However, in many situations, there are many more conditions. The **get** construct can be used in such cases. The syntax of the construct is as follows.

Syntax

```
<dictionary name>.get('<value to be searched>', 'default
    value>')
```

Illustration 4.5 demonstrates the use of the get construct.

Illustration 4.5:

This illustration has a directory containing the names of the books and the corresponding year. The statements that follow, finds the year of publication, for a given name. If the name is not found, the string (given as the second argument, in get) "Bad choice" is displayed.

Program

```
hbbooks = {'programming in C#': 2014, 'Algorithms': 2015,
    'Python': 2016}
print(hbbooks.get('Programming in C#', 'Bad Choice'))
print(hbbooks.get('Algorithms', 'Bad Choice'))
print(hbbooks.get('Python', 'Bad Choice'))
print(hbbooks.get('Theory Theory, all the way', 'Bad Choice'))
```

Output:

```
Bad Choice
2015
```

```
2016
Bad Choice
>>>
```

Note that in the first case, the "P" of "Programming" is capital. Hence, "Bad Choice" is displayed. In the second and third cases, the get function can find the requisite value. In the last case, the value is not found; hence, the second argument of the **get** construct appears. Also, note that the later part is similar to the default of the C-type **switch** statement. The flow diagram given in Figure 4.4 shows a program having many branches.

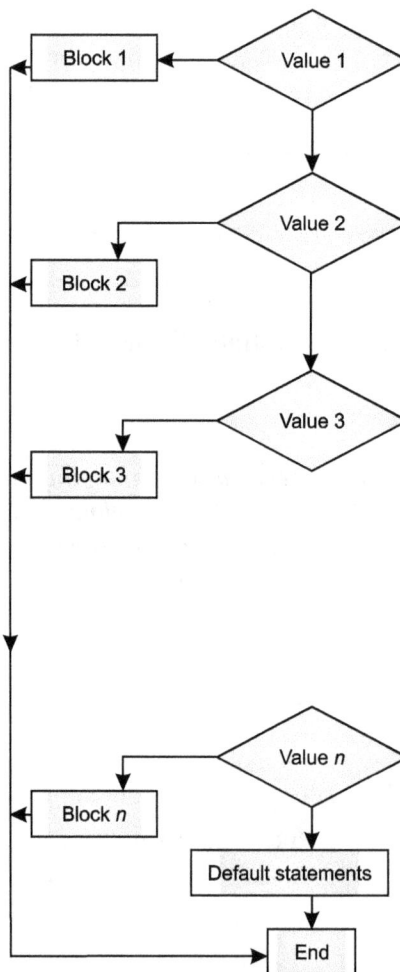

FIGURE 4.4 A program having multiple conditional statements.

Observation

In Python, dictionaries and lists form an integral part of the language basics. The **get** construct implements the concept of conditional selection. Notably, this construct greatly reduces the problems of dealing with situations where mapping is required and hence is important.

4.7 EXAMPLES

The "**if**" condition is also used for input validation. The following program asks the user to enter a character and checks whether its ASCII **value is greater a certain value**.

Illustration 4.6:

Ask the user to enter a number and check whether its ASCII value is greater than 80.

Program:

```
inp = input('Enter a character :')
if ord(inp) > 80:
    print('ASCII value is greater than 80')
else:
    print('ASCII value is less than 80')
```

Output 1:

```
Enter a character: A
ASCII value is less than 80
```

Output 2:

```
Enter a character: Z
ASCII value is greater than 80
```

The construct can also be used to find the value of a multi-valued function. For example, consider the following function.

$$\begin{cases} f(x) = x^2 + 5x + 3, & \text{if } x > 2 \\ \qquad\qquad x + 3, & \text{if } x < 2 \end{cases}$$

The following example asks the user to enter the value of *x* and calculates the value of the function as per the given value of *x*.

Illustration 4.7:

Implement the above function and find the values of the function f(x) above at x = 2 and x = 4.

Program:

```
f(x) = x^2 + 5x + 3 , if x > 2
     = x + 3 , if x <= 2
"""
x = int (input('Enter the value of x\t:'))
if x > 2:
    f = ((pow(x,2)) + (5*x) + 3)
else:
    f = x + 3
print('Value of function f(x) = %d' % f )
```

Output:

```
Enter the value of x   :4
Value of function f(x) = 39
Enter the value of x   :1
Value of function f(x) = 4
```

The "if-else" construct, as stated earlier, can be used to find the outcome based on certain conditions. For example, two lines are parallel if the ratio of the coefficients of *x*'s is the same as that of those of *y*'s. That is, if the equations are

$a_1x + b_1y + c_1 = 0$ and $a_2x + b_2y + c_2 = 0$. Then the condition of the lines being parallel is

$$\frac{a_1}{a_2} = \frac{b_1}{b_2}$$

The following program checks whether two lines are parallel or not.

Illustration 4.8:

Ask the user to enter the coefficients of $a_1x + b_1y + c_1 = 0$ and $a_2x + b_2y + c_2 = 0$ and find whether the two lines depicted by the above equations are parallel or not?

Program:

```
print('Enter Coefficients of the first equation [ a1x + b1y + c1 =
    0 ]\n')
r1 = input('Enter the value of a1: ')
a1 = int (r1)
r1 = input('Enter the value of b1: ')
b1 = int (r1)
r1 = input('Enter the value of c1: ')
c1 = int (r1)
print('Enter Coefficients of second equation [ a2x + b2y + c2 = 0
    ]\n')
r1 = input('Enter the value of a2: ')
a2 = int (r1)
r1 = input('Enter the value of b2: ')
b2 = int (r1)
r1 = input('Enter the value of c2: ')
c2 = int (r1)
if (a1/a2) == (b1/b2):
    print('Lines are parallel')
else:
    print('Lines are not parallel')
```

Output:

```
Enter Coefficients of the first equation [ a1x + b1y + c1 = 0 ]
Enter the value of a1: 2
Enter the value of b1: 3
Enter the value of c1: 4
Enter Coefficients of second equation [ a2x + b2y + c2 = 0 ]
Enter the value of a2: 4
Enter the value of b2: 6
Enter the value of c2: 7
Lines are parallel
>>>
```

The above program can be extended to find whether the lines are intersecting or overlapping. Two lines intersect if the following condition is true.

$a_1x + b_1y + c_1 = 0$ and $a_2x + b_2y + c_2 = 0$. Then the lines intersect if

$$\frac{a_1}{a_2} \neq \frac{b_1}{b_2}$$

And the two lines overlap if

$$\frac{a_1}{a_2} = \frac{b_1}{b_2} = \frac{c_1}{c_2}$$

The following flowchart shows the flow of control of the program (Figure 4.5).

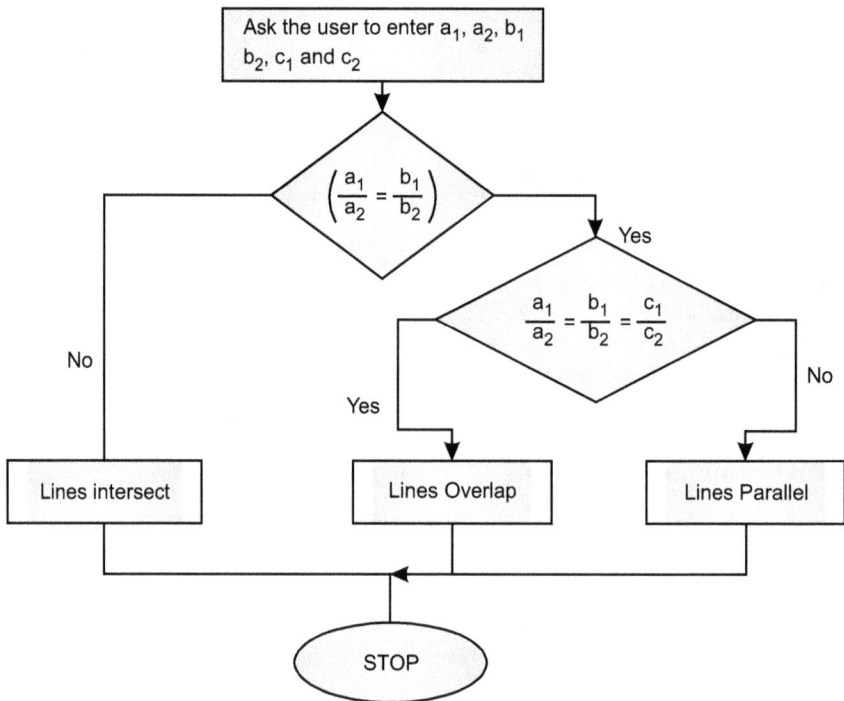

FIGURE 4.5 Checking whether lines are parallel, overlapping, or they intersect.

The following program implements the logic.

Illustration 4.9:

Ask the user to enter the values of a1, a2, b1, b2, c1, and c2 and find whether the lines are parallel, they overlap, or they intersect.

Program:

```
print('Enter Coefficients of the first equation [ a1x + b1y + c1 =
    0 ]\n')
r1 = input('Enter the value of a1: ')
a1 = int (r1)
r1 = input('Enter the value of b1: ')
b1 = int (r1)
r1 = input('Enter the value of c1: ')
c1 = int (r1)
print('Enter Coefficients of second equation [ a2x + b2y + c2 = 0
    ]\n')
r1 = input('Enter the value of a2: ')
a2 = int (r1)
r1 = input('Enter the value of b2: ')
b2 = int (r1)
r1 = input('Enter the value of c2: ')
c2 = int (r1)

if ((a1/a2) == (b1/b2))&((a1/a2)==(c1/c2)):
    print('Lines overlap')
elif (a1/a2)==(b1/b2):
    print('Lenes are parallel')
else:
    print('Lines intersect')
```

Output:

```
Enter Coefficients of the first equation [ a1x + b1y + c1 = 0 ]

Enter the value of a1: 2
Enter the value of b1: 3
Enter the value of c1: 4
```

```
Enter Coefficients of second equation [ a2x + b2y + c2 = 0 ]
Enter the value of a2: 1
Enter the value of b2: 2
Enter the value of c2: 3
Lines intersect
>>>
```

4.8   SUMMARY

As stated in the first chapter, we write a program with some purpose. The purpose to be accomplished by a program generally requires making decisions. This decision-making capacity also empowers a programmer to write code that requires branching. Moreover, many problems, as explained in the chapter, can be easily solved using **"if-else"** constructs. Python greatly reduces the unnecessary clutter, as against C or a C++ program. In Python code, there is hardly a need for braces or for that matter handling obvious conditions. Python also provides us with a switch-like construct to handle multiple conditions. This chapter discusses the basics of conditional statements and presents ample illustrations to clarify things. Conditional statements are used everywhere; from a basic programs to decision support systems and expert systems. The reader is required to go through the points to remember and implement the problems given in the exercise for a better understanding. One must also understand that conditional statements are the first step toward programming. However, understanding conditional statements, though essential, is just the beginning. Your journey of becoming a programmer has just started.

GLOSSARY

1. 'if' construct

```
if    <test condition>:
        <block if the test condition is true>
```

2. 'if else' construct

```
if    <test condition>
        <block if the test condition is true>
else:
        <block if the test condition is not true>
    ...
```

3. 'If else ladder':

```
if    <test condition>:
       <block if the test condition is true>
elif <test 2>:
       <second block>
elif <test 3>:
       <third block>
else:
       <block if the test condition is true>
```

POINTS TO REMEMBER

- The '**if**' statement implements conditional branching.
- The test condition is a Boolean expression that results in a **true** or a **false**.
- The block of "**if**" executes if the test condition is **true**.
- The **else** part executes if the test condition is **false**.
- Multiple branches can be implemented using **if-elif** ladder.
- Any number of **if-else** can be nested.
- A ternary if can be implemented in Python.
- Logical operators can be used in implementing conditional statements.

EXERCISES

Multiple Choice Questions

What will be the output of the following snippets?

1. What will be the output of the following?

```
if 28:
     print('Hi')
else:
     print('Bye')
```

(a) Hi

(b) Bye

(c) None of the above

(d) The above snippet will not compile

2.
```
a = 5
b = 7
c = 9
if a>b:
    if b>c:
        print(b)
    else:
        print(c)
else:
    if b>c:
        print(c)
    else:
        print(b)
```
(a) 7   **(b)** 9

(c) 34   **(d)** None of the following

3.
```
a = 34
b = 7
c = 9
if a>b:
    if b>c:
        print(b)
    else:
        print(c)
else:
    if b>c:
        print(c)
    else:
        print(b)
```
(a) 7   **(b)** 9

(c) None of the above   **(d)** The code will not compile

4.
```
a = int(input('First number\t:'))
b = int(input('Second number\t'))
c = int(input('Third number\t:'))
if ((a>b) & (a>c)):
        print(a)
```

```
elif ((b>a) &(b>c)):
        print(b)
else:
        print(c)
```

(a) The greatest of the three numbers entered by the user

(b) The smallest of the three numbers entered by the user

(c) None

(d) The code will not compile

5. ```
n = int(input('Enter a three digit number\t:'))
if (n%10)==(n/100):
    print('Hi')
else:
    print('Bye')
    # The three digit number entered by the user is 453
```

**(a)** Hi            **(b)** Bye

**(c)** None of the above      **(d)** The code will not compile

6. In the above question if the number entered is 545, what would be the answer?

**(a)** Hi            **(b)** Bye

**(c)** None of the above      **(d)** The code will not compile

7. ```
hb1 = ['Programming in C#','Oxford University Press', 2014]
hb2 = ['Algorithms', 'Oxford University Press', 2015]
if hb1[1]==hb2[1]:
    print('Same')
else:
    print('Different')
```

(a) Same   **(b)** Different

(c) No output   **(d)** The code would not compile

8. ```
hb1 = ['Programming in C#','Oxford University Press', 2014]
hb2 = ['Algorithms', 'Oxford University Press', 2015]
if (hb1[0][3]==hb2[0][3]):
    print('Same')
```
```

```
else:
    print('Different')
```

(a) Same   **(b)** Different

(c) No output   **(d)** The code will not compile

9. In the snippet, given in question 8, the following changes are made. What will be the output?

```
hb1 = ['Programming in C#','Oxford University Press', 2014]
hb2 = ['Algorithms', 'Oxford University Press', 2015]
if (str(hb1[0][3])==str(hb2[0][3])):
    print('Same')
else:
    print('Different')
```

(a) Same   **(b)** Different

(c) No output   **(d)** The code will not compile

10. Finally, the code in question number 8 is changed to the following. What will be the output?

```
hb1 = ['Programming in C#','Oxford University Press', 2014]
hb2 = ['Algorithms', 'Oxford University Press', 2015]
if (char(hb1[0][3])==char(hb2[0][3])):
    print('Same')
else:
    print('Different')
```

(a) Same   **(b)** Different

(c) No output   **(d)** The code will not compile.

Programming Exercises

1. Ask the user to enter a number and find the number obtained by reversing the order of the digits.

2. Ask the user to enter a four-digit number and check whether the sum of the first and the last digits is the same as the sum of the second and the third digit.

3. In the above question, if the answer is true, then obtain a number in which the second and the third digit are one more than that in the given number.

   Example: Number 5342, the sum of the first and the last digit = 7, that of the second and the third digit = 7. New number: 5452

4. Ask the user to enter the concentration of hydrogen ions in a given solution (C) and find the PH of the solution using the following formula.

   $$PH = \log_{10}C$$

5. If the PH is <7, then the solution is deemed as acidic, else it is deemed as basic. Find the solution whose hydrogen ion concentration is entered by the user, is acidic or basic?

6. In the above question, find whether the solution is neutral? (A solution is neutral if its pH is 7)

7. The centripetal force acting on a body (mass m), moving with a velocity v, in a circle of radius r, is given by the formula mv^2/r. The gravitational force on the body is given by the formula $(GmM)/R^2$, where m and M are the masses of the body and Earth, and R is the radius of the Earth. Ask the user to enter the requisite data and find whether the two forces are equal or not.

8. Ask the user to enter his salary and calculate the TADA, which is 10% of the salary; the HRA, which is 20% of the salary and the gross income, which is the sum total of the salary, TADA, and the HRA.

9. In the above question, find whether the net salary is greater than INR 3,00,000.

10. Use the Tax Slab of the current year to find the tax on the above income (question number 8), assuming that the savings are INR 1,00,000.

11. Find whether a number entered by the user is divisible by 3 and 13.

12. Find whether the number entered by the user is a perfect square.

13. Ask the user to enter a string and find the alphanumeric characters from the string.

14. In the above question, find the digits in the strings.

15. In question number 11, find all the components of the string which are not digits or alphabets.

5

LOOPING

Objectives

After reading this chapter, the reader should be able to

- Understand the importance and use of loops
- Appreciate the importance of the **while** and **for**
- Use **range**
- Process list of lists
- Understand nesting of loops

5.1   INTRODUCTION

Consider an example of writing the multiples of a given number (from 1 to 10). Writing this requires writing, say "$n \times$ " followed by "i" (i varying from 1 to n) and then the result of calculation (i.e., $n \times 1$, $n \times 2$, and so on). Many such situations require us to repeat a task multiple times. This repetition can be used to calculate a function's value, print a pattern, or simply repeat something. This chapter discusses loops and iterations, which is an integral part of procedural programming. Looping means repeating a set of statements till a condition is true. The number of times, this set is repeated, depends on the test condition. Also, what must be repeated must be chalked out with due deliberation. In general, repeating a block requires the following (Figure 5.1).

Deciding what is to be repeated: the set of statements

The test condition or the number of times the set of statements is to be repeated

Special cases wherein the loop breaks (or continues, escaping certain statements)

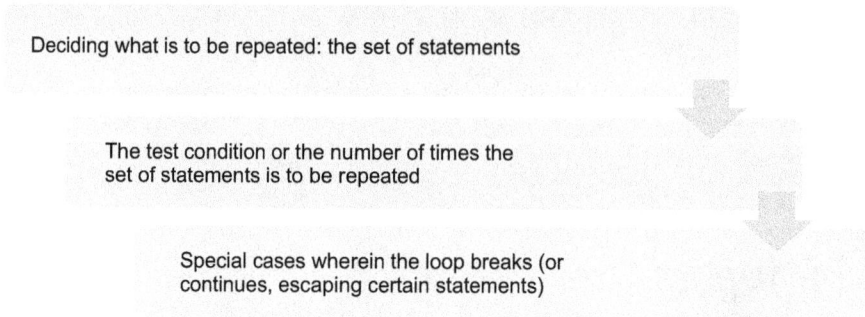

FIGURE 5.1 Looping.

Python provides two types of loops: **for** and **while** (Figure 5.2).

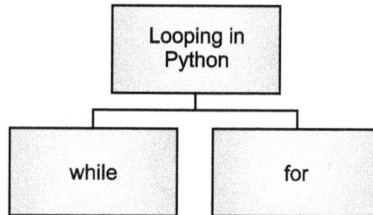

Looping in Python

while

for

FIGURE 5.2 Loops in Python.

While loop is one of the most general constructs in any programming language. If one comes from a C background, he must be equipped with the above construct. **While** loop retains most of its features in Python as well, however, there are notable differences also.

The **while** loop repeats a block, identified by indentation, till the test condition remains true. Then, as we will see in the following discussion, one can come out of the loop using **break** and **continue**. Also, if the loop repeats as per the test condition, the **else** condition executes. This is an additional feature in Python.

The use of **for** in Python is a bit different as compared to C-like languages. The **for** construct, in Python, is generally used for lists, tuples, arrays, etc. The chapter introduces **range,** which would help the programmer to select a value from a given range. The reader is advised to go through the discussion of lists and tuples presented in Chapter 3 of this book, before starting with **for** loop.

The chapter has been organized as follows. Section 5.2 of this chapter presents the basics of **while** loop. Section 5.3 uses looping to create patterns. Section 5.4 introduces the concept of nesting and the processing of lists and tuples using **for** loops. The last section concludes.

5.2   WHILE

In Python, the **while** loop is the most commonly used construct for repeating a task over and over again. The task is repeated till the test condition remains true, after which the loop ends, and if the exit occurs without a **break,** then the **else** part of the construct executes. The syntax of the loop is as follows.

Syntax

```
while test:
        ...
        ...
else:
        ...
```

It may be stated here that the indentation determines the body of the loop. This is the reason why one must be extremely careful in indentation. Also, the **else** part, which is an addition in Python is optional. In order to understand the concept, let us go through the following Illustrations.

Illustration 5.1:

Ask the user to enter a number and calculate its factorial.

Solution:

The factorial of a number n is defined as follows.

$$factorial = 1 \times 2 \times 3 \times \ldots \times n$$

That is, the factorial of a number, **n,** is the product of **n** terms starting from 1. To calculate the factorial of a given number, first of all, the user is asked to input a number. The number is then converted into integer. This is followed by the initialization of **"factorial"** by 1. Then a **while** loop successively multiplies **i** to **"factorial"** (note that after each iteration, the value of

i increases by 1). The following program calculates the factorial of a number entered by the user.

Program:

```
n = input('Enter number whose factorial is required')#ask user
    to enter number
m = int(n)#convert the input to an integer
factorial = 1#initialize
i=1# counter
while i<=m:
    factorial =factorial*i
i=i+1
print('\factorial of '+str(m)+' is '+str(factorial))
```

Output:

```
Enter number whose factorial is required6
Factorial of 6 is 720
```

Illustration 5.2:

Ask the user to enter two numbers "a" and "b" and calculate "a" to the power of "b."

Solution:

"a" raised to the power "b" can be defined as follows.

$$\text{power} = a \times a \times a \times ... \times a$$

$$(b \; times)$$

To calculate the power, first of all, the user is asked to input two numbers. The numbers are then converted into integers. This is followed by the initialization of **"power"** by 1. Then a **while** loop successively multiplies **"a"** to **"power"** (note that after each iteration the value of **i** increases by 1). The following program implements the logic.

Program:

```
a = int(input('Enter the first number'))
b = int(input('Enter the second number'))
```

```
power=1
i = 1
while i<=b:
    power = power*a
    i=i+1
else:
    print(str(a)+'to the power of '+str(b)+' is '+str(power))
```

Output:

```
Enter the first number4
Enter the second number5
4 to the power of 5 is 1024
```

Illustration 5.3:

An Arithmetic Progression is obtained by adding the common difference "d" to the first term "a," successively. The ith term of the Arithmetic progression is given by the following formula.

$$T(i) = a + (i - 1) \times d$$

Ask the user to enter the value of "a," "d," and "n" (the number if terms), and find all the terms of the AP. Also, find the sum of all the terms.

Solution:

The following program asks the user to enter the values of **"a," "d**," and **"n."** The input is then converted into integers.

Since all the terms are to be calculated, this evaluation is done inside a loop. The **"sum"** is initialized to **0** and the terms are added to **"sum,"** in each iteration.

Program:

```
a = int(input('Enter the first term of the Arithmetic
    Progression\t:'))
d = int(input('Enter the common difference\t:'))
n = int(input('Enter the number of terms\t:'))
i = 1
sum = 0#initialize
```

```
while i<=n:
    term = a +(i-1)*d
    print('The '+str(i)+'th term is '+str(term))
    sum = sum + term
    i=i+1
else:
    print('The sum of '+str(n)+' terms is\t:'+str(sum))
```

Output:

```
RUN C:/Users/ACER ASPIRE/AppData/Local/Programs/Python/
    Python35-32/Tools/scripts/AP.py
Enter the first term of the Arithmetic Progression:5
Enter the common difference:6
Enter the number of terms  :7
The 1th term is 5
The 2th term is 11
The 3th term is 17
The 4th term is 23
The 5th term is 29
The 6th term is 35
The 7th term is 41
The sum of 7 terms is  :161
```

Illustration 5.4:

The Geometric Progression is obtained by multiplying the common ratio **"r"** to the first term **"a,"** successively. The ith term of the progression is given by the following formula.

$$T(i) = a \times r^{i-1}$$

Ask the user to enter the value of **"a,"** **"r,"** *and* **"n"** *(the number if terms), and find all the terms of the GP. Also, find the sum of all the terms.*

Solution:

The following program asks the user to enter the values of **"a,"** **"r,"** and **"n."** Since all the terms are to be calculated, this evaluation is done inside a loop. The **"sum"** is initialized to 0 and the terms are added to **"sum,"** in each iteration.

Program:

```
a = int(input('Enter the first term of the Geometric
    Progression\t:'))
r = int(input('Enter the common ratio\t:'))
n = int(input('Enter the number of terms\t:'))
i = 1
sum = 0#initialize
while i<=n:
    term = a * (r**(i-1))
    print('The '+str(i)+'th term is '+str(term))
    sum = sum + term
    i=i+1
else:
    print('The sum of '+str(n)+' terms is\t:'+str(sum))
```

Output:

```
Enter the first term of the Arithmetic Progression:5
Enter the common ratio:3
Enter the number of terms  :5
The 1th term is 5
The 2th term is 15
The 3th term is 45
The 4th term is 135
The 5th term is 405
The sum of 5 terms is  :605
```

5.3   PATTERNS

Have you ever wondered why quizzes and riddles are integral to any intelligence test? The following incident would help the reader to understand the importance of patterns. During the Second World War, the Britons were striving hard to break Enigma, the machine used by the Germans for encrypting their messages. The army somehow got Alan Turing, who was never, in his lifetime recognized, for the above task. He wanted a team to help him, for which he conducted an exam. In the exam, he asked the candidates to solve the given puzzles in a given time. This incident underlines the importance of comprehending patterns. What happens thereafter is a history. Decoding pattern, solving puzzles helps to judge the intellect of a person. This is much more important as compared to learning a formula. This section presents the design of patterns using loops to help the reader understand the concept of nesting. Moreover, this book also intends to inculcate the problem-solving approach in the reader. Therefore, this section becomes all the more important.

The following illustrations show how to assign values to the counters of the inner and the outer loops to carry out the given task. The patterns, as such, may not be very useful. However, doing the following programs would help the reader to comprehend the concept of nesting. Therefore, the methodology of making a pattern has been explained in each of the following illustrations.

Illustration 5.5:

Ask the user to enter the number of rows and write a program to generate the following pattern in Python.

```
*

*  *

*  *  *

*  *  *  *
```

Solution:

The number of rows **n**, would determine the value of the counter (from 0 to **n**). The value of i denote the row number, in the following program. In each row, the number of stars is equal to the row number. The values of **j,** in each iteration, denote the number of stars in each row. This loop is, therefore, nested. Also, note that after the inner loop ends a new line is printed using the **print()** function.

Program:

```
>>>
n = input('Enter the number of rows')
m = int(n)
k=1
for i in range(m):
    for j in range(1, i+2):
    print ('*', end=" ")
    print ()
```

Output:

```
Enter the number of rows 5
*
* *
* * *
* * * *
```

Illustration 5.6

Ask the user to enter the number of rows and write a program to generate the following pattern in Python.

```
1
2 2
3 3 3
4 4 4 4
```

Solution:

The number of rows would determine the value of the counter **i,** (from 0 to **n**). The value of **i** denote the row numbers. In each row, the number of elements is equal to the row number. The values of **j,** in each iteration, denote the number of elements in each row. This loop is, therefore, nested. The value of **i +1** is then printed. Also, note that after the inner loop ends, a new line is printed using the **print()** function.

Program:

```
>>>
n = input('Enter the number of rows')
m = int(n)
k=1
for i in range(m):
    for j in range(1, i+2):
        print(i+1, end=" ")
      print()
```

Output:

```
Enter the number of rows5
1
2 2
```

```
3 3 3
4 4 4 4
5 5 5 5 5
```

Illustration 5.7:

Ask the user to enter the number of rows and write a program to generate the following pattern in Python.

```
1
1 2
1 2 3
1 2 3 4
```

Solution:

The number of rows, entered by the user, would determine the value of **i** (from 0 to **n**). The value of **i** denote the row number. In each row, the number of elements is equal to the row number. The values of **j,** in each iteration, denote the number of elements in each row. This loop is, therefore, nested. The value of **j + 1** is then printed. Also, note that after the inner loop ends, a new line is printed using the **print()** function.

Program:

```
>>>
n = input('Enter the number of rows')
m = int(n)
k=1
for i in range(m):
        for j in range(1, i+2):
            print(j+1, end=" ")
            print()
```

Output:

```
Enter the number of rows5
2
2 3
2 3 4
2 3 4 5
2 3 4 5 6
```

Illustration 5.8:

Ask the user to enter the number of rows and write a program to generate the following pattern in Python.

```
1
2 3
4 5 6
7 8 9 10
```

Solution:

The value of **i** denote the row number in the following program. In each row, the number of elements is equal to the row number. The values of **i,** in each iteration, would denote the number of elements in each row. This loop is, therefore, nested. The value of **k** is then printed, which starts from 1 and increments in each iteration. Also, note that after the inner loop ends, a new line is printed using the **print()** function.

Program:

```
>>>
n = input('Enter the number of rows')
m = int(n)
k=1
for i in range(m):
    for j in range(1, i+2):
        print(k, end=" ")
        k=k+1
    print()
```

Output:

```
Enter the number of rows7
1
2 3
4 5 6
7 8 9 10
11 12 13 14 15
16 17 18 19 20 21
22 23 24 25 26 27 28
```

Illustration 5.9:

Ask the user to enter the number of rows and write a program to generate the following pattern in Python.

```
     *
    ***
   *****
  *******
 *********
***********
```

Solution:

The value of **i** denotes the row number in the following program. In each row, the number of stars is equal to the row number. The values of **k,** in each iteration, denote the number of stars in each row, which ranges from 0 to **(2*i +1).** This loop is, therefore, nested. The leading spaces are governed by the value of **j,** which ranges from 0 to **(m - i -1).** This is because if the value of **i** is 0, the number of spaces should be 4 (if the value of n is 5). In case the value of **i** is 1, the number of spaces should be 3 and so on. Also, note that after the inner loop ends, a new line is printed using the **print()** function.

Program:

```python
.n = input('Enter the number of rows')
m = int(n)
for  i in range(m):
    for j in range(0, (m-i-1)):
        print(' ', end="")
    for k in range(0, 2*i+1):
        print('*',end="")
    print()
```

Output:

```
Enter the number of rows6
     *
    ***
   *****
  *******
 *********
***********
```

5.4 NESTING AND APPLICATIONS OF LOOPS IN LISTS

Nested loops can be used to generate matrices. In order to do this, the inner loop is designed to govern the rows and the outer to govern each element of a particular row. The following Illustration shows the generation of a matrix having **i**[th] element given by the following formula.

$$a_{i,j} = 2 \times (i + j)^2$$

Note that in the following illustration, two loops have been used. The outer loop runs **n** times, where **n** is the number of rows, and the inner loop runs **m** times, where **m** is the number of columns. The number of columns can be perceived as the number of elements in each row.

The inner loop has one statement, which calculates the element. At the end of each iteration (of the outer loop), a new line is printed using the **print()** function.

Illustration 5.10:

Generate a n × m, matrix, wherein each element (a$_{ij}$), is given by

$$a_{i,j} = 2 \times (i + j)^2$$

Program:

```
n = int(input('Enter the number of rows'))
m = int(input('Enter the number of columns'))
for i in range (n):
    for j in range(m):
        element = 5*(i+j)*(i+j)
        print(element, sep=' ', end= ' ')
print()
```

Output:

```
Enter the number of rows3
Enter the number of columns3
0 5 20
5 20 45
20 45 80
```

It may be noted that in the following chapters, this nesting is used to deal with most of the operations of matrices. As a matter of fact, addition and subtraction of two matrices requires two levels of nesting, whereas multiplication of two matrices requires three levels of nesting.

Illustration 5.11:

Handling list of lists: *Note that in the following code the first list's second element is itself a list. It's first element can be accessed by writing hb[0][1] and the first letter of the first element of the nested list would be hb[0][1][0].*

Program:

```
hb=["Programming in C#",["Oxford University Press", 2015]]
rm=["SE is everything",["Obscure Publishers", 2015]]
authors=[hb, rm]
print(authors)
print("List:\n"+str(authors[0])+"\n"+str(authors[1])+"\n")
print("Name of books\n"+str(authors[0][0])+"\n"+str(authors[1]
    [0])+"\n")
print("Details of the books\n"+str(authors[0][1])+"\n"+str(au-
    thors[1][1])+"\n")
print("\nLevel 3 Publisher 1\t:"+str(authors[0][1][0]))
```

Output:

```
[[['Programming in C#', ['Oxford University Press', 2015]], ['SE
    is everything', ['Obscure Publishers', 2015]]]
List:
['Programming in C#', ['Oxford University Press', 2015]]
['SE is everything', ['Obscure Publishers', 2015]]
Name of books
Programming in C#
SE is everything
Details of the books
['Oxford University Press', 2015]
['Obscure Publishers', 2015]
Level 3 Publisher 1    :Oxford University Press
```

The following two Illustrations handle the list of lists using nested loops. Kindly note the output and the corresponding mappings.

Illustration 5.12:

Handling list of lists using loops: *The elements of nested lists can also be dealt with using nested loops as shown in this illustration.*

Program:

```
hb=["Programming in C#",["Oxford University Press", 2015]]
rm=["SE is everything",["Obscure Publishers", 2015]]
authors=[hb, rm]
print(authors)
for i in range(len(authors)):
    for j in range(len(authors[i])):
        print(str(i)+" "+str(j)+" "+str(authors[i][j])+"\n")
    print()
```

Output:

```
[['Programming in C#', ['Oxford University Press', 2015]], ['SE
    is everything', ['Obscure Publishers', 2015]]]
0 0 Programming in C#

0 1 ['Oxford University Press', 2015]
1 0 SE is everything
1 1 ['Obscure Publishers', 2015]
```

Illustration 5.13:

Processing nested lists: *Another Illustration of the use of loops in processing nested lists. The user is expected to observe the output and comprehend it.*

Program:

```
hb=["Programming in C#",["Oxford University Press", 2015]]
rm=["SE is everything",["Obscure Publishers", 2015]]
authors=[hb, rm]
print(authors)
for i in range(len(authors)):
    for j in range(len(authors[i])):
        for k in range(len(authors[i][j])):
```

```
        print(str(i)+" "+str(j)+" "+str(k)+" "+str(authors[i][j]
        [k])+"\n")
print()
```

Output:

```
RUN  C:/Users/ACER  ASPIRE/AppData/Local/Programs/Python/
    Python35-32/Tools/scripts/listfor.py
[[['Programming in C#', ['Oxford University Press', 2015]], ['SE
    is everything', ['Obscure Publishers', 2015]]]
0 0 0 P
0 0 1 r
0 0 2 o
0 0 3 g
0 0 4 r
0 0 5 a
0 0 6 m
0 0 7 m
0 0 8 i
0 0 9 n
0 0 10 g
0 0 11
0 0 12 i
0 0 13 n
0 0 14
0 0 15 C
0 0 16 #
0 1 0 Oxford University Press
0 1 1 2015
1 0 0 S
1 0 1 E
1 0 2
1 0 3 i
1 0 4 s
1 0 5
1 0 6 e
1 0 7 v
1 0 8 e
```

```
1 0 9 r
1 0 10 y
1 0 11 t
1 0 12 h
1 0 13 i
1 0 14 n
1 0 15 g
1 1 0 Obscure Publishers
1 1 1 2015
```

5.5   CONCLUSION

Repeating a task is an immensely important job. This is needed, in various situations, to accomplish different tasks. This chapter introduced the two most important looping constructs in Python and demonstrated the use of looping constructs by taking simple examples. Having a loop within a loop is called nesting. The nesting of loops has been explained using patterns and a list of lists. The following chapters briefly revisit one of the constructs and compare the use of iterators and generators. The reader is expected to solve the problems at the chapter's end for better understanding. It may be stated that, Python provides us with other constructs which would greatly simplify program writing. As of now, try various permutations and combinations, observe the outputs and learn.

GLOSSARY

1.   Looping means repeating a task, a certain number of times.

2.   Syntax of for loop

```
for i in range(n):
        ...
        ...
```

OR

```
for i in range(n, m):
        ...
        ...
```

OR

```
for i in range (_, _,...)
        . . .
        . . .
        . . .
```

3. Syntax of while loop

```
while <test condition>:
        . . .
```

POINTS TO REMEMBER

▧ Looping in Python can be implemented using **while** and **for.**

▧ **"while"** is the most common looping construct in Python.

▧ The statements in the **while** block execute till the test condition remains true.

▧ The else part executes if the loop ends without a break.

▧ **"for"** can be used for all the purposes for which a **"while"** is used.

▧ **"for"** is generally used for processing lists, tuples, matrices etc.

▧ **range** (n) means values from 0 to (n-1).

▧ **range** (m, n) means all the values from m to (n-1).

▧ A loop can be nested in a loop.

▧ There can be any number of nestings, though this is undesirable.

EXERCISES

Multiple Choice Questions

1. What will be the output of the following?

```
a=8
i=1
while a:
     print(a)
     i=i+1
     a=a-i
print(i)
```

(a) 8, 6, 3

(b) 8, 6, 3, -1

(c) 8, 6, 3, -1, ...

(d) None of the above

2.
```
a=8
  i=1
  while a:
      print(a)
      i=i+1
      a=a/2
  print(i)
```
(a) 8, 4, 2, 1 **(b)** 8, 4, 2, 1, 0

(c) 8, 4, 2, 1, 0.5 **(d)** Infinite loop

3. How many times the following loop executes?
```
n = int(input('Enter number'))
i = n
while (i>0):
    print(n)
i=i+1
    n = int(n/2)
print(i)
#The value of n entered by the user is 10
```
(a) 4 **(b)** 5

(c) Infinite **(d)** The code will not compile

4. Which loop can be used when the number of iterations are not known?

(a) while **(b)** for

(c) both **(d)** None of the above

5. How many levels of nesting are possible in for?

(a) 2 **(b)** 3

(c) Both **(d)** The code will not compile

6.
```
n = int(input('Enter number'))
for i in (0,7):
    print('i is '+str(i))
    i = i+1;
else:
    print('bye')
```

How many values would be printed?

(a) 2   **(b)** 3

(c) 6   **(d)** None of the above

7.
```
n = int(input('Enter number'))
    for i in range(n, 1, -1):
        for j in range(i):
            print(i, j)
    #value entered by the user is 5
```

(a) (5, 0), (5, 1), ...(2, 1)   **(b)** (5, 1), (5,2),...(2, 0)

(c) (0,1), (0,2), ...(5, 2)   **(d)** None of the above

8. In order to print the elements of a given matrix which of the following is essential?

(a) Nested loops   **(b)** Single loop

(c) if-else   **(d)** None of the above

9. What is meant by range (5)?

(a) Integers from 0 to 4   **(b)** Integers from 0 to 5

(c) Integers from 1 to 4   **(d)** Integers from 1 to 5

10. What is meant by range (3, 8)?

(a) 3, 4, 5, 6, 7, 8   **(b)** 3, 4, 5, 6, 7

(c) 1, 2, 4, 5, 6, 7, 8   **(d)** 8, 8, 8

Programming Exercises

1. Ask the user to enter a number and find whether it is a prime number.

2. Ask the user to enter a number and find all its factors.

   Example: If number = 30, then factors are 2, 3, and 5.

3. Find whether the number entered by the user is a perfect square?

4. Ask the user to enter two numbers and find the lowest common multiple.

   Example: If numbers are 30 and 20, then LCM is 60, as both 20 and 30 are factors of 60

5. Ask the user to enter two numbers and find the highest common factor.

   Example: If numbers are 30 and 20, the HCF is 10

6. Find the mean of numbers entered by the user.

$$\text{Mean} = \frac{x_1 + x_2 + x_3 + \ldots + x_n}{n}$$

7. Find the variance and standard deviation of the numbers entered by the user.

8. Ask the user to enter the values of a and b and find a^{b^a}.

9. Find the common factor of n numbers entered by a user.

10. Ask the user to enter three numbers and find all possible permutations of the numbers.

11. In the above question, what happens if we have four numbers in place of three?

12. Can the above logic be extended for n numbers?

13. Ask the user to enter n numbers and find the minimum of the numbers without using arrays.

14. Ask the user to enter n numbers and find the maximum of the numbers without using arrays.

15. Create a list of authors, in which the record of each author is itself a list consisting of the name of the book, publisher, year of publication, ISSN, and the city. Now process the list using **for** loop.

6

FUNCTIONS

Objectives

After reading this chapter, the reader should be able to

- Appreciate the importance of modular programming
- Understand the components and types of functions
- Implement Linear Search using functions
- Understand the concept of scope of a variable
- Understand and use recursion

6.1 INTRODUCTION

If one has to perform a bigger task, then it is advisable to divide it into smaller, manageable tasks. This division has many advantages, discussed in the following sections. The units of program, which can be called on as it is basis, take some input, process it and may generate some output are referred to as functions.

> Functions are units which perform a particular task, take some input and may give some output.

This concept is the soul of procedural programming. The readers familiar with C (or for that matter C++, JAVA, C#, etc.) must be familiar with the idea and use of functions. However, a brief discussion on the features and advantages of functions follows in the next section.

This chapter introduces the concept of functions. The chapter has been organized as follows. The next section briefly explains the features of a function; the third section explains the basic terminology, and the fourth section explains the definition and invocation of a function. The fifth section presents a brief discussion on various types of functions. The sixth section illustrates the concept by taking the example of linear search. The seventh section discusses the scope of a variable, and the eighth section presents recursion. The last section concludes.

6.2 FEATURES OF A FUNCTION

As discussed earlier, functions form the basis of procedural programming. One of the most obvious advantages of using functions is the division of a program into smaller parts. This section briefly discusses the advantages of functions.

6.2.1 Modular Programming

A good program is divided into small parts, in such a way that a particular part perform some specific task. The clubbing together of similar functions gives rise to modular programming.

6.2.2 Reusability of Code

A function can be called many times. This spares the programmer from the horror of rewriting the same code again, which in turn can reduce the length of the program.

6.2.3 Manageability

Dividing a bigger task into smaller functions makes the program manageable. It becomes easy to locate bugs and leads to increased reliability. Also, it becomes easy to carry out local optimizations in a function. To summarize, manageability leads to the following.

6.2.3.1 Easy debugging

In order to understand why creating functions would make debugging easy, let us consider White Box Testing. This type of testing, which uses code for testing, requires elicitation of paths and crafting test cases catering to them. It is easy to effectively analyse smaller functions rather than the whole task.

6.2.3.2 Efficient

It is essential to make code efficient both in terms of time and memory. As a matter of fact, even in C's compiler, most of the code optimization is attributed to the developer rather than the compiler.

The above factors point to the fact that dividing the task into functions is a good practice. It may be noted here that even Object-Oriented Programming relies on functions for implementing the behavior of a class.

6.3   BASIC TERMINOLOGY

The importance of functions in procedural programming has already been discussed in the previous section. This section briefly introduces the terminology of functions and presents the syntax, which would form the foundation stone of the discussion that follows.

6.3.1 Name of a Function

Function can have any legal literal name. For example, **sum1** is a valid function name, as it satisfies all the constrains on the name (also discussed in Section 6.4). It may be stated here that in general a class can have more than one functions having same name but different parameters. This is referred to as overloading.

6.3.2 Arguments

The arguments of a function denote the input given to a function. A function can have any number of arguments. As a matter of fact, it is possible that a function may not have any argument.

6.3.3 Return Value

A function may or may not return a value. The beauty of Python lies in not specifying the return type and hence using the same functions for returning various data types.

In Python, a function can be made in the command prompt. This implies that unlike C (or for that matter C++, JAVA, or C#) a function need not be a part of a program. Moreover, the return type, as described in this section, need not to be mentioned. This inculcates flexibility in the procedures.

6.4 DEFINITION AND INVOCATION

This section discusses how to define a function and call a function that has already been defined. The definition of a function depicts the behavior. The task to be performed by the function is contained in its definition. In the discussion that follows, the components of a function have been explained in detail.

The invocation of a function means calling a function. As is explained in Section 6.8: a function can also be called within itself. This is referred to as recursion.

It may also be noted that a function is defined only once. However, it can be called any number of times.

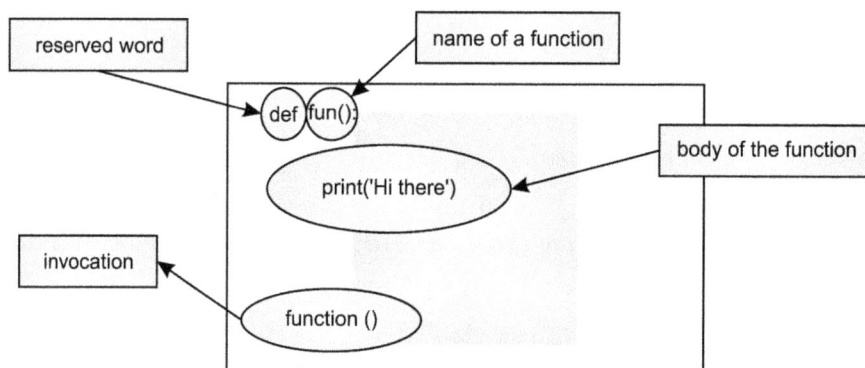

FIGURE 6.1 Example of a function.

The definition of a function contains the following:

Name of a function: The name of a function should be any valid identifier. The name of a function should be meaningful and if possible, convey the task to be performed by the function.

Parameter: The list of parameters (separated by commas) is given in the parenthesis following the name of the function. The parameters are basically the input to the function. A function may have any number of parameters.

Body of the function: The body of the function contains the code that implements the task to be performed by the function.

Figure 6.1 shows the name of the function (**fun**), the list of parameters in the parenthesis following the name of the function (in this case there are no parameters), and the body of the function.

It may also be noted that the closing parenthesis containing the parameters is followed by a colon. The body of a function starts with a proper indentation.

The invocation of a function can be at any place after the definition. However, exceptions to this premise are found in the case of recursion.

The syntax of a function is depicted in Figure 6.2.

Syntax:

def <name of the function>(list

of parameters):

       <body>

FIGURE 6.2 Syntax of a function.

6.4.1 Working

Consider a function which multiplies two numbers passed as parameters.

```
def product(num1, num2):
    prod= num1*num2
    print('The product of the numbers is \t:'+str(prod))
```

The name of this function is **product**. It takes two arguments as input (**num1** and **num2**), calculates the product, and displays the results.

```
The function can be invoked as follows.
num1=int(input('Enter the first number\t:'))
num2=int(input('Enter the second number\t:'))
print('Calling the function...')
product(num1, num2)
print('Back to the calling function');
```

Here, calling **product** shifts the control to the function, inside which the product is calculated and the result is displayed. The control then comes back to the calling function (Figure 6.3).

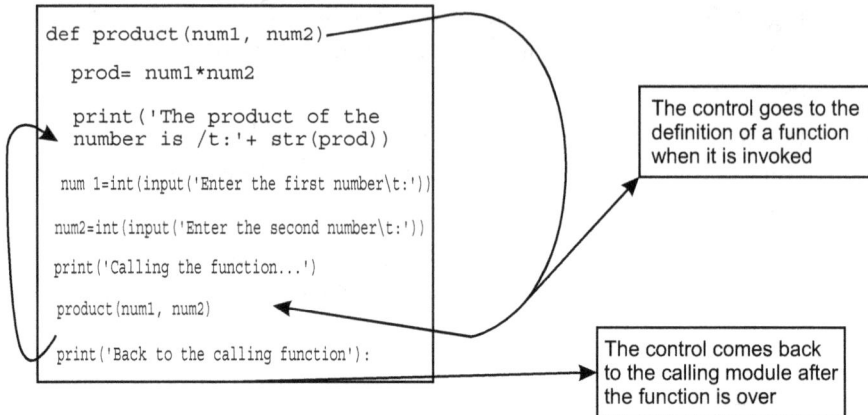

FIGURE 6.3 Calling a function.

A function can be called any number of times. The following example shows a function which does not take any input and does not return anything. The function called, just prints the lines of **Ecclesiastes**. The following listing shows the function and the output of the program follows.

Illustration 6.1:

Basic Function

Listing:

```
def Ecclesiastes_3():
    print('To everything there is a season\nA time for every
    purpose under Heaven')
    print('A time to be born\nand a time to die\nA time to
    plant\nand a time to reap')
    print('A time to kill\nand a time to heal\nA time to break
    down\nand a time to build up')
    print('A time to cast away stones\nand a time to gather
    stones\nA time to embrace\nand a time to refrain')
    print('A time to gain\nand a time to lose\nA time to keep\
    nand a time to cast away')
    print('A time of love\nand a time of hate\nA time of war\
    nand a time of peace')
print('Calling function\n')
Ecclesiastes_3()
```

```
print('Calling function again\n')
Ecclesiastes_3()
>>>
```

Output:

```
Calling function

To everything there is a season
A time for every purpose under Heaven
A time to be born
and a time to die
A time to plant
and a time to reap
A time to kill
and a time to heal
A time to break down
and a time to build up
A time to cast away stones
and a time to gather stones
A time to embrace
and a time to refrain
A time to gain
and a time to lose
A time to keep
and a time to cast away
A time of love
and a time of hate
A time of war
and a time of peace
```

6.5 TYPES OF FUNCTION

Based on the parameters and the return type, functions can be divided into the following categories. The first type of function does not take any parameter nor returns anything. The program given in Illustration 6.2 examples one such function.

The second type of function takes parameter but do not return anything. The second function in Illustration 6.2, exemplifies such function. The third type of function takes parameters and returns output. The example that follows, adds two numbers using functions. The task has been accomplished in three different ways, in the first function (**sum1**) the input is taken inside the function and the result is displayed using a print statement, which is also present inside the function.

The second function takes the two numbers as input (via parameters), adds then, and prints the result inside the function itself. The third function (**sum3**) takes two parameters and returns the sum.

Illustration 6.2:

Write a program to add two numbers, using functions. Craft three functions, one which does not take any parameters and does not return anything. The second function should take parameters and not return anything. The third function should take two numbers as parameters and should return the sum.

Program:

```
def sum1():
    num1=int(input('Enter the first number\t:'))
    num2=int(input('Enter the second number\t:'))
    sum= num1+num2
    print('The sum of the numbers is \t:'+str(sum))
def sum2(num1, num2):
    sum= num1+num2
    print('The sum of the numbers is \t:'+str(sum))

def sum3(num1, num2):
    sum= num1+num2
    return(sum)
print('Calling the first function...')
sum1()
num1=int(input('Enter the first number\t:'))
num2=int(input('Enter the second number\t:'))
print('Calling the second function...')
sum2(num1, num2)
print('Calling the third function...')
```

```
result=sum3(num1, num2)
print(result)
```

Output:

```
RUN  C:/Users/ACER ASPIRE/AppData/Local/Programs/Python/
    Python35-32/Tools/scripts/sum_of_numbers.py
Calling the first function...
Enter the first number  :3
Enter the second number     :4
The sum of the numbers is  :7
Enter the first number  :2
Enter the second number     :1
Calling the second function...
The sum of the numbers is  :3
Calling the third function...
3
```

6.5.1 Arguments: Types of Arguments

In Python, unlike C, while defining a function, the types of arguments are not specified. This has the advantage of giving different types of arguments to the same function. For example, in the function that follows, the first invocation passes an integer value in the function. The function adds the two numbers. In the case of the second invocation, the addition operator concatenates the strings, passed as a parameters.

Illustration 6.3:

Arguments

Listing 1:

```
def sum1(num1, num2):
    return (num1+num2)
    sum1(3,2)
    sum1('hi', 'there')
```

Output:

```
5
'hithere'
```

Listing 2:

```
def sum1(num1, num2):
    return (num1+num2)
print('Calling function with integer arguments\t: Result:
    '+str(sum1(2,3)))
print('Calling the function with string arguments\t: Result:
    '+sum1('this',' world'))
```

Output:

```
Calling function with integer arguments     :Result: 5
Calling the function with string arguments   :Result: this world
```

6.6   IMPLEMENTING SEARCH

This section demonstrates one of the most important use of the topics discussed so far: **Searching**. In the search problem, if the element is present in a given list, then its position should be printed, otherwise the message "Not Found" should be displayed. There are two major strategies to accomplish this task. They are linear search and binary search. In linear search, the elements are iterated one by one. If the required element is found, the position of the element is printed. The absence of an element can be judged using a **flag**.

The algorithm has been implemented in Illustration 6.4.

Illustration 6.4:

Write a program to implement linear search.

Solution:

Code:

```
def search(L, item):
    flag=0
    for i in L:
        if i==item:
            flag=1
            print('Position ',i)
        if flag==0:
            print('Not found')
L =[1, 2, 5, 9, 10]
search(L, 5)
search(L, 3)
```

Output:

```
    Position  5
Not found
```

The above search strategy works well. However, the complexity of this algorithm is O(n). There is another strategy of search called binary search. In binary search, the input list must be sorted. The algorithm checks whether the item to be searched is present at the first position, the last position, or at the middle position. If the requisite element is not present at any of these positions and it is less than the middle element, then the left part of the list becomes the input of the procedure; else the right part of the list becomes the input to the procedure. The reader is advised to implement binary search. The complexity of binary search is O (log n).

6.7   SCOPE

The scope of a variable in Python, is the part of the program wherein its value is legal or valid. It may be stated here that, though, Python allows global variable, the value of a local variable must be assigned before being referred. Illustration 6.5 exemplifies the concept. The illustration has three listings. In the first listing the value of **"a"** has been assigned outside the function as well as inside the function. This leads to problem as a variable cannot be referenced before being assigned.

In the second case this contention is resolved. Finally, the last listing shows that global variables are very much allowed in Python, for some strange reason. As an active programmer, I firmly believe that should not have been allowed and there are multiple reasons of not allowing global variables in a programming language.

Illustration 6.5:

Scope of a variable

Listing 1:

Code:
```
# Note that a = 1 does not hold when function is called
a = 1
def fun1():
    print(a)
```

```
    a=7
    print(a)

def fun2():
    print(a)
    a=3
    print(a)

fun1()
fun2()
```

Output:

```
Traceback (most recent call last):
  File "C:/Python/Functions/scope.py", line 12, in <module>
    fun1()
  File "C:/Python/Functions/scope.py", line 3, in fun1
    print(a)
UnboundLocalError: local variable 'a' referenced before
    assignment
```

Listing 2:

Code:

```
a = 1
def fun1():
    a=1
    print(a)
    a=7
    print(a)

def fun2():
    a=1
    print(a)
    a=3
    print(a)

fun1()
fun2()
```

Output:

```
1
7
1
3
Also, note that had 'a' been not assigned in the functions, the
    global value would have sufficed.
```

Listing 3:

Code:

```
a = 1
def fun1():
    print(a)
  def fun2():
    print(a)
fun1()
fun2()
```

Output:

```
1
1
```

6.8   RECURSION

A function can also be called within itself. Calling a function in itself is referred to as recursion. The concept is used to accomplish many tasks easily and intuitively. For example, consider the following series:

1, 1, 2, 3, 5, 8, 13, ...

Note that each term of this series is the sum of the previous two terms, the first and the second term being 1 and 1, respectively. This series is referred to as Fibonacci series. The series is due to the famous rabbit problem, which has been described as follows.

6.8.1 Rabbit Problem

Initially, there is a single pair of rabbits. This pair of rabbits do not breed for the first two months, after which they generate a pair of rabbits every month.

This way, there is a single pair of rabbit for the first two months, after which the series become 2, 3, 5, 8, 13, and so on. The formation of the series is depicted in Table 6.1. R0 refers to the first pair, R01 refers to the pair generated from R0, in the third month. Likewise, R02 is the pair generated from R0 in the fourth month.

TABLE 6.1 Fibonacci series.

Month	Pair of rabbits	Number of pair
1	R0	1
2	R0	1
3	R0 -> R01	2
4	R0 -> R01, R02	3
5	R0 -> R01 (->R010), R02, R03	5
6	R0 -> R01 (->R010, R011), R02 (->R020), R03, R04	8

Note that in the above series, each term is the sum of the preceding two terms. The series can be mathematically represented as follows.

$$\text{fib}(n) = \begin{cases} 1, \text{ for } n = 1 \\ 1, \text{ for } n = 2 \\ \text{fib}(n-1) + \text{fib}(n-2) \end{cases}$$

Illustration 6.6 depicts the implementation of Fibonacci series, using recursion.

Illustration 6.6:

Ask the user to enter the value of n and find the nth Fibonacci term.

Solution:

Code:
```
def fib(n):
    if n==1:
        return 1
    elif n==2:
```

```
    return 1
  else
    return (fib(n-1) + fib(n-2))
n=input('Enter the number\t:')
f=fib(n)
print('The nth fib term is ',str(f))
```

Output:

```
Enter the number  :5
The nth fib term is 5
```

Note that the calculation of Fibonacci uses the Fibonacci term calculated earlier. For example, the calculation of the 5th Fibonacci term requires the following calculations. fib(5) requires fib(4) and fib(3); fib(4) requires fib(3) and fib(2) and fib(3) requires fib(2) and fib(1) (Figure 6.4).

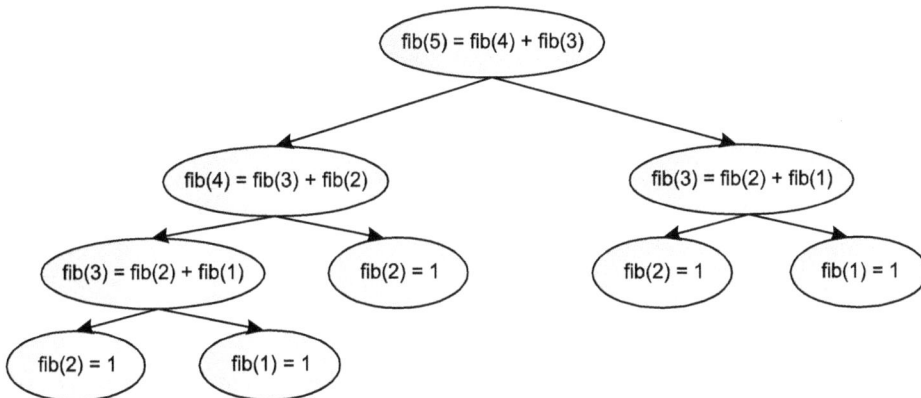

FIGURE 6.4 Calculation of the fifth Fibonacci term.

The next example calculates the factorial of a number using recursion. The factorial of a number n (positive and integer) is the product of all the integers from 1 to n. That is

$n! = 1 \times 2 \times 3 \times \ldots \times n$

Note that since $(n-1)! = 1 \times 2 \times 3 \times \ldots \times (n-1)$

Therefore, $n! = n \times (n-1)!$

Also factorial of 1 is 1, that is $1! = 1$, which can be used as the base case while implementing factorial using recursion. The program has been depicted in Illustration 6.7.

Illustration 6.7:

Ask the user to enter the value of n and calculate the factorial of n.

Solution:

Code:

```python
def fac(n):
    if n==1:
        return 1;
    else:
        return(n*fac(n-1))
n = int(input('Enter the number\t:'))
factorial = fac(n)
print('Factorial of ',n, ' is ', factorial)
```

Output:

```
Enter the number   :5
Factorial of  5  is  120
```

The power of a number raised to another number can also be calculated using recursion. Since $a^b = a \times a^{b-1}$ that is power(a, b) = a*power(a, b - 1). Also, $a^1 = a$ that is, power(a, 1) = 1. The above logic has been implemented in the illustration that follows.

Illustration 6.8:

Ask the user to enter the values of a and b and calculate a to the power of b, using recursion.

Program:

Code:

```python
def power(a , b):
    if b==1:
        return a
    else:
        return (a*power(a, b-1))
a = int(input('Enter the first number\t:'))
b = int(input('Enter the second number\t:'))
p = power(a,b)
print(a, ' to the power of ',b,' is ', p)
```

Output:

```
Enter the first number  :3
Enter the second number    :4
3  to the power of  4  is  81
```

6.8.2 Disadvantages of Using Recursion

In spite of the fact that recursion makes things easy and helps to accomplish some of the tasks intuitively, there is a flip side. Consider the first illustration. Though, the program calculates the nth Fibonacci term, the complexity of the procedure is too high ($O(\phi^n)$, where Φ is gold number). The same task can be accomplished in linear time using a paradigm called dynamic programming.

Similarly, the recursive procedures in Divide and Conquer also require huge time. In addition to the above problem, there is another flip side. Recursion requires a lot of memory. Though, a portion of the memory is reserved for stacks, a recursive procedure may eat up all the available memory.

6.9 CONCLUSION

This chapter introduced the concept of functions. The idea of dividing the given program into various parts is central to manageability. The chapter forms the foundation stone of the chapters that follow. It may also be stated that functions implement the behavior of a class; therefore, before moving to the Object-Oriented Paradigms, one must be familiar with functions and procedures.

The concept of recursion is also central to the implementations, which involve the ideas of Divide and Conquer and that of Dynamic Programming. So, one must also be equipped with the power of recursion and should be able to use the concept to solve problems, if possible.

GLOSSARY

- **Function:** Functions accomplish a particular task. They help in making a program manageable.
- **Argument:** Arguments are the values passed in a function.
- **Recursion:** A function may call itself. It is referred to as recursion.

POINTS TO REMEMBER

- A function can have any number of arguments.
- A function may return a maximum of one value.
- A function may not even return a value.
- A function may call itself.
- A function can be called any number of times.

EXERCISES

Multiple Choice Questions

1. Which of the following keyword is used to define a function?

    (a) Def

    (b) Define

    (c) Definition

    (d) None of the above

2. The values passed in a function are called

    (a) Arguments

    (b) Return values

    (c) Yield

    (d) None of the above

3. A recursive function is one that calls

    (a) Itself

    (b) Other function

    (c) The main function

    (d) None of the above

4. Which of the following should be present in a recursive function?

    (a) Initial values

    (b) Final values

    (c) Both

    (d) None of the above

5. Which of the following can be accomplished using recursion?

    (a) Binary search

    (b) Fibonacci series

    (c) Power

    (d) All of the above

6. Which of the following is allowed in a function?

    (a) If

    (b) While

    (c) Calling a function

    (d) None of the above

7. Which types of functions are supported in Python?

    (a) Build in   **(b)** User defined

    (c) Both   **(d)** None of the above

8. Which of the following is true?

    (a) A function helps in dividing a program in small parts

    (b) A function can be called any number of times

    (c) Both

    (d) None of the above

9. Which of the following is true?

    (a) One can have a function that called any number of functions

    (b) Only a limited number of functions can be called from a function

    (c) Nested functions are not allowed in Python

    (d) Nested functions are allowed only in certain conditions

10. Nested functions incorporates the concept of

    (a) Stack   **(b)** Queue

    (c) Linked List   **(d)** None of the above

Programming Exercises

1. Write a function that calculates the mean of numbers entered by the user.

2. Write a function that calculates the mode of numbers entered by the user.

3. Write a function that calculates the median of numbers entered by the user.

4. Write a function that calculates the standard deviation of the numbers entered by a user.

5. Write a function that finds the maximum of the numbers from a given list.

6. Write a function that finds the minimum of the numbers from a given list.

7. Write a function that finds the second maximum of the numbers from a given list.

8. Write a function that finds the maximum of three numbers entered by the user.

9. Write a function that converts the temperature in Celsius to that in Fahrenheit.

10. Write a function that searches an element from a given list.

11. Write a function that sorts a given list.

12. White a function that takes two lists as input and returns the merged list.

13. Write a function that finds all the factors of a given number.

14. Write the function that finds common factors of two given numbers.

15. Write a function that returns a number obtained by reversing the order of digits of a given number.

Questions Based on Recursion

Use recursion to solve the following problems

1. Find the sum of two given numbers.

2. Find the product of two given numbers.

3. Given two numbers, find first number to the power of the second.

4. Given two numbers, find the greatest common divisor of the numbers.

5. Given two numbers, find the least common multiples of the numbers.

6. Generate n Fibonacci terms.

7. In a series, the first three terms are 1, 1, and 1; the ith term is obtained using the following formula

$$f(i) = 2 \times f(i-1) + 3 \times f(i-2)$$

Write a function to generate n terms of the sequence.

8. Find the element in a given sorted list.

9. Find the maximum from a given list.

10. Reverse the order of digits of a given number.

Theory

1. What are the advantages of using functions in a program?

2. What is a function? What are the components of a function?

3. What is the importance of parameter and return type in a function? Can a function have more than one return values?

4. What is recursion? Which data structure is used internally while implementing recursion?

5. What are the disadvantages of recursion?

Extra Questions

1. What will be the output of the following program?

```
def fun1(n):
    if n==1:
        return 1
    else:
        return (3*fun1(n-1)+2*fun1(n))
fun1(2)
```

(a) 1

(b) 5

(c) 3

(d) Maximum iteration depth reached

2. What will be the output of the following?

```
def fun1(n):
    if n==1:
        return 1
    elif n==2:
        return 2
    else:
        return (3*fun1(n-1)+2*fun1(n))
fun1(5)
```

(a) 5 **(b)** 27

(c) Maximum iteration depth reached **(d)** None of the above

3. What will be the output of the following?

```
def fun1(n):
    if n==1:
        return 1
    elif n==2:
        return 2
    else:
        return (3*fun1(n-1)+2*fun1(n-2))
print(fun1(5))
```

(a) 5

(b) 100

(c) 25

(d) Maximum iteration depth reached

4. What will be the output of the following?

```
def fun1(n):
    if n==1:
        return 1
    elif n==2:
        return 2
    else:
        return (3*fun1(n-1)+2*fun1(n-2))
```

```
for i in range(10):
    print(fun1(i), end=' ')
```

(a) 1 2 8 28 100 356 1268 4516 16084

(b) 1 3 5 7 9 11 13 15

(c) Maximum iteration depth reached

(d) None of the above

5. What will be the output of the following?

```
def fun1(n):
    if n==1:
        return 1
    elif n==2:
        return 2
    else:
        return (3*fun1(n-1)+2*fun1(n-2))
for i in range(1, 10, 1):
    print(fun1(i), end=' ')
```

(a) 1 2 8 28 100 356 1268 4516 16084

(b) 1 3 5 7 9 11 13 15

(c) Maximum iteration depth reached

(d) None of the above

6. What will be the output of the following?

def _main_():

    print('I am in main')

    fun1()

    print('I am back in main')

def fun1():

    print('I am in fun1')

    fun2()

    print('I am back in fun1')

def fun2():

    print('I am in fun 2')

      main()am in fun 2')

      >>>

(a) I am in main

   I am in fun1

   I am in fun 2

   I am back in fun1

   I am back in main

(b) Reverse of the above

(c) None of the above

(d) The program does not execute

7. Conceptually which data structure is implemented in the above program?

   (a) Stack

   (b) Queue

   (c) Graph

   (d) Tree

8. Which technique is implemented in the following code?

```python
def search(L, item):
    flag=0
    for i in L:
        if i==item:
            flag=1
            print('Position ',i)
    if flag==0:
        print('Not found')
L =[1, 2, 5, 9, 10]
search(L, 5)
search(L, 3)
```

   (a) Linear search

   (b) Binary search

   (c) None of the above

   (d) The code does not execute

9. What is the complexity of the above?

   (a) O (n)

   (b) O (n^2)

   (c) O (log n)

   (d) None of the above

10. Which is better linear search or Binary search?

   (a) Linear

   (b) Binary

   (c) Both are equally good

   (d) depends on the input list.

7

FILE HANDLING

Objectives

After reading this chapter, the reader should be able to

- Understand the importance of File Handling
- Appreciate the mechanism of File Handling in Python
- Learn various file access modes
- Understand various functions for File Handling in Python
- Implement the concepts studied in the chapter

7.1 INTRODUCTION

The data types and control structures discussed so far would help us to accomplish many simple tasks. The problem, though, is that we have not been able to store the data or the results obtained for future use. Moreover, at times the results produced by a program are voluminous. In such cases it becomes difficult to store data in the memory or even to read the data. File handling can help us to handle the above situations.

The reader would also appreciate the fact that the main memory is volatile. The data produced by a program, therefore, cannot be used for future endeavours. Many times it is required to store the data for use in future. For example, if one develops a student management system, it will be useful only if the data stored can be retrieved as and when required.

While storing data, the format of the data should be taken care of. At the programmer's level, however, the data can be stored in files or in databases. Databases store and manage related data. The ease of retrieval, the security and flexibility make database one of the most important topics in Computer Science. The concept of databases, their usage and related issues constitute a dedicated subject. This chapter, however, only concentrates on file handling.

A file can be perceived as set of records, where each record has some fields. The fields, in turn, store data. Files, as discussed later, can have many formats. The chapter, though, concentrates on binary and text files. The two formats differ in the how the end of a line is represented, the representation of the end of the file and in the storage of standard data types. A file may have certain permissions associated with it. For example, one may not have the write permissions for a file which is to be used by the operating system. For that matter, a user may not even have the read permissions for such files. Such constraints are kept in mind while writing programs for file handling.

Python provides many functions to carry out the operations related to file handling. The creation of a file, writing data to a file, reading the data, appending something to the file and standard directory operations have been discussed in this chapter. Moreover, to make the things interesting, the use of the above operations in encryption has also been discussed.

The chapter has been organized as follows. The second section discusses the general file handling mechanism. The third section discusses the **open()** function and the various modes in which a file can be opened. The fourth section discusses the functions for reading and writing to the file. The section also introduces the functions to get and set the position of a cursor in a file. The fifth section briefly discusses the Command Line Arguments, and the last section, concludes.

7.2 THE FILE HANDLING MECHANISM

In Python, files are accessed using the file objects. As a matter of fact, the file objects help us to access not just normal disk files but can help us to accomplish many additional tasks, involving other kinds of files.

The file handling mechanism in Python is simple. The file needs to be opened first. As in, the file is hooked to an Object [1]. This is done with the help of the **open()** function. The function takes the name of the file and the mode as its arguments. In fact, the function can have three arguments. These arguments are discussed in the next section. The open function returns an object of the file. The object then uses the library functions to read the file, write into it or append it. Finally, the memory space occupied by the object is freed using the **close()** function. The mechanism has been depicted in the following figure (Figure 7.1).

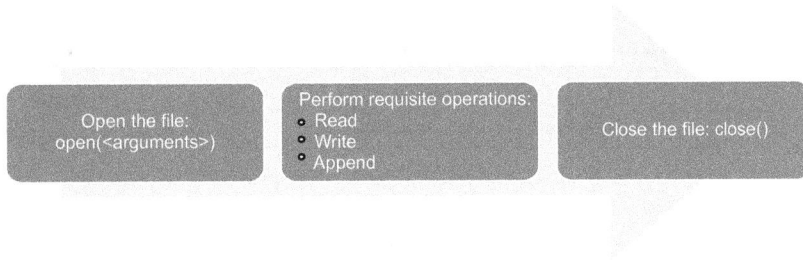

FIGURE 7.1 File handling in Python.

Having discussed the mechanism of handling a file, in Python, let us move on to the file access modes and the **open** function in Python.

7.3   THE OPEN FUNCTION AND FILE ACCESS MODES

The files are accessed using the object created with the help of the **open()** function. Note that this returns a file object or a file like object. This abstraction is helpful for considering file as interface for communication. Therefore, this communication can be perceived as transfer of bytes since a file can be considered as a sequence of bytes.

So, to be able to do input /output to/from a file, the **open()** function is needed. If the file is opened successfully, the file Object is returned. If the file is not opened successfully, the **IOERROR** exception is raised.

The **open** function takes three arguments. The first argument is the name of the file, the second the mode in which the file is to be opened and the third indicates the buffer string. As a matter of fact, the third would rarely be used. The first argument is a string of characters, which is either a valid filename or a path. The path can be relative or absolute. The access mode is the mode in which the file would be opened (Figure 7.2). The various modes have been presented in Figure 7.3. These open the file in read, write or append mode. In the read mode ("*r*"), the file is opened, if it exists. The write mode ("*w*") opens the file for writing. If the file already exists, the existing contents of the file are truncated. The append mode("*a*") opens the file for writing but does not truncate the existing contents. In this mode, if the file does not exist, it would be created.

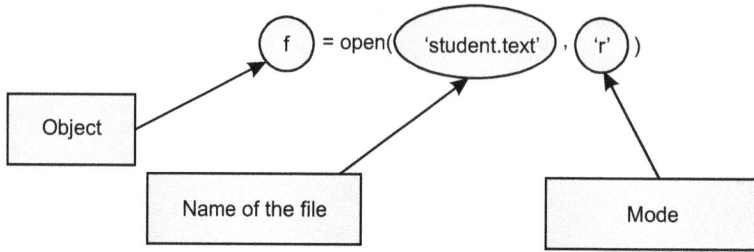

FIGURE 7.2 (a) The open function.

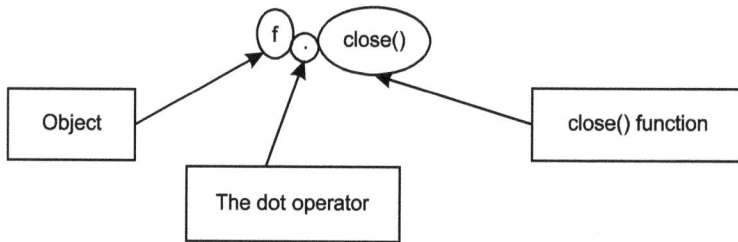

FIGURE 7.2 (b) The close function.

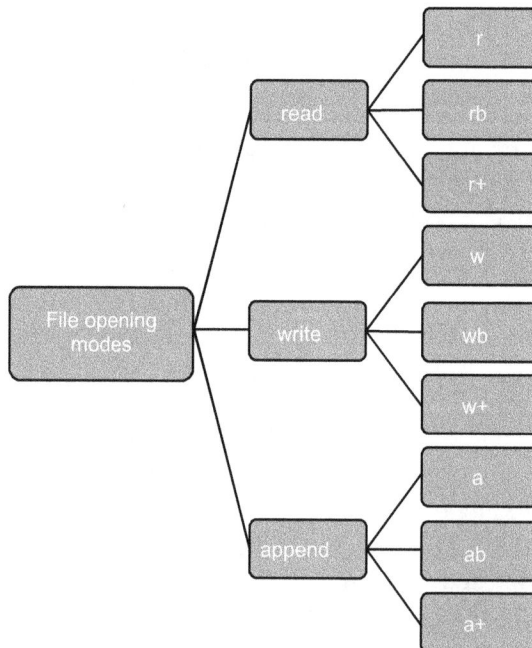

FIGURE 7.3 File opening modes in Python.

The modes can be suffixed with a letter "*b*" indicating the binary access. The "+" suffix can be used to grant the read and write access to the file. Table 7.1 presents the various modes and the corresponding operations that can be performed.

TABLE 7.1 Access modes for file.

File Mode	Operations
r	Reading from a file
w	Write to a file; creates the file if it does not exist; truncate the file if it already exists.
a	Append to the file; if the file does not exist creates the file
r+	Open for read and write
w+	**w** for both read and write
a+	**a** for both read and write
rb	Read a binary file
wb	Write mode for a binary file
ab	Append mode for a binary file
rb+	**r+** for a binary file
wb+	**w+** for a binary file
ab+	**a+** for a binary file

7.4 PYTHON FUNCTIONS FOR FILE HANDLING

Python provides various library functions to carry out the standard tasks. The functions help us to read from a file, write to a file, and to append something in the existing file. Moreover, Python also provides the programmer with functions to take the cursor to a particular location, or to read from a given location.

7.4.1 The Essential Ones

The use of these functions has been briefly explained in this section. The reader is expected to experiment with the functions in order to get a clear insight.

The read() function

The function reads bytes in a string. It may take an integer argument indicating the number of bytes to read. If the argument is –1, the files must be read to the end. Also, if no argument is given, the default is taken as –1.

TIP!   *read() is same as read(-1)*

If the content of the file is larger than the memory then only the content which can fit into the memory would be read. Moreover, when the read operation ends, a "(an empty string) is returned.

readline() and readlines()

The **readline()** method is used to read a line, till the newline character is read. It may be stated here that the newline character is retained in the string that is returned. The **readlines()** method reads all the lines from a given file and returns a list of strings.

write() and writelines()

The **write()** method writes the string in a given file. The method is complementary to the **read()** method. The **writelines()** method write a list of strings to the file.

TIP!   *There is no **writeline()** method in Python 3.x*

seek()

The **seek** function takes the cursor to the stated position in the given file. The position is decided with respect to the offset given. The offset can be 0, 1 or 2. "0" indicates the beginning of the file. The value "1" indicates the current position and the value 2 indicates the "end of the file."

tell()

The **tell()** function is complementary to the **seek()** function. The function returns the position of the cursor.

close()

The **close()** function closes the file. The object should be assigned to another file after it is closed. Though Python closes a file after a program finishes (see Garbage Collection), it is advisable to close the file when the required task is accomplished. The repercussions of not closing the file can be observed at the most unexpected times.

fileno()

The **fileno()** function returns a descriptor for the file. For example, the descriptor of the file named "Textfile.txt," in the following snippet is 3.

```
>>> f=open('Textfile.txt')
>>>f.fileno()
3
```

7.4.2 The OS Methods

The methods which deal with the issues related to operating system help the programmer to create a generic program. The methods also spare the programmer from the horror of dealing with the uncanny formatting details. For example, the end of a line is represented by different character sets in different operating systems. In Unix, a newline is indicated by "\n," In MAC the newline character is "\r" in DOS it is "\r\n." Similarly, file separator Unix is "/", whereas that in windows is "\" and that in MAC is ":". There inconsistencies make the life of a programmer miserable. This is the reason why a consistent approach is needed to handle such situations. Table 7.2 presents the names and functions of some of the most important **OS** methods.

TABLE 7.2 OS methods.

OS method	Function
linesep	string used to separate lines in a file
sep	used to separate file pathname components
Pathsep	delimit a set of file pathnames
Curdir	current directory
Pardir	parent directory

7.4.3 Miscellaneous Functions and File Attributes

Except for the functions stated above, the **flush** and **isatty** are also used to make a program more robust.

flush(): The flush function flushes the internal buffer.

isatty(): The function returns a "1," if the file is a tty-like device

For more such functions, the reader may refer to the references at the end of this book.

File attributes

It may also be stated here that the file attributes help the programmer to see the state of a file and its features like the name and the mode. Table 7.3 presents some of the most important file attributes.

TABLE 7.3 File attributes.

File attribute	Importance
file.closed	1 if file is closed, 0 otherwise
file.mode	access mode
file.name	name of the file

The following illustration demonstrates the use of the above attributes.

Illustration 7.1:

Open a file called "Textfile.txt" in the read mode. Check the name of the file, its mode and find whether it is closed using the file attributes.

Solution:

Code:

```
f=open('Textfile.txt','r')
print('Name of the file\t:',f.name)
print('Mode\t:',f.mode)
print('File closed?\t:',f.closed)
f.close()
print('Mode\t:',f.mode)
print('File closed?\t:',f.closed)
```

Output:

```
Name of the file   : Textfile.txt
Mode : r
File closed?  : False
Mode : r
File closed?  : True
```

7.5   COMMAND LINE ARGUMENTS

If the compiler knows the name of the script, then the name of the script along with the additional arguments are stored in a list called **argv.** The **argv** variable is in the **sys** module. The arguments along with the name of the script are called the command line arguments. It may be noted here that even the name of the script is the part of the list. As matter of fact, the name of the script is the first element of the list. The rest of the arguments are stored in the succeeding locations of the list. The **argv** can be accessed by importing the **sys** module. The following illustration demonstrates the use of the **argv** variable.

Illustration 7.2:

Display the number of the command line arguments and the individual arguments.

Solution:

Code:

```
import sys
print('The number of arguments',len(sys.argv))
print('Arguments\n')
for x in sys.argv:
    print('Argument\t:',x)
```

Output:

```
The number of arguments 1
Arguments
Argument : C:/Python/file handling/commandLine.py
```

The following example presents the bubble sort which takes the numbers entered at the command line as the input.

Illustration 7.3:

Sort the numbers (using bubble sort) entered as the command line arguments.

Solution:

Code:

```
import sys
def sort(L):
```

```
            i=0;
        while(i<(len(L)-1)):
            print('\nIteration\t:',i,'\n');
            j=0
            flag=0
            while(j<(len(L)-i-1)):
                if(L[j]<L[j+1]):
                    flag=1
                    temp=L[j]
                    L[j]=L[j+1]
                    L[j+1]=temp
                #print(L[j],end=' ')
                j=j+1
            print(L)
            if(flag==0):
                break
            i=i+1
        return(L)
L=[]
for x in sys.argv:
    L.append(x)
print('Before sorting\t:',L)
print(sort(L))
```

7.6 IMPLEMENTATION AND ILLUSTRATIONS

Having seen the mechanism of file handling, the functions and the attributes, let us now have a look at the usage of the above functions and the implanta-tion of an interesting task. We will begin with writing something to a file (say "TextFile.txt"), after opening the file in the write mode. The **open** function, in this case, would have two parameters: name of the file("TextFile.txt") and the mode ("w"). Also, the file needs to be closed. Note that the **write** function returns the number of bytes written in the file.

```
>>> f = open('TextFile.txt','w')
>>> f.write('Hi there\nHow are you?')
21
>>>f.close()
```

The **read** function reads the bytes of the given file. The **read** function, as stated earlier, may not take any argument. This implies reading file till the end. The read text can be stored in a string ("text").

```
>>> text=f.read()
>>> text
'Hi there\nHow are you?'
>>>f.close()
>>>
```

A file can be renamed using the **rename** function of **os**. The **rename** function takes two arguments: the first being the name of the original file and the second being the new name of the file. In the following snippet, a file called "TextFile.txt" is renamed to "TextFile1.txt" and read into "str" using the open function.

```
>>> import os
>>> os.rename('TextFile.txt','TextFile1.txt')
>>> f=open('TextFile1.txt','r')
>>> str=f.read()
>>> str
'Hi thereHow are you'
>>>
```

Writing a list of string in a file

As stated earlier, a list of strings can be written into a file using the **writelines**()function. The use of the function has been illustrated as follows. In the following snippet, the lines entered by the user are put into a list, L, and this list is then written into the file f.

Illustration 7.4:

Write a program to ask the user to enter lines of text. The user should be able to enter any number of lines. In order to stop, he must enter "\e." The lines should be appended to an empty list (say L). This list should then be written to a file called lines.txt. The program should then read the lines of lines.txt.

Solution:

Code:

```
print('Enter text, press \'\\e\' to exit')
L=[]
```

```
i=1
in1=input('Line number'+str(i)+'\t:')
while(in1 !='\e'):
    L.append(in1)
    i=i+1
    in1=input('Line number'+str(i)+'\t:')
print(L)
f=open('Lines.txt','w')
f.writelines(L)
f.close()
f=open('lines.txt','r')
for l in f.readline():
    print(l, end=' ')
f.close()>>>
```

Output:

```
Enter text, press '\e' to exit
Line number1  :Hi there
Line number2  :How are you
Line number3  :I am good
Line number4  :\e
['Hi there', 'How are you', 'I am good']
Hi there How are you I am good >>>
```

Reading n characters and the seek() function

The use of the **read(n)** function, which reads the first "n" characters of the file has been demonstrated in the following illustration (Illustration 7.5). Note that, the **tell** function tells the position of the cursor. The **seek()** function takes two parameters, the first being the offset and the second position. For example, **seek(0, 0)** positions the cursor at the first position from the beginning.

Illustration 7.5:

Open a file TextFile.txt and write a few lines in it. Now open the file in the read mode and read the first 15 characters from the file. Then read the next five characters. In each step show the position of the cursor in the file. Now, go back to the first position in the file and read 20 characters from the file.

Solution:

Code:

```
f=open('TextFile.txt','w')
f.writelines(['Hi there', 'How are you'])
f.close()
f = open('TextFile.txt', 'r+')
str = f.read(15)
print('String str\t: ', str)
pos = f.tell()
print('Current position\t:', pos)
str1=f.read(5)
print('Str1\t:',str1)
pos = f.seek(0, 0)
print('Current position\t:',pos)
str = f.read(20);
print('Again read String is : ', str)
f.close()
```

Output:

```
String str    :   Hi thereHow are
Current position  : 15
Str1 :   you
Current position  : 0
Again read String is :   Hi thereHow are you
```

Creating directories and navigating between them

One can also create directories in Python, using the **mkdir()** function. The function takes name of the directory as one of the arguments. The reader is advised to go through the references at the end of this book for a detailed description of this function. The **chdir()** function changes the current directory and the **getpwd()** function prints the name(along with the path) of the current working directory. The use of these functions has been demonstrated as follows.

```
'>>> import os
>>> os.mkdir('PythonDirectory')
>>> os.chdir('PythonDirectory')
>>> os.getcwd()
'C:\\Python\\file handling\\PythonDirectory'
>>>
```

An example of encryption

The following illustration uses the **ord(c)** function which prints the **ASCII** value of the character "**c**" and that of the **chr(n)** function, which returns a character corresponding to the ASCII value **n.**

Illustration 7.6:

Write "Hi there how are you" in a file called "TextFile.txt." Now, read characters from the file, one by one and write the character obtained by adding k(entered by the user) to the ASCII value of the character. Also, decrypt the string in the second file by subtracting "k" from the ASCII values of the characters in the second file.

Solution:

Code:

```
f=open('TextFile.txt','w')
f.write('Hi there how are you')
f.close()
k=int(input('Enter a number'))
f =open('TextFile.txt','r')
f1=open('TextFile1.txt','w')
for s in f.read():
    for c in s:
        print('Character ',c,' Ascii value\t:',ord(c))
        f1.write(str(chr(ord(c)+k)))

f1.close()
print((open('TextFile1.txt').read()))
f1 =open('TextFile1.txt','r')
f2=open('TextFile2.txt','w')
for s in f1.read():
    for c in s:
        print('Character ',c,' Ascii value\t:',ord(c))
        f2.write(str(chr(ord(c)-k)))

f2.close()
print((open('TextFile2.txt').read()))
```

Output:

```
Enter a number4
Character  H  Ascii value  : 72
Character  i  Ascii value  : 105
Character     Ascii value  : 32
Character  t  Ascii value  : 116
Character  h  Ascii value  : 104
Character  e  Ascii value  : 101
Character  r  Ascii value  : 114
Character  e  Ascii value  : 101
Character     Ascii value  : 32
Character  h  Ascii value  : 104
Character  o  Ascii value  : 111
Character  w  Ascii value  : 119
Character     Ascii value  : 32
Character  a  Ascii value  : 97
Character  r  Ascii value  : 114
Character  e  Ascii value  : 101
Character     Ascii value  : 32
Character  y  Ascii value  : 121
Character  o  Ascii value  : 111
Character  u  Ascii value  : 117
Lm$xlivi$ls{$evi$}sy
Character  L  Ascii value  : 76
Character  m  Ascii value  : 109
Character  $  Ascii value  : 36
Character  x  Ascii value  : 120
Character  l  Ascii value  : 108
Character  i  Ascii value  : 105
Character  v  Ascii value  : 118
Character  i  Ascii value  : 105
Character  $  Ascii value  : 36
Character  l  Ascii value  : 108
Character  s  Ascii value  : 115
Character  {  Ascii value  : 123
Character  $  Ascii value  : 36
Character  e  Ascii value  : 101
Character  v  Ascii value  : 118
```

```
Character  i  Ascii value  : 105
Character  $  Ascii value  : 36
Character  }  Ascii value  : 125
Character  s  Ascii value  : 115
Character  y  Ascii value  : 121
Hi there how are you
>>>
```

Copying the contents of one file to another

In order to copy the content of a file to another, the first file is opened in the read mode and the second file is opened in the write mode. The lines of the first file are then read one by one (using the read function) and written to the other file (using the write function). The program follows.

Illustration 7.7:

Copy the contents of one file to another.

Solution:

Code:

```
f1=open('source.txt','r')
f2=open('dest.txt','w')
char=f1.read()
print(char)
f2.write(char)
f1.close()
f2.close()
```

Illustration 7.8:

Write a program to count the number of words in a file.

Solution:

The number of words in a file can be calculated by initializing the 'n' (=1) and reading one line at a time, splitting the line into words and successively incrementing the value of **n**.

Code:

```
fname = 'source.txt'
n = 0
with open(fname, 'r') as f:
```

```
for line in f:
    w = line.split()
    n += len(w)
print('Word count\t:',n)
```

7.7 CONCLUSION

File handing provides the user with the power of persistence. The user must be equipped with the knowhow of the file access modes, the **open(), close()** functions and the functions which help in reading a file and writing to it. This chapter briefly explains the most essential functions used for file handling in Python. This chapter also introduces the user to the OS methods and the essential file attributes to help the user achieve the task at hand. The chapter also includes ample illustrations and explanation, to make the concept clear in the simplest manner.

POINTS TO REMEMBER

- The **open** function takes three arguments
- The mode of opening file decides the tasks that can be accomplished
- The file should be closed after the required task has been completed
- The **seek** method helps to move the cursor within a file
- The value of **file.closed** is 1 after the **close** function has been called
- The **file.name** attribute prints the name of the file
- The **file.mode** attribute gives the file access mode
- The **os.getpwd** function returns the present working directory
- The **os.chdir** function changes the directory

EXERCISES

Multiple Choice Questions

1. Which of the following is a solid argument for using file handling?

    (a) It is not possible to store all data produced by the program in the main memory

    (b) It is used for persistent storage

    (c) Both

    (d) None of the above

2. In which of the format the end of the line is denoted by "\n" and "\r"?

   (a) Text   **(b)** Binary

   (c) Both   **(d)** None of the above

3. To be able to use a file it must be opened. The reason for doing so is

   (a) To allocate memory to the object so formed

   (b) To specify the access mode

   (c) To specify the offset (optional)

   (d) All of the above

4. In f =open('abc.txt', 'r'), the offset is

   (a) 0 from the beginning   **(b)** 0 from the end

   (c) Random   **(d)** None of the above

5. How many arguments does the open function take?

   (a) 1   **(b)** 2

   (c) 3   **(d)** None of the above

6. The file must be closed if it is opened in which of the following mode?

   (a) r   **(b)** w

   (c) both   **(d)** None of the above

7. If the file is not opened successfully, which of the following exception is raised?

   (a) File not found   **(b)** IOERROR

   (c) IO   **(d)** None of the above

8. In f =open('abc.txt', 'w'), if the file "abc.txt" does not exist, then

   (a) IOERROR is raised   **(b)** The program does not compile

   (c) A new file is created   **(d)** None of the above

9. Which suffix is used for opening a binary file

(**a**) b   (**b**) bin

(**c**) ab   (**d**) None of the above

10. The + suffix allows

(**a**) Read   (**b**) Read and write

(**c**) Read or write   (**d**) None of the above

11. How many file access modes are there in Python?

(**a**) 3   (**b**) 6

(**c**) 9   (**d**) 12

12. The integer argument in the **read**() function denotes the number of bytes to be read, if no argument is given, which of the following is the default argument?

(**a**) -1   (**b**) 0

(**c**) len(file)   (**d**) None of the above

13. To read all the lines in a file, which of the following function can be used?

(**a**) readline()   (**b**) readlines()

(**c**) both   (**d**) None of the above

14. Which of the following methods can be used to write a list of strings in a file?

(**a**) writeline()   (**b**) writelines()

(**c**) write()   (**d**) None of the above

15. Which of the following argument in the seek() function denotes the end of the file?

(**a**) 1   (**b**) 2

(**c**) 0   (**d**) None of the above

16. Which function returns the descriptor of the file?

    (a) fileno()   **(b)** filedisp()

    (c) descriptor()   **(d)** None of the above

17. The linesep function is used to find which of the following?

    (a) The new line   **(b)** The end of the file

    (c) The current directory   **(d)** None of the above

18. Which of the following is not a file attribute?

    (a) closed   **(b)** opened

    (c) name   **(d)** softspace

19. In which of the following variable, the command line argument is saved?

    (a) argv   **(b)** argc

    (c) Both   **(d)** None of the above

20. Which of the following function helps to create a directory?

    (a) os.mkdir()   **(b)** os.chdir()

    (c) os.getpwd()   **(d)** None of the above

21. Which of the following function helps to change the current directory?

    (a) os.mkdir()   **(b)** os.chdir()

    (c) os.getpwd()   **(d)** None of the above

22. Which of the following function helps to print the name of the current directory?

    (a) os.mkdir()   **(b)** os.chdir()

    (c) os.getpwd()   **(d)** None of the above

23. Which function is used to find the ASCII value of a character?

    (a) ascii   **(b)** ord

    (c) chord   **(d)** None of the above

24. Which of the following is not a file access mode in Python?

    (a) a   **(b)** ab

    (c) ab+   **(d)** abc

25. Which of the following is incorrect?

    (a) f =open('file.txt')   **(b)** f =open('file.txt','r')

    (c) f =open('file.txt','r',0)   **(d)** None of the above is incorrect

Theory

1. What is the importance of file handling? Explain the mechanism of file handling in Python.

2. Explain various file access modes.

3. Explain the signature and usage of the following functions

    a. open   b. close

    c. read   d. write

    e. readline   f. readlines

    g. writeline   h. seek

4. What are file attributes? Explain the file attributes provided by python.

5. Briefly explain the usage of the following os functions in Python

    a. mkdir

    b. chdir

    c. getpwd

Programming Exercises

1. Write a program to copy the contents of a file to another.

2. Write a program to capitalize the first character of each word in a file.

3. Write a program to find the ASCII value of each character in a file.

4. Write a program to find the frequency of each character in a file.

5. Write a program to find all occurrences of a word, entered by the user, in a given file.

6. Write a program to replace a given character with another in a file.

7. Write a program to replace a given word with another, in a given file.

8. Write a program to find the frequency of a given word in a file.

9. Write a program to find the word used minimum number of times in a given file.

10. Write a program and change the name of a file to the name entered by the user.

11. Write a program to create a directory and then create a new file in it.

12. Write a program to print the name, number of characters and number of spaces in a file.

13. Write a program to convert the characters of a given file to binary format.

14. Write a program to find the words starting with a vowel from a given file.

15. Write a program to implement any substitution cipher on the text of a given file.

8

LISTS, TUPLE, AND DICTIONARY

Objectives

After reading this chapter, the reader should be able to
- Understand the difference between list, tuple, and dictionary
- Understand slicing and indexing
- Understand strings in Python
- Understand the in-built functions of list, tuple, dictionaries

8.1 INTRODUCTION

Python provides us with lists, tuples, and dictionaries, all of which have become synonym with ease of programming and can be used in diverse applications. In Python, a list can contain different types of elements and is mutable whereas a tuple can contain different types of elements and is immutable. In a dictionary, the items can be accessed using strings as indices (Figure 8.1).

Dictionaries are collections in which each value is associated with a key. This key-value pair constitutes an item in a dictionary. In order to understand the idea, consider the following example. In the 1980s, the phone numbers of the people in a city could be found using a telephone directory. The directory had phone numbers of people of the city in the lexicographic order (of names). So, using one's name, their phone number could be found. That is, the index of value (phone number in this case) was the name of the person. Python dictionary uses the same idea. However, a directory could have a name twice but in a dictionary each key is unique.

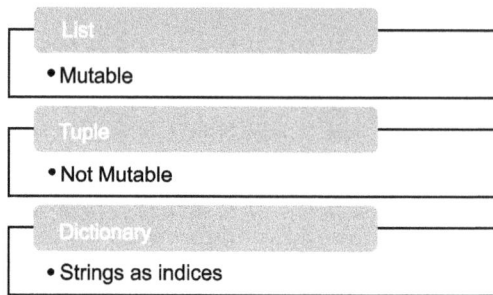

FIGURE 8.1 Lists, tuples, and dictionaries.

The following chapter discusses strings. Strings are the sequence of characters. These data structures are used to store text. For example, if one wants to store the name of a person, or for that matter, his address, then strings are the most appropriate data structures. As a matter of fact, the knowledge of strings is essential in the development of many applications like word processors and parser.

This chapter has been organized as follows. The second section discusses lists, the third discusses tuples and the next section discusses dictionaries. The last section concludes.

8.2 LISTS

In Python, a list is a sequence object. A list can be an empty list ([]) It may contain any number of elements, which may or may not be of the same type. Also, a list can contain different types of elements, including a list. Moreover, a list inherits many properties of a string. However, unlike strings, they are mutable and may also contain different types of elements.

In the following snippet, **L1** is an empty list, **L2** contains 2, 4, 8, 16, 32, and 64. **L3** contains an integer (2), a string ("Harsh"), another integer (3), another string ("Manan"), a double (4.7865), and another string ("Ali"). **L4** is a list of lists. It contains three lists, each containing different number of elements.

Code: Creation

```
#Empty List
L1=[]
#Homogeneous list
L2=[2,4,8,16,32,64]
```

```
#Heterogenous list
L3=[2, 'Harsh', 3,'Manan', 4.7865,'Ali']
#List of Lists
L4=[[1,2],[2,4,8],[3,9,27,81]]
print('First List',L1,'\nSecond List',L2,'\nThird List',L3)
```

Output:

```
First List []
Second List [2, 4, 8, 16, 32, 64]
Third List [2, 'Harsh', 3, 'Manan', 4.7865, 'Ali']
```

8.2.1 Accessing Elements: Indexing and Slicing

The elements of a list can be accessed using indexing. The index of the first element of the list is 0, that of the second element is 1, and so on. The negative index denotes the elements from the end. For example, **L[-1]** denotes the last element of the list **L**, likewise, **L[-2]** denotes the second last element of a list (Figure 8.2).

```
L = [1, 21, 5, 7, 9, 11]
L[0] is 1
L[-1] is 11
L[3] is 7
```

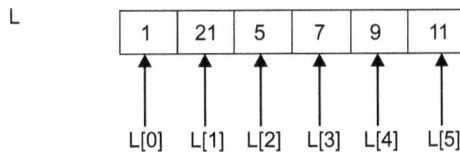

FIGURE 8.2 Indexing in Python Lists. Note that **L[0]** denotes the first element, **L[1]** denotes the second element and **L[-1]** (that is **L[5]**) denotes the last element.

Slicing is generally used to take out a sublist from a given list. In the above case

L[:2] is a sublist having all the elements of **L** starting from the first element till the element at the first index (not the element at index 2) that is [1, 3].

L[2:] is a sublist having all the elements of **L** starting from the element at the second index till the last element that is [5, 7, 9 , 11].

L[1:4] is a sublist having all the elements of **L** starting from the element at the first index till the element at the third index that is [3, 5, 7].

8.2.2 Mutability

It may be stated that a list is mutable. That is, one can update an element of the list. In order to understand this, consider the following snippet. The following code sets the second element of the list as 675. Note that an element of a list can be accessed by a square bracket.

Code: Mutability

```
#Mutable
L2=[2,4,8,16,32,64]
L2[1]=675
#The second element of the list L2 becomes 675
print(L2)
```

Output:

```
[2, 675, 8, 16, 32, 64]
```

8.2.3 Operators

Python provides four operators for lists. These are: +, *, **in** and **not in**. The + operator in **L1+L2** concatenates two lists **L1** and **L2**. The * operator in **L1*n** repeats the elements of the list **L1**, **n** times. The **in** operator checks if a given element is available in the list. The **not in** operator checks if a given element is not available in the list. In order to understand the operators, consider the following code. The list **L9** is the concatenation of **L2** and **L8**. The list **L10** has elements of **L2** repeated thrice. Note that we cannot use the – operator with a list.

Code: Operator +, *, in and not in

```
#Using + operator
L8= [90, 80, 70]
#Concatenate two lists
L9=L2+L8
print(L9)
#Using * operator
num=3
#repeat the elements 'num' times
L10=L2*num
```

```
print(L10)
#- Illegal
L11=L2-L8
print(L11)
```

Output:

```
[2, 675, 8, 16, 32, 64, 90, 80, 70]
[2, 675, 8, 16, 32, 64, 2, 675, 8, 16, 32, 64, 2, 675, 8, 16,
    32, 64]
TypeError Traceback (most recent call last) <ipython-input-14-
    7ccfbd405905> in <module>()1
#- Illegal----> 2L11=L2-L8 3 print(L9)
TypeError: unsupported operand type(s) for -: 'list' and 'list'
```

8.2.4 Traversal

A list is a sequence. Each element in a list can be accessed using its index in square brackets. The elements can also be accessed using loops and iterators. Though any loop can be used to accomplish the said task, **for** loop's use make things easy. The **for** loop can be used to traverse a list as shown in the following code which stores each item of the list in **i** and prints them. In the next code, nested loops are used to process lists of lists. The last code processes a list having different types of elements.

Code: Print each element in a list

```
L2=[2, 675, 8, 16, 32, 64]
#Print all elements
for i in L2:
    print(i)
```

Output:

```
2
675
8
16
32
64
```

Code: Processing list of lists using nested loops

```
L4=[[1,2],[2,4,8],[3,9,27,81]]
#Print all the lists in list L4
for i in L4:
    print(i)
```

Output:

```
[1, 2]
[2, 4, 8]
[3, 9, 27, 81]
```

Code: Processing all elements of a list of lists using nested loops

```
L4=[[1,2],[2,4,8],[3,9,27,81]]
#All the elements of the lists inside the list L4 are accessed
    and printed using nested loops
for i in L4:
    print('List', end='\t')
    for j in i:
      print(j, end=' ')
    print('')
```

Output:

```
List 1 2
List 2 4 8
List 3 9 27 81
```

Code: Processing a list containing different types of elements

```
L3=[2, 'Harsh', 3,'Manan', 4.7865,'Ali']
#The different types of elements are printed using for loop
for i in L3:
    print(i)
```

Output:

```
2
Harsh
3
Manan
4.7865
Ali
```

The above discussion revealed the methods of creation of a list, its processing, slicing, and indexing. Having seen the basics, let us now perform slightly more complicated tasks by using the predefined functions of lists.

8.2.5 Functions

The functions associated with a list are shown in Table 8.1. The table enlists some of the most important methods used for lists and their explanation. Note that the following tasks can also be accomplished without using the functions. However, these functions help us to accomplish many tasks efficiently and effectively. The code that follows demonstrates the use of these functions.

TABLE 8.1 Functions of List.

Function	Explanation
list.append(item)	The append method adds "item" at the end of the list.
list.extend(item)	The extend method adds the "item" at the end of the list.
list.insert(index, item)	The insert function inserts "item" at "index."
list.remove(item)	The remove function removes the first instance of "item" from the list.
list.pop(index)	The pop function removes the item at the given index in the list. Also, pop() removes the last element from the list.
list.clear()	The clear method removes all the elements from the list.
list.index(item)	Return zero-based index in the list of the first item whose value is equal to item.
list.count(item)	The count function counts the number of instances of "item" in the list.
list.sort()	The sort function sorts the items of the list. An additional argument reverse=True sorts the list in descending order.
list.reverse()	The reverse function reverses the order of the list.
list.copy()	The copy function returns a shallow copy of the list.

The above functions have been used as follows. Note that the input is prefixed with >> and the statements without >> shows the output.

```
#Functions
>>L13=[1,2,3]
>>L14=[7,8,9,10]
>>L14.extend(L13)
```

```
>>print(L14)
[7, 8, 9, 10, 1, 2, 3]
>>L14.remove(10)
>>print(L14)
[7, 8, 9, 1, 2, 3]
>>L14.insert(2,787)
>>print(L14)
[7, 8, 787, 3, 9, 1, 2, 3]
>>L14.index(3)
3
>>L14.count(3)
2
>>L14.pop()
3
>>L14.reverse()
>>print(L14)
[2, 1, 9, 3, 787, 8, 7]
>>L14.sort()
print(L14)
[1, 2, 3, 7, 8, 9,`787]
```

8.3 TUPLE

A tuple can be empty or can also contain any number of elements. It may also contain different types of elements like strings, integers, float, doubles can even contain lists, and can be a combination of the above. For example, in the following code, **T2** contains an integer 2, a string "Harsh," two more integers 5 and 67. **T3** is a tuple consisting of three different tuples – (1,2), (4,5), and (7,8). **T4** contains three different lists and **T5** consists of a tuple having elements 3 and 4, a list with elements 4, 5, and 6, and a string, "Harsh."

Code:

```
#Create tuple
T1=(2,3,4,5,6,7,8,9,) #Tuple homogeneous
print(T1)
T2=(2,'harsh',5,67) #Tuple heterogeneous
print(T2)
```

```
T3=((1,2),(4,5),(7,8)) #Tuple of tuples
print(T3)
T4=([1,2],[-4],[8,9,10]) #Tuple of lists
print(T4)
T5=((3,4),[4,5,6],'Harsh') #Tuple consisting a tuple, list and a
    string
print(T5)
```

Output:

```
(2, 3, 4, 5, 6, 7, 8, 9)
(2, 'harsh', 5, 67)
((1, 2), (4, 5), (7, 8))
([1, 2], [-4], [8, 9, 10])
((3, 4), [4, 5, 6], 'Harsh')
```

8.3.1 Accessing Elements of a Tuple

The elements of a tuple can be accessed using indexing. The index of the first element of the tuple is 0, that of the second element is 1, and so on. The negative index denotes the elements from the end. For example, **T[-1]** denotes the last element of the list **T**, likewise, **T[-2]** denotes the second last element of a list (Figure 8.3).

```
T = (1, 3, 5, 7, 9, 11)
T[0] is 1
T[-1] is 11
T[3] is 7
```

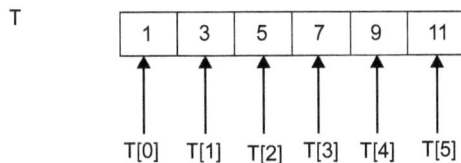

FIGURE 8.3 Indexing in Python tuple. Note that **T[0]** denotes the first element, **T[1]** denotes the second element and **T[-1]** (that is **T[5]**) denotes the lest element.

Slicing is generally used to take out a subtuple from a given tuple. In the above case

T[:2] is a sublist having all the elements of **T** starting from the first element till the element at the first index (not the element at index 2) that is (1, 3).

T[2:] is a sublist having all the elements of **T** starting from the element at the second index (the third element as the list is a zero-based index) till the last element that is (5, 7, 9, 11).

T[1:4] is a sublist having all the elements of **T** starting from the element at the first index till the element at the third index that is (3, 5, 7).

8.3.2 Nonmutability

Unlike a list, a tuple is nonmutable as one cannot update any element of the tuple. On setting the element of a tuple to a particular number we get the following error.

```
# Non-Mutability
T1[2]=3
```

```
-----------------------------------------------------------------
TypeError                                    Traceback (most recent
    call last)
<ipython-input-12-bc71ae9aae00> in <module>()
----> 1 T1[2]=3
TypeError: 'tuple' object does not support item
```

8.3.3 Operators

Python provides four operators for tuples. These are: +, *, **in**, and **not in**. The + operator in **T1+T2** concatenates two tuples **T1** and **T2**. The * operator in **T1*n** repeats the elements of the list **T1**, **n** times. The **in** operator checks if a given element is available in the tuple. The **not in** operator checks if a given element is not available in the tuple. In order to understand the operators, consider the following code. The list **T9** is the concatenation of **T2** and **T8**. The list **T10** has elements of **T2** repeated thrice. Note that we cannot use the – operator with a tuple.

Code: Operator +, *, in and not in

```
T2=(90, 80, 70)
#Concatenate two tuples
T9=T2+T8
print(T9)
#Using * operator
num=3
```

```
#repeat the elements 'num' times
T10=T2*num
print(T10)
#- Illegal
T11=T2-T8
print(T11)
```

Output:

```
(2, 675, 8, 16, 32, 64, 90, 80, 70)
(2, 675, 8, 16, 32, 64, 2, 675, 8, 16, 32, 64, 2, 675, 8, 16,
    32, 64)
TypeError Traceback (most recent call last) <ipython-input-14-
    7ccfbd405905> in <module>()1
#- Illegal----> 2L11=L2-L8 3 print(L9)
TypeError: unsupported operand type(s) for -: 'list' and 'list'
```

8.3.4 Traversal

The **for** loop can be used to traverse a tuple in the same way as a list. In order to understand the usage, consider the following code which stores each item of the tuple in **i** and print them.

Code:

```
#Display
def display(T):
    for i in T:
print(i, end=',')
print()
T=(2,3,4,5,6,7,8,10,)
display(T)
```

Output:

```
2,3,4,5,6,7,8,10,
```

The above discussion revealed the methods of creation of a tuple, its processing, slicing, and indexing. Having seen the basics, let us now perform slightly more complicated tasks like displaying a tuple entered by the user and finding the maximum and the minimum elements from a tuple.

Illustration 8.1:

*Ask the user to enter a tuple and find the element having the maximum value from the tuple, without using the **max** function.*

Solution:

In the following code, the variable **maxVal** would store the maximum element of the tuple. Initially, it stores the first element of the tuple. As we traverse, if we can find an element which is greater than **maxVal**. The value of **maxVal** would be updated by the greater value found in the tuple. At the end of the program, the value of **maxVal** is printed.

Program:

```
#Find maximum
def max(T):
    maxVal=T[0]
    for i in T:
        if i>maxVal:
            maxVal=i
    return(maxVal)
T=(2,13,41,455,678,7,8,10,)
m=max(T)
print(m)
```

Output:

```
678
```

8.3.5 Functions

Table 8.2 shows the various functions for tuples. These functions can be used to carry out many of the tasks easily and efficiently. The illustration that follows shows the use of these functions.

TABLE 8.2 Functions pertaining to Tuple.

Function	Description
len(T)	The function returns the number of elements in the tuple T
max(T)	The function returns the element with the maximum value
min(T)	The function returns the element with the minimum value
tuple(S)	This function converts the list (S) into a tuple

Illustration 8.2:

Write a program that asks the user to enter two tuples and checks the following:

(*a*) Compare the two tuples.
(*b*) Count the number of elements in the two tuples.
(*c*) Returns the element with the maximum value in the first tuple.
(*d*) Returns the element with the minimum value in the second list.
(*e*) Convert a list into a tuple.

The above tasks should be accomplished using the methods of the functions.

Solution:

Table 8.2 shows the various functions associated with tuples. The program to accomplish the given task follows.

Program:

```
print('For the first tuple')
x1=int(input('Enter the first number'))
y1=int(input('Enter the second number'))
T1=(x1,y1)
print('For the second tuple')
x2=int(input('Enter the first number'))
y2=int(input('Enter the second number'))
T2=(x2,y2)
l=len(T1)
print('Length of the first tuple',l)
m=max(T1)
print('Maximum in the first tuple',m)
m=min(T1)
print('Minimum in the first tuple',m)
if(T1 == T2):
print('The two tuples are same')
else:
print('They are not same')
L=[2,4,3,1]
T=tuple(L)
print(T)
```

Output:

```
For the first tuple
Enter the first number1
Enter the second number2
For the second tuple
Enter the first number1
Enter the second number2
Length of the first tuple 2
Maximum in the first tuple 2
Minimum in the first tuple 1
The two tuples are same
(2,4,3,1)
```

8.4 ASSOCIATE ARRAYS AND DICTIONARIES

An associate array is a data structure having a collection of pairs of keys and corresponding values. The keys in an associative array are unique. This is because each value is accessed using the corresponding key. The presence of more than one key with the same value would lead to ambiguity. In an associate array, one can insert items in the collection, remove items, update them, and look for a particular value associated with the key.

Dictionaries in Python are also a kind of associate arrays. They contain key-value pairs, where a key can be any nonmutable data structure like a string, tuple, etc. In Python, one can create a dictionary using curly braces. The elements of a dictionary are separated by a comma and each element has two parts: a key and a value. The key, in a dictionary, can be any immutable object e.g., a string. The value corresponding to a particular key is written by putting a colon after the key. For example, the following statement creates a dictionary called **PersonAge**. The first element of the dictionary has "Harsh" as a key and the value corresponding to this key is 100. The second element of the dictionary has "Rohan" as its key and the value corresponding to this key is 21 and third element of the dictionary has "Tarush" as key and the value corresponding to this key is 21. The **print** function can be used to print the dictionary. The dictionary has been shown in Figure 8.4 and the items and corresponding keys in Table 8.3.

```
PersonAge={'Harsh':100,'Rohan':21,'Tarush':20}
print(PersonAge)
```

Output:

```
{'Harsh': 100, 'Rohan': 21, 'Tarush': 20}
```

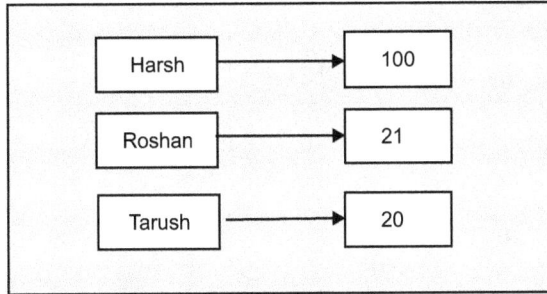

PersonAge

FIGURE 8.4 Dictionary associates a key with value.

TABLE 8.3 Dictionary Person Age.

Key	Value	Item
"Harsh"	100	("Harsh," 100)
"Rohan"	21	("Rohan," 21)
"Tarush"	20	("Tarush," 20)

8.4.1 Displaying Elements of a Dictionary

The items of a dictionary can be seen by using the **items()** function. The following statement prints the items of the dictionary **PersonAge**.

```
PersonAge.items()
```

Output:

```
dict_items([('Harsh', 100), ('Rohan', 21), ('Tarush', 20)])
```

Syntax

<name of the dictionary>.item()

The value corresponding to a particular key can be seen by using the key as the index. For example, the value corresponding to the key "Harsh" can be seen as follows:

```
PersonAge['Harsh']
```

Output:

```
100
```

Syntax: Indexing in dictionary

<name of the dictionary>[<key>]

Dictionaries in Python come with many build-in functions for displaying the keys, values, or the complete item. Some of them are presented in Table 8.4.

TABLE 8.4 Functions for displaying keys, values, and items of a dictionary.

Name of the function	Explanation
Keys()	Displays the list of keys of the given dictionary
values()	Displays the list of values of the given dictionary
Items()	Displays the items of the given dictionary

The following code shows how to create a dictionary of 5 elements. Note that each string is enclosed in a single quote and the value corresponding to a given key is separated using a colon.

```
BookPages = {'Programming in C': 250, 'Programming in C#': 450,
    'Python for beginners': 400, 'Physics': 100, 'Chemistry':
    120}
```

The **keys** function will display the keys of the dictionary.

```
BookPages.keys()
```

Output:

```
dict_keys(['Programming in C', 'Programming in C#', 'Python for
    beginners', 'Physics', 'Chemistry'])
```

Syntax

<name of the dictionary>.keys()

The **values** function will display the values of the dictionary.

```
BookPages.values()
```

Output:

```
dict_values([250, 450, 400, 100, 120])
```

Syntax

<name of the dictionary>.values()

8.4.2 Some Important Functions of Dictionaries

This section discusses some of the important functions for dealing with dictionaries. The meaning of the function along with the usage has been presented in this section.

8.4.2.1 *The **len** function returns the number of **elements** in a given dictionary.*

Example

```
len(BookPages)
```

Output:

```
5
```

8.4.2.2 *The **max** function returns the key with **maximum** value. If the key is a string, then the value in the lexicographic ordering would be returned.*

Example:

```
max(BookPages)
```

Output

```
'Python for beginners'
```

8.4.2.3 *The **min** function returns the key with **minimum** value. If the key is a string, then the value in the lexicographic ordering would be returned.*

Example:

```
min(BookPages)
```

Output:

```
'Chemistry'
```

8.4.2.4 *The **sorted** function would sort the **elements** of a given dictionary by their keys. If the keys are strings then lexicographic ordering would be followed.*

```
sorted(BookPages)
```

Output:

```
['Chemistry', 'Physics', 'Programming in C', 'Programming in
    C#', 'Python for beginners']
```

8.4.2.5 *The **pop** function takes **out** the element with the given key from the dictionary.*

```
BookPages.pop('Physics')
```

Output:

```
100
```

Note that after calling the **pop** function, the item with the key "Physics" has been removed from the dictionary.

```
BookPages
```
```
{'Chemistry': 120, 'Programming in C': 250, 'Programming in C#':
    450, 'Python for beginners': 400}
```

8.4.3 Input from the User

The following illustration shows how to ask the user to input the values of the dictionary. The program asks the user to enter the name of the book and the corresponding number of pages.

A dictionary called **BookPages** is initialized to {}. This is followed by asking the user to enter the number of elements. In each iteration, the key and the corresponding value is taken and inserted into the dictionary.

Input

```
BookPages={}
n=int(input('Enter the number of pages\t:'))
for i in range(n):
    Book=input('Enter the name of the book\t:')
    pages=int(input('Enter the number of pages\t:'))
    BookPages[Book]=pages
print(BookPages)
```

Output:

```
Enter the number of pages :5
Enter the name of the book :Programming in C
```

```
Enter the number of pages :250
Enter the name of the book :Programming in C#
Enter the number of pages :450
Enter the name of the book :Python for beginners
Enter the number of pages :400
Enter the name of the book :Physics
Enter the number of pages :100
Enter the name of the book :Chemistry
Enter the number of pages :120
{'Programming in C': 250, 'Programming in C#': 450, 'Python for
    beginners': 400, 'Physics': 100, 'Chemistry': 120}
```

8.5   CONCLUSION

Lists, tuples, and dictionaries are the most important objects in Python. The student must be equipped with the power of these objects before proceeding any further. The following chapters make extensive use of the concepts studied in this chapter; therefore, the reader is advised to spend considerable time in attempting the exercises that follow.

GLOSSARY

List: A Python list is a sequence object. It may contain any number of elements, which may or may not be of the same type. They are mutable.

Tuples: A tuple can contain different types of elements and are not mutable.

Dictionary: A dictionary maps the index (which may be a string) to a value.

POINTS TO REMEMBER

- Lists are mutable.
- A list may contain different types of elements.
- Tuples are not mutable.
- Dictionary maps a key to a value.
- Strings in Python are nonmutable.

EXERCISES

Multiple Choice Questions

1. In Python a list is

    (a) Mutable

    (b) Can contain different types of elements

    (c) Both

    (d) None of the above

2. A list in Python

    (a) Is 0 index based

    (b) Can be accessed by indexing

    (c) Negative indices can be used in indexing

    (d) All of the above

3. L_names=["Harsh," "Amit," "Sahil," "Viresh"]

    L_names.sort()

    print(L_names)

    (a) ["Amit," "Harsh," "Sahil," "Viresh"]

    (b) ["Amit," "Sahil," "Viresh"]

    (c) ["Harsh," "Amit," "Sahil," "Viresh"]

    (d) Error as sort function cannot be applied on a list having strings as its elements.

4. min(L_names)

    (a) "Harsh"   **(b)** "Amit"

    (c) "Viresh"   **(d)** None of the above

5. L_names[2:3]

    (a) ["Sahil"]   **(b)** ["Sahil," "Raven"]

    (c) ["Harsh," "Sahil"]   **(d)** None of the above

6. Which function is used to delete an element from a list?

   (a) Clear

   (b) Pop

   (c) Append

   (d) None of the above

7. Can a tuple have a list as its element?

   (a) Yes

   (b) No

   (c) Depending upon the situation

   (d) None of the above

8. Which of the following is used to delete a tuple?

   (a) del

   (b) clear

   (c) Both

   (d) None of the above

9. Which function can be used to convert a list to a tuple?

   (a) tuple

   (b) to_tuple

   (c) list

   (d) None of the above

10. Which of the following can be the index in a dictionary?

   (a) String

   (b) List

   (c) Tuple

   (d) None of the above

11. Which of the following cannot be the index in a dictionary?

   (a) String

   (b) List

   (c) Character

   (d) None of the above

12. Which of the following is used to take out a particular element of a dictionary?

   (a) Clear

   (b) Pop

   (c) Push

   (d) None of the above

Theory

1. What is a list? Is it mutable?

2. Explain indexing and slicing in lists.

3. Explain the following functions vis-à-vis Python lists.

    (a) append() **(b)** extend()

    (c) insert() **(d)** remove()

    (e) pop() **(f)** clear()

    (g) index() **(h)** count()

    (i) sort() **(j)** reverse()

    (k) copy()

4. Explain the purpose of the following operators vis-à-vis lists.

    (a) + **(b)** *

    (c) in **(d)** not in

5. Explain the various functions of a dictionary. Is a dictionary the same as a tuple?

6. Explain the following functions:

    (a) isanum() **(b)** isalpha()

    (c) isdecimal() **(d)** isdigit()

    (e) isidentifier() **(f)** islower()

    (f) isupper() **(h)** swapcase()

    (i) isspace() **(j)** lstrip()

    (k) rstrip() **(l)** replace()

    (m) join()

Programming Exercises

1. Ask the user to enter the names of his friends in a list.

    (a) Sort the above list.

    (b) Can we apply the **min** function to the above list?

    (c) Create a sublist from the above list having elements at the odd positions.

(d) Does the list contain any duplicate elements?

(e) From the above list find the name having the maximum number of vowels.

2. Ask the user to enter the x and y coordinates of n points and find the distances between all the possible pairs.

3. Write a program to reverse a string.

4. Write a program to find the sum of ASCII values of the characters of a given string.

5. Write a program to find a particular substring in a given string.

6. Write a program to split a given text into tokens.

7. Write a program to check which of the tokens obtained in the above question are in the title case.

8. Write a program to check how many alphanumeric strings are there in the tokens obtained in question 6.

9. Write a program to check how many alpha strings are there in the tokens obtained in question 6.

10. Write a program to check how many numeric strings are there in the tokens obtained in question 6.

9

ITERATIONS, GENERATORS, AND COMPREHENSIONS

Objectives

After reading this chapter, the reader should be able to

- Understand the use and application of iterators
- Use iterators to produce sequences
- Use generators to generate sequences
- Understand and use list comprehensions

9.1 INTRODUCTION

Python is powerful due to the presence of lists, strings, tuples, dictionary, and files. However, one should be able to efficiently access and manipulate the elements of these objects to accomplish a given task. Though, **for** loop can be used to access and manipulate the elements of these objects, Python comes with better option namely iterators. Iterators help us to achieve the said goal efficiently and effectively. One can also define an iterable object, in Python. Generators, in Python, facilitates the dynamic generation of lists and sequences. This chapter also introduce comprehensions, which provide an elegant way to craft lists, tuples, and sets.

The chapter has been organized as follows. The second section of this chapter revisits **for**. The iterators have been introduced in the third section of this chapter. The fourth section explains how to define an iterable object. The generators have been introduced and explained in the fifth section of this chapter. The sixth section of this chapter deals with the comprehensions.

The chapter assumes importance as it forms the foundation of many of the difficult tasks presented in the following chapters. Also, the knowledge of these would make the day to day tasks easy and spare the programmer from the horror of writing longer codes.

9.2 THE POWER OF "FOR"

A **for** loop can be used to iterate through a list, tuple, string, or a dictionary. This section gives a brief description of the **for** loop for the above iterable objects. Let us start with the syntax of **for**. Note that, in the following code, an element from L is extracted one by one and, in each iteration, i stores the object.

Syntax:

```
for i in L:
    #do something
```

L is list, string, tuple or dictionary

When one writes **"i in L,"** where **L** is a list, **i** points to the first element of the list and as the iteration progresses, **i** points to the second element, the third element, and so on. These elements can be, then, independently manipulated. The concept has been exemplified in Illustration 9.1. The illustration shows the manipulation of a list using the **for** loop. In the illustration, the given list contains a set of numbers, some of them positive and some negative. The negative numbers are appended to a list called **N**, whereas the positive numbers are appended in a list called **P**.

Illustration 9.1:

From a given list, put all the positive numbers in one list and negative numbers in the other list.

Solution:

Create two lists **P** and **N**. Initialize both of them to []. Now check each number in the list. If the number is positive, put it in P and if the number is negative put it in N.

Program:

```
L= [1, 2, 5, 7, -1, 3, -6, 7]
P=[]
N=[]
for num in L:
  if(num>0):
     P.append(num)
  elif (num<0):
     N.append(num)
print('The list of positive numbers \t:',P)
print('The list of negative numbers \t:',N)
>>>
```

Output:

```
The list of positive numbers   : [1, 2, 5, 7, 3, 7]
The list of negative numbers   : [-1, -6]
>>>
```

A **for** loop can also be used to manipulate strings. When one writes **"i in str,"** where **str** is a string, **i** points to the first character of the string and as the iteration progresses, **i** points to the second character, the third character and so on. These characters can be, then, independently manipulated. The concept has been exemplified in Illustration 9.2.

Illustration 9.2:

Ask the user to enter a string and put all the vowels of the string in one string and the consonants in the other string.

Solution:

Create two strings: **str1** and **str2**. Initialize both to "". Now, check each character in the string, if it is a vowel, concatenate it to with **str1** otherwise concatenate it to with **str2**.

Program:

```
string =input('Enter a string\t:')
str1=""
str2=""
for i in string:
```

```
  if((i =='a')|(i=='e')|(i=='i')|(i=='o')|(i=='u')):
    str1=str1+str(i)
  else :
    str2=str2+str(i)
print('The string containing the vowels is '+str1)
print('The string containing consonants '+str2)
>>
```

Similarly, a **for** loop can be used to iterate through a tuple and keys of a dictionary as shown in Illustrations 9.3 and 9.4.

Illustration 9.3:

*This illustration demonstrates the use of **for** for Iterating through a tuple.*

Solution:

```
T=(1, 2, 3)
for i in T:
 print(i)
print(T)
>>>
```

Output:

```
1
2
3
(1, 2, 3)
>>>
```

Illustration 9.4:

*This illustration demonstrates the use of **for** for iterating through a dictionary.*

Solution:

```
Dictionary={'Programming in C#': 499, 'Algorithms Analysis and
    Design':599}
print(Dictionary)
for i in Dictionary:
 print(i)

>>>
```

Output:

```
{'Programming in C#': 499, 'Algorithms Analysis and Design':
    599}
Programming in C#
Algorithms Analysis and Design
>>>
```

9.3 ITERATOR

The above tasks can also be accomplished using iterators. The **"iter"** function returns the iterator of the object passed as an argument. The itertaor can be used to manipulate lists, strings, tuples, files, and dictionary, in the same way as a **for** loop. However, the use of iterator ensures flexibility and additional power to a programmer. This would be established in the following section.

An iterator can be set on a list using the following:

```
<name of the iterator> = iter(<name of the List>)
```

The iterator can move to the next element, using the **__next__**() method. An iterator, as stated earlier, can iterate through any iterable object, including list, tuple, string, or a directory. When there are no more elements then a **StopIteration** exception is raised.

The following illustration shows the manipulation of a list using iterators. In the illustration, the given list contains a set of numbers, some of them positive, and some negative. The negative numbers are appended to a list called **N**, whereas the positive numbers are appended in a list called **P**. The same problem was solved using the **for** loop in Illustration 9.1.

Illustration 9.5:

Through using iterators, the following program puts the positive and negative numbers of a list into two separate lists and raises error at the end of the program.

Solution:

```
L = [ 1,2,3,-4,-5,-6]
P = []
N = []
t = iter(L)
```

```
try:
    while True:
     x = t.__next__()
     if x >= 0:
         P.append(x)
     else:
         N.append(x)
except StopIteration:
    print( 'original List- ' , L , '\nList containing the
    poisitive numbers- ', P , '\nList containing the negative
    numbers- ', N )
    raise StopIteration
```

The next example deals with a string. The iterator is set to the first element of the string and is then set to the second element, third element, and so on. The following illustration solves the problem given in Illustration 9.2 using iterators.

Illustration 9.6:

The program uses iterators to separate the vowels and consonants of a given string and raises error at the end of the program.

Solution:

The **vow** and **cons** strings are initialized to "" and each character of the given list is checked. If the character is a consonant it is concatenated to **cons**, otherwise it is concatenated to vow.

```
s = 'colour'
t = iter(s)
vow = ''
cons = ''
try:
    while True:
     x = t.__next__()
     if x in ['a','e','i','o','u']:
         vow += x
     else:
         cons += x
```

```
except StopIteration:
    print( 'String - ' + s + '\nVowels - ' + vow + '\nConsonents
    - ' + cons )
    raise StopIteration
```

A little more complex example of iterators has been shown in the following illustration. The illustration adds the corresponding elements of the two lists and then sorts the concatenated list.

Illustration 9.7:

Add the corresponding elements of two given lists and sort the final list.

Solution:

```
#The program concatenates two lists into one by iterating over
    individual elements of the lists using the list function and
    then sorts the concatenated list.
L1 = [ 3, 6, 1, 8, 5]
L2 = [ 7, 4, 6, 2, 9]
i1 = iter(L1)
i2 = iter(L2)
i3 = sorted( list(i1) + list(i2) )
print( 'List1 - ', i1 , '\nList2 - ', 12 , '\nSortedCombn - ', 13 )
```

9.4 DEFINING AN ITERABLE OBJECT

One can define ones' own class, in which **__init__**, **__iter__** and **__next__** can be defined as per the requirement. The **init** function initializes the variables of the class, the **iter** defines the mechanism of iterations and the next method implements the mechanism its jump to the next item.

Illustration 9.8:

*This illustration demonstrates the use of **Iterator**.*

Solution:

```
class yrange:
 def _init_(self, n):
    self.a = int(input('Enter the first term\t:'))
    self.d=int(input('Enter the common differnce\t:'))
```

```
        self.i=self.a
        self.n=n
    def _iter_(self):
        return self
    def _next_(self):
      if self.i<self.n:
        i=self.i
        self.i = self.i + self.d
        return i
      else:
        raise StopIteration()
y=yrange
y._init_(y, 8)
print(y)
print(y._next_(y))
print(y._next_(y))
print(y._next_(y))
>>>
```

Output:

```
Enter the first term    :1
Enter the common differnce  :2
<class '_main_.yrange'>
1
3
5
>>>
```

9.5   GENERATORS

Generators are functions that generate the requisite sequence. However, there is an inherent difference between a normal function and a generator. In a generator, the values are generated as and when we proceed. So, if one comes back to the function after a particular value is generated, then and instead of starting from the beginning the function starts from the point where we let off.

The task seems difficult but has an advantage. The concept can help the programmer to generate lists containing the desired sequences. For example, if one wants to dynamically generate a list containing the terms of an arithmetic progression, in which each term is "d" more than the first term, generators come to the rescue. Similarly, the sequences like geometric progression, Fibonacci series etcetera can be easily generated using generator.

Python comes with **yield**, which helps to start from the point where we let of. This is markedly different from **return** used in normal functions which does not save the state where we let off. If the function having **return** is called again it starts all over again.

The following illustration exemplifies the use of generators to produce simple sequences like arithmetic progression, geometric progression, Fibonacci series, etc.

Illustration 9.9:

Write a generator to produce arithmetic progression, where in the first term, the common difference and the number of terms is entered by the user.

Solution:

```
def arithmetic_progression(a, d, n):
 i=1
 while i<=n:
    yield (a+(i-1)*d)
    i+=1
a=int(input('Enter the first term of the arithmetic
    progression\t:'))
d=int(input('Enter the common differ nce of the arithmetic
    progression\t:'))
n=int(input('Enter the number of terms of the arithmetic
    progression\t:'))
ap = arithmetic_progression(a, d, n)
print(ap)
for i in ap:
 print(i)
>>>
```

Output:

```
Enter the first term of the arithmetic progression:3
Enter the common difference of the arithmetic progression  :5
Enter the number of terms of the arithmetic progression    :8
<generator object arithmetic_progression at 0×031C2DE0>
3
8
13
18
23
28
33
38
>>>
```

Illustration 9.10:

Write a generator to produce Geometric Progression, where in the first term, the common ratio and the number of terms is entered by the user.

Solution:

```
def geometric_progression(a, r, n):
  i=1;
 while i<=n:
    yield(a*pow(a, i-1))
    i+=1


a=int(input('Enter the first term of the geometric
    progression\t:'))
r=int(input('Enter the common ratio of the geometric
    progression\t:'))
n=int(input('Enter the number of terms of the geometric
    progression\t:'))
gp=geometric_progression(a, r, n)
for i in gp:
 print(i)
>>>
```

Output:

```
Enter the first term of the geometric progression  :3
Enter the common ratio of the geometric progression  :4
Enter the number of terms of the geometric progression  :7
3
9
27
81
243
729
2187
>>>
```

Illustration 9.11:

Write a generator to produce Fibonacci series.

Solution:

```
def fib(n):
  a=[]
  if n==1:
    a[0]=1
    yield 1
   elif n==2:
    a[1]=1
    yield 1
   else:
    a[0]=1
    a[1]=1
    i=2
    while i<=n:
        a[i]=a[i-1]+a[i-2]
        yield (a[i])
n=int(input('Enter the number of terms\t:'))
fibList=fib(n)
for i in fibList:
 print(i)
```

It may be stated here that the value of **i** increments as in the next iteration. The value does not change before or after the **yield**. In order to understand the concept, let us go through the following illustration.

Illustration 9.12:

*This illustration demonstrates the effect of **yield** on the value of the counter.*

Solution:

The reader is expected to note the change in the value after and before yield.

Program:

```
def demo():
 print ('Start')
 for i in range(20):
    print('Value of i before yield\t:',i)
     yield i
    print('Value of i after yield\t:',i)
 print('End')
a=demo()
for i in a:
 print (i)
```

Output:

```
Start
Value of i before yield  : 0
0
Value of i after yield  : 0
Value of i before yield  : 1
1
Value of i after yield  : 1
Value of i before yield   : 2
2
Value of i after yield  : 2
Value of i before yield   : 3
3
Value of i after yield  : 3
```

```
Value of i before yield   : 4
4
Value of i after yield : 4
Value of i before yield   : 5
5
Value of i after yield : 5
Value of i before yield   : 6
6
Value of i after yield : 6
Value of i before yield   : 7
7
Value of i after yield : 7
Value of i before yield :   : 8
8
Value of i after yield : 8
Value of i before yield   : 9
9
Value of i after yield : 9
Value of i before yield   : 10
10
Value of i after yield : 10
Value of i before yield   : 11
11
Value of i after yield : 11
Value of i before yield   : 12
12
Value of i after yield : 12
Value of i before yield   : 13
13
Value of i after yield : 13
Value of i before yield   : 14
14
Value of i after yield   : 14
Value of i before yield : 15
15
Value of i after yield : 15
Value of i before yield   : 16
```

```
16
Value of i after yield  : 16
Value of i before yield   : 17
17
Value of i after yield  : 17
Value of i before yield   : 18
18
Value of i after yield  : 18
Value of i before yield   : 19
19
Value of i after yield  : 19
End
>>>
```

9.6   COMPREHENSIONS

The aim of a programming language is to make things easy for a programmer. A task, though, can be performed in many ways but the one which requires least coding is the most appealing to a coder. Python has many features which facilitate programming. Comprehensions are one of them. Comprehensions allow sequences to be built from other sequences. Comprehensions can be used for lists, dictionary and set comprehension. In the earlier version of Python (Python 2.0) only list comprehensions were allowed. However, in the newer versions comprehensions can also be used with dictionary and sets also.

The following illustration explains the use of comprehensions to generate lists in various cases.

- The **range(n)** function generates numbers up to n. The first comprehension generates the list of numbers which are cubes of all the numbers generated by the range function.
- The second comprehension works in the same way but generates 3 to the power of x.
- The third comprehension generates a list having numbers generated by the range (n) function, which are multiple of 5.
- In the fourth comprehension the comprehension takes the words of sentence "Winter is coming" and generates a list containing the word in caps, in running and the length of the word.
- Comprehensions can also be used to generate lists, satisfying a given condition.

Illustration 9.13:

Generate the following lists using comprehensions.

- x^3, x from 0 to 9
- 3^x, x from 2 to 10
- All the multiples of 5 from the previous list
- The caps, running version and the length of each word in the sentence "Winter is coming"

Solution:

```
L1 = [x**3 for x in range(10)]
print(L1)
L2 = [3**x for x in range(2, 10, 1)]
print(L2)
L3 = [x for x in L2 if x%5==0]
print(L3)
String = "Winter is coming".split()
print(String)
String_cases=[[w.upper(), w.lower(), len(w)] for w in String]
for i in String_cases:
 print(i)
list1 = [1, '4', 9, 'a', 0, 4]
square_int = [ x**2 for x in list1 if type(x)==int]
print(square_int)
>>>
```

Output:

```
[0, 1, 8, 27, 64, 125, 216, 343, 512, 729]
[9, 27, 81, 243, 729, 2187, 6561, 19683]
[]
['Winter', 'is', 'coming']
['WINTER', 'winter', 6]
['IS', 'is', 2]
['COMING', 'coming', 7]
[1, 81, 0, 16]
>>>
```

A comprehension contains the input sequence along with the expression that represents the members. A comprehension may also have an optional predicate expression.

In order to understand the concept, let us consider one more illustration in which a list of temperatures in Celsius is given and the corresponding list containing the temperatures in Kelvin are to generated. The temperatures in Celsius and Kelvin are related as follows.

$$\text{Kelvin}(T) = \text{Celsius}(T) + 273.16$$

Illustration 9.14:

Given a list containing temperatures in Celsius, generate a list containing temperatures in Kelvin.

Solution:

The list **L_kelvin**, is a list where in each element is 273.16 more than the corresponding element in **L_cel**. Note that the task has been accomplished in the definition of the list **L_Kelvin** itself.

Program:

```
L_Cel = [21.2, 56.6, 89.2, 90,1, 78.1]
L_Kelvin = [x +273.16 for x in L_Cel]
print('The output list')
for i in L_Kelvin:
 print(i)
```

Output:

```
The output list
294.36
329.76000000000005
362.36
363.16
274.16
351.26
>>>
```

Another important application of comprehension is to generate the Cartesian product of two sets.

The cross product of two sets, A and B, is a set containing tuples of the form (x, y), where x belongs to the set A and y belongs to the set B. Illustration 9.15 implements the program.

Illustration 9.15:

Find the Cartesian product of two given sets.

Solution:

```
A= ['a', 'b', 'c']
B= [1, 2, 3, 4]
AXB = [(x, y) for x in A for y in B]
for i in AXB:
 print(i)
>>>
```

Output:

```
('a', 1)
('a', 2)
('a', 3)
('a', 4)
('b', 1)
('b', 2)
('b', 3)
('b', 4)
('c', 1)
('c', 2)
('c', 3)
('c', 4)
>>>
```

The above program is important because the concept of relations and hence functions, in mathematics, originates from the cross product. As a matter of fact, any subset of A × B is a relation from A to B. There are four types of relations in mathematics: one to one, one to many, many to one, and many to many. Out of these relations one to one and many to one are referred to as functions.

9.7 CONCLUSION

The chapter explained the use of **for** for iterating over a list, string, tuple, or a dictionary. It may be stated here that in C or C++, **for** is generally used for the same purpose as **while**. However, in Python, **for** can be used to visit each element individually. Note that this can also be done in JAVA or C#. In order to define an iteratble object, **__iter__** and **__next__** needs to be defined for the class. The reader is also expected to take note of the fact that **yield** and **return** perform different tasks in Python. The use of these two has been demonstrated in the illustrations. Finally, while defining a list, each element can be crafted as per the need of the question. The chapter, though easy, becomes important in the light of excessive use of these techniques in machine learning and pattern recognition tasks.

GLOSSARY

- Iterator takes an iterable object and helps to traverse the object
- **_next_():** It produces the next value of the iterable object
- **_iter_():** It helps in iteration

POINTS TO REMEMBER

- The **for** statement can be used for looping over a list, string, tuple, file, and dictionary.
- **Iter** takes an object and returns corresponding iterator.
- **_next_** gives us the next element.
- **Built-in** functions like list etc. accept iterators as arguments.
- A generator produces sequence of results.
- **Yield** is used when many values are to be produced from a function/generator.

EXERCISES

Multiple Choice Questions

1. Which of the following can be an argument in _iter()_

    (a) string   **(b)** tuple

    (c) list   **(d)** dictionary   **(e)** all of the above

2. The **iter** takes which type of object?
   - **(a)** iterable
   - **(b)** Any object
   - **(c)** Comprehension
   - **(d)** Generator

3. What is the function of **_next()_**?
   - **(a)** To produce next object of the iteration
   - **(b)** To produce a new iteration
   - **(c)** To iterate through a generator
   - **(d)** None of the above

4. Which of the following transfers the control to the calling function?
   - **(a)** return
   - **(b)** yield
   - **(c)** both
   - **(d)** one of the above

5. Which of the following does not transfer the control to the calling function?
   - **(a)** return
   - **(b)** yield
   - **(c)** both
   - **(d)** one of the above

6. Which of the following is essentially used in generators?
   - **(a)** yield
   - **(b)** return
   - **(c)** both
   - **(d)** none of the above

7. Which of the following is true?
   - **(a)** One can use interators with generators
   - **(b)** One can use iterators with list
   - **(c)** One can use iterators with comprehensions
   - **(d)** All of the above

8. Which of the following can be iterated using a for loop
   - **(a)** string
   - **(b)** list
   - **(c)** tuple
   - **(d)** all of the above

9. Which of the following can be iterated using **a for** loop

   (a) string   **(b)** comprehension

   (c) file   **(d)** all of the above

10. Which of the following behaves in the same manner as the combination of _iter()_ and _next()_

    (a) for   **(b)** if

    (c) both   **(d)** none of the above

Theory

1. Explain the iteration protocol in Python.

2. What is the function of a Generator?

3. What is the difference between yield and return?

4. What are list comprehensions? Explain how comprehensions help in evading the use of loops.

5. Explore some of the iteration tools in Python.

Programming Exercises

(For references regarding AP, GP, HP, primes refer to the references given at the end of this book)

1. Write a generator that produces the terms of arithmetic progression.

2. For the above question write the corresponding iterator class.

3. Write a generator that produces the terms of a geometrical progression.

4. For the above question write the corresponding iterator class.

5. Write a generator that produces the terms of a harmonic progression.

6. For the above question write the corresponding iterator class.

7. Write a generator that produces all the prime numbers up to a given number.

8. For the above question write the corresponding iterator class.

9. Write a generator that produces all the Fibonacci numbers up to n.

10. For the above question write the corresponding iterator class.

11. Write a generator that produces all the Armstrong numbers up to n.

12. For the above question write the corresponding iterator class.

13. Write a generator that produces Pythagoras triples in the range (1, 20).

14. For the above question write the corresponding iterator class.

15. Write a generator that produces all the multiples of 6 up to the given number.

16. For the above question write the corresponding iterator class.

17. Write a list comprehension that produces all the numbers which are multiple of 2 or 5.

18. Write a list comprehension that converts a list containing the temperature in degree to that in Fahrenheit.

19. Write a list comprehension that produces all the prime numbers.

20. Write a list comprehension that produces all the numbers which leave remainder 1 when divided by 5.

21. Write a list comprehension that produces all the vowels of a given string.

22. Write a list comprehension that produces the fourth power of numbers of a given list.

23. Write a list comprehension that produces the absolute powers of numbers in a given list.

STRINGS

Objectives

After reading this chapter, the reader should be able to

- Understand the concept and importance of strings
- Understand various string operators
- Learn about the built-in functions to manipulate strings

10.1 INTRODUCTION

Strings are the sequence of characters. These data structures are used to store text. For example, if one wants to store the name of a person, or for that matter, his address, then strings are the most appropriate data structures. As a matter of fact, the knowledge of strings is essential in the development of many applications like word processor and parser.

Strings, in Python, can be enclosed in single quotes or double quotes, or even in triple quotes. Though, there is no difference between a string enclosed in single quotes or double quotes. That is "harsh" is same as "harsh." Triple quotes are generally used in special cases as discussed later in this chapter. Strings in Python, come with wide variety of operators and in-build functions.

The chapter examines various aspects of strings, like nonmutability, traversal, operators, and in-build functions. One of the most prominent difference between a string and a list is nonmutability. Once a value is given to a string, one cannot change the value of a character present at a particular position. For the users familiar with "C," "C++," "C#," or "Java," the operators discussed in the chapter, notably * would be a pleasant surprise. Moreover, Python provides many build-in functions to help the programmers to handle strings.

This chapter examines the above issues and exemplifies them. The chapter has been organized as follows. The second section of the chapter explores the use of standard **for** and **while** loops in strings. The third section deals with the operators that can be used with strings. The in-built functions used for accomplishing various tasks have been dealt with in the fourth section. The fifth section deals with the concept of indexing and slicing. The sixth section deals with the features of strings and the last section concludes.

10.2 LOOPS REVISED

The traversal of a strings has already been discussed in the previous chapter. A brief description of the topic has been presented in this section.

As stated earlier, strings are iterable objects, therefore standard loops (read **for** and **while**) can be used to iterate though them. The **for** loop helps to iterate through each character by storing the character in some variable. The following illustration depicts the use of a **for** loop to iterate a string.

Illustration 10.1:

Ask the user to enter a string and print each character using a for loop.

Listing:

```
str1= input('Enter a string\t:')
for i in str1:
    print('Character \t:',i)>>>
```

Output:

```
================= RUN C:/Python/String/str2.py =================
Enter a string     :harsh
Character          : h
Character          : a
Character          : r
Character          : s
Character          : h
```

The above procedure can also help us to find the length of string. Note that there is a built-in function to accomplish this task. However, the purpose here is to be able to use the **for** loop in order to imitate the **len** function. In the following illustration, a variable called length is initialized to 0, and is incremented as we proceed.

Illustration 10.2:

Ask the user to enter a string and find its length.

Listing:

```
name=input('Enter your name\t');
length=0
for i in name:
    length=length +1
print('The length of ',name,' is ',length)
```

Output:

```
Enter your name    harsh
The length of  harsh  is  5
>>>
```

The ability to handle each character individually, in a string, makes tasks like basic cryptography manageable. In order to understand the concept, consider the following example. The example that follows, displaces the characters two positions to the right. This is referred to as transposition. The next example shifts the characters by "k" positions, where "k" is entered by the user.

Illustration 10.3:

Ask the user to enter a string and displace each character by two positions to the right.

Solution:

```
str1=input('Enter the string\t:')
i=0
str2=""
while i<len(str1):
    str2[i]=str1[(i+2)%len(str1)]
print(str2)
```

Illustration 10.4:

Ask the user to enter a string and displace each character by k positions to the right.

Solution:

```
str1=input('Enter the string\t:')
k=int(input('Enter the value of k\t:'))
i=0
str2=""
while i<len(str1):
    str2+=str1[(i+k)%len(str1)]
    print(str2)
    i+=1
print(str2)
>>>
```

Output:

```
============= RUN C:/Python/String/transposition.py ============
Enter the string  :harsh
Enter the value of k   :4
h
hh
hha
hhar
hhars
hhars
>>>
```

Substitution means replacing a symbol with some other symbol. This replacement would result in the formation of another string by a given string. Loops can be used for substitution. The example that follows implements one of the most basic substitutions. Here, each character is replaced by a character obtained by adding two to the ASCII value of the character and finding the requisite character.

Illustration 10.5:

Ask the user to enter a string. Replace each character by that obtained by adding two to the ASCII value of that character.

Solution:

```
str1=input('Enter the string\t:')
k=int(input('Enter the value of k\t:'))
i=0
```

```
str2=""
while i<len(str1):
    str2+=str((ascii(str1[i])+k))
    print(str2)
    i+=1
print(str2)
```

10.3 STRING OPERATORS

Python provides the programmer with a wide variety of extremely useful operators to manipulate strings. These operators help a user to perform involved tasks with ease and efficiency. Here, it may be stated that the replication and membership operators make Python stand apart from its counterparts. This section briefly introduces and exemplifies the operators.

10.3.1 The Concatenation Operator (+)

The concatenation operator takes two strings and produces a concatenated string. The operator acts on values as well as variables. In the example that follows, the result generated by applying the concatenation operators have been stored in variables called **result1** and **str2**.

```
name=input('Enter your name\t:')
result1 = 'Hi'+' there'
print(result1)
str1='Hello'
str2=str1 +' '+name
print(str2)
```

Output:

```
>>>
=============== RUN C:/Python/String/operator1.py ==============
Enter your name    :Harsh
Hi there
Hello Harsh
>>>
```

10.3.2 The Replication Operator (*)

The replication operator in Python, replicates the strings as many times as the first operand. The operator operates on two operands: the first being a number and the second being a string. The result is a string in which the input string is repeated as many times as the first argument. In the example that follows, the result has been stored in a variable called **result1**.

```
name=input('Enter your name\t:')
print('Hi', ' ', name)
str1=input('Enter a string\t:')
num=int(input('Enter a number\t:'))
result1=num*str1
print(result1)
```

Output:

```
>>>
=============== RUN C:/Python/String/operator2.py ===============
Enter your name    :harsh
Hi    harsh
Enter a string     :abc
Enter a number     :4
abcabcabcabc
>>>
```

10.3.3 The Membership Operator

The membership operator checks whether a given string is in a given list or not. The operator returns a **True**, if the first string is a part of the given list, otherwise it returns a **False**.

```
>>> 'Hari' in ['Har', 'Hari', 'Hai']
True
>>>
>>> 'Hari' in ['Har', 'hari', 'Hai']
False
>>>
```

It may be noted here that this operator is also used in manipulating iterations. The reader is advised to revisit the last chapter, for a detailed discussion regarding the use of **"in"** in for. It may also be noted that the operator can also be used in tuples. In the listing that follows, the string **"Hari"** is present in the given tuple and hence **"True"** is returned.

```
>>> 'Hari' in ('Hari', 'Har')
True
>>>
```

The reader may also note that corresponding to the **"in"** operator, there is a **"not in"** operator which works in the exactly opposite manner vis-a-vis **"in."**

A string in Python can span over many lines. This can be accomplished by putting a "\" at the end of the line. For example, **str2** is "Harsh Bhasin Author Delhi." However, it has been written in three lines, using the "\" character.

```
>>> str2="'Harsh Bhasin\
Author\
Delhi'"
>>> str2
"'Harsh BhasinAuthorDelhi'"
```

10.4 IN-BUILT FUNCTIONS

This section presents some of the most common functions used to manipulate strings in Python. It may be stated here that, though all the following tasks can be done, without the predefined functions, with a varying degree of ease, the presence of these functions help the programmer to do the task easily and efficiently. Moreover, when one designs and implements one's version of a function, the implementation may not be efficient in terms of time or space or both. However, while implementing these predefined functions in Python, the issues related to memory and times have already been dealt with. Let us, now, have a look at the names, meanings, and usage of the predefined functions in Python.

10.4.1 len()

Usage:

```
>>>len(<string>)
```

Explanation:

The function returns the number of characters in a string. For example, if a variable called **str1** stores "Harsh Bhasin," then the length of the string can be calculated by writing **len(str1)**. Note that, the space between "Harsh" and "Bhasin" has also been taken into account while calculating the length of the string. To summarize, the function takes a string argument and returns an integer, which is the length of the string.

Example (s):

```
str1 ='Harsh Bhasin'
len(str1)
```

Output:

```
12
```

Code:

```
len('Harsh Bhasin')
```

Output:

```
12
```

Code:

```
len('')
```

Output:

```
0
```

10.4.2 Capitalize()

Usage:

```
capitalize()
```

Explanation:

The function capitalizes the first character of the string. Note that only the first character would be capitalized. If one wants to capitalize the first characters of all the words in the string the **title()** function can be used.

Example (s):

```
str2='harsh bhasin'
str2
```

Output:

```
'harsh bhasin'
```

Code:

```
str2.capitalize()
```

Output:

```
'Harsh bhasin'
```

10.4.3 Find()

Usage:

```
<name of the string>.find(<parameter(s)>)
```

Explanation:

The location of a given substring in a given string can be found by using the function **"find."** Also, if the location of a substring after a particular position (and before a particular index) is to be determined, then three arguments can be passed to the function: the substring, initial index, and the final index. The following examples depict the usage of the function.

Example(s):

```
str2.find('ha')
```

Output:

```
0
```

Code:

```
str2.find('ha',3,len(str2))
```

Output:

```
7
```

10.4.4 Count

Usage:

```
<name of the string>.count(<parameter(s)>)
```

Explanation:

The number of occurrences of a particular substring can be found by the **count** function. The function takes three arguments: the substring, the initial index, and the final index. The following examples show the usage of the function.

Example(s):

```
str3.count('ha',0,len(str3))
```

Output:

```
1
```

Code:

```
str3.count('ka',0,len(str3))
```

Output:

```
0
```

10.4.5 endswith()

```
<name of the string>.endswith(<parameter(s)>)
```

Explanation:

One can determine if a string ends with a particular substring. This can be done using the **endswith()** function. The function returns a **True** if the given string ends with the given substring, otherwise it returns a **False**.

Example(s):

```
str3.endswith('n')
```

Output:

```
True
```

10.4.6 encode

Usage:

```
<name of the string>.encode(<parameter(s)>)
```

Explanation:

Python provides a function called **encode** to encode a given string in various formats. It takes two arguments: encoding=<value> and errors=<value>. The encoding can be one of the many encodings (refer to *https://www.tutorials point.com/python/string_encode.htm*). The following examples demonstrate the use of this function.

Example(s):

```
>>>str3.encode(encoding='utf32',errors='strict')
b'\xff\xfe\x00\x00H\x00\x00\x00A\x00\x00\x00R\x00\x00\x00S\x00\
    x00\x00H\x00\x00\x00\x00\x00\x00b\x00\x00\x00h\x00\x00\x00a\
    x00\x00\x00s\x00\x00\x00i\x00\x00\x00n\x00\x00\x00\x00'
```

10.4.7 decode

The **decode** function is complimentary to the **encode** function in Python. The reader is advised to visit the References given at the end of the book for detailed discussion.

```
Str3. decode
```

Usage:

```
<name of the string>.decode(<parameter(s)>)
```

Explanation:

This function is complementary to the **encode** function. It returns the decoded string.

10.4.8 Miscellaneous Functions

Except for the functions discussed above, there are some more functions to accomplish assorted tasks. The following list presents some of the functions and the code that follows presents an example of the usage.

List:

1. isanum()
2. isalpha()
3. isdecimal()
4. isdigit()
5. isidentifier()
6. islower()
7. isupper()

8. swapcase()

9. isspace()

10. lstrip()

11. rstrip()

12. replace()

13. join()

Explanation:

The contents of a given string can be checked using the following functions. The **isalnum()** function checks if the given string is alphanumeric. The other functions like **isalpha()** and **isdecimal()** also check the type of contents in a given string.

Whether a given string contains only digits can be checked using the **isdigit()** function. Similarly, whether a given string is an identifier can be checked using the **isidentifier()** function. The **islower()** function checks if the given string contains only the lower case characters and the **isupper()** function checks if the given string contains only the upper case characters. The **swapcase()** functions, swaps the case of the given string, as in converts the upper case to lower and lower to upper. The presence of spaces(only) can be checked using the **isspace()** function. Extra spaces can be removed from the left and the right hand side by using the **lstrip()** and **rstrip()** functions. The **replace()** function replaces the instances of the first argument by the string in the second argument. The **split** function, splits the given strings into tokens. The following illustration depicts the use of this function for splitting the string into constituent words. The function of **join()** is exactly the opposite as that of **split**.

Example(s):

```
str3.isalnum()
```

Output:

```
False
str3.isalpha()
```

```
False
>>>
str3.isdecimal()
```

```
False
>>>
str3.isdigit()
```

```
False
>>>
str3.isidentifier()
```

```
False
>>>
str3.islower()
```

```
False
>>>
str3.isnumeric()
```

```
False
>>>
str3.replace('h','p')
```

```
'HARSH bhasin'
>>>
```

Illustration 10.6:

*A string **str4** contains a sentence "I am a good boy." Split the string into tokens and also display each token using a **for** loop.*

Solution:

```
>>> str4='I am a good boy'
>>> str4.split()
['I', 'am', 'a', 'good', 'boy']
>>>
>>> for i in str4.split():
    print('Token\t:',i)
```

Output:

```
Token    : I
Token    : am
Token    : a
Token    : good
Token    : boy
```

10.5 CONCLUSION

In C and C++, strings are character arrays. They are a special type of arrays with a "\0" character at the end. Strings, in C, come with a set of built-in functions. However, there were two major issues. First, string is not an independent data type in C or C++, second is mutability of strings. In Python the importance of strings has been duly recognized by creating an object type. Moreover, strings, in Python, are nonmutable. Strings come with a wide range of built-in functions. Also, there are useful operators, to help the programmer accomplish a given task easily and efficiently. This chapter introduces the concept, operators, and functions of strings. However, the reader is expected to complete the end-chapter exercise to be able to understand and use strings.

GLOSSARY

String: Strings are the sequence of characters. These data structures are used to store text.

POINTS TO REMEMBER

- Strings in Python are nonmutable.
- The negative index denotes the characters from the right hand.
- Strings are iterable objects.

EXERCISES

Multiple Choice Questions

1. Which of the following is true?

    (a) A string in Python is iterable

    (b) A string in Python is not iterable

    (c) Iterability of a string depends upon the situation

    (d) None of the above

2. Is a string, in Python, mutable?

    (a) No   **(b)** Yes

    (c) Depends on the situation   **(d)** None of the above

3. If str1="Hari," what is the output of print(str1[4])

    (a) i   **(b)** \0

    (c) Exception is raised   **(d)** None of the above

4. If str1="Hari," what is the output of print(str1[-3])

    (a) "a"   **(b)** "H"

    (c) Exception is raised   **(d)** None of the above

5. What is the output of "Hari"=="hari"

    (a) True   **(b)** False

    (c) An exception is raised   **(d)** None of the above

6. What is the output of "a"< >"A"

    (a) True   **(b)** False

    (c) Exception is raised   **(d)** None of the above

7. What is the output of "567">"989"

    (a) True   **(b)** False

    (c) An exception is raised   **(d)** None of the above

8. Which of the following helps to find the ASCII value of "C"?

    (a) ord("C")   **(b)** chr("C")

    (c) both   **(d)** None of the above

9. Which of the following helps to find the character represented by ASCII value 67?

   (a) ord(67) (b) chr(67)

   (c) Both (d) None of the above

10. What are "in" and "not in" in Python?

    (a) Relational operators (b) Membership operators

    (c) Concatenation operator (d) None of the above

11. What is the output of "A" + "B"

    (a) "A + B" (b) "AB"

    (c) 131 (d) None of the above

12. What is the output of 3*"A"

    (a) "3A"

    (b) Character corresponding to the ASCII value 65X3

    (c) "AAA"

    (d) None of the above

13. Which function capitalizes the first character of a given string?

    (a) Capitilize() (b) Titlecase()

    (c) Toupper() (d) None of the above

14. The find() function, in Python takes

    (a) 1 argument (b) 3 arguments

    (c) Both (d) None of the above

15. If str1="hari," then what would be the output of str1.asalnum()?

    (a) True (b) False

    (c) Exception is raised (d) None of the above

16. If str1="hari3," then what would be the output of str1.asalnum()?

    (a) True (b) False

    (c) Exception is raised (d) None of the above

17. If str1="hari feb," then what would be the output of str1.asalnum()?

(a) True   **(b)** False

(c) Exception is raised   **(d)** None of the above

18. If str1="123h," then what would be the output of str1.digit()?

(a) True   **(b)** False

(c) Exception is raised   **(d)** None of the above

19. Which function checks whether all the characters in a given string are in lower case?

(a) lower()   **(b)** islower()

(c) istitle()   **(d)** None of the above

20. Which function checks whether all the characters in a given string are in upper case?

(a) upper()   **(b)** isupper()

(c) istitle()   **(d)** None of the above

21. Which function removes the whitespaces from the right hand of a given string?

(a) rstrip()   **(b)** strip()

(c) lstrip()   **(d)** None of the above

22. Which of the following functions convert a given string into a list of words?

(a) split()   **(b)** break()

(c) breakup()   **(d)** None of the above

23. Which of the following helps in breaking a string into two substrings of desirable length?

(a) slicing   **(b)** splitting

(c) both   **(d)** None of the above

24. Which of the following functions combines the strings given as the argument?

(a) split   **(b)** join

(c) slice   **(d)** None of the above

25. Which of the following is illegal in Python (assume that str1 is a string, having initial value "hari")?

    (a) str1= "Harsh"   **(b)** str1[0]= "t"

    (c) str1[0]=str[2]   **(d)** None of the above

Theory

1. Write a program to reverse a string.

2. Write a program to encode a string in UTF format.

3. Write a program to find the sum of ASCII values of the characters of a given string.

4. Write a program to find a particular substring in a given string.

5. Write a program to split a given text into tokens.

6. Write a program to check which of the tokens obtained in the above question are Keywords of (C).

7. Write a program to check how many alphanumeric strings are there in the tokens obtained in question 6.

8. Write a program to check how many alpha strings are there in the tokens obtained in question 6.

9. Write a program to check how many numeric strings are there in the tokens obtained in question 6.

10. Write a program to convert a string entered by a user to that obtained by adding "k" to each character's ASCII value.

11. Implement the first phase of complier design (for "C"). (Please refer to the Bibliography for a brief overview of compiler design).

12. In the above question design deterministic finite acceptors for the keywords of C.

SECTION III
OBJECT-ORIENTED PROGRAMMING

Having learned the fundamentals of Python, and procedural programming, let us move toward the Hogwarts of Programming: object-oriented programming (OOP). This section has five chapters. Chapter 11 presents the fundamentals of OOP, its importance, need, and comparison with other paradigms.

Class is a real or a conceptual entity, having importance to the problem at hand. Chapter 12 discusses the concepts of class and related topics like constructors and the foundation stone of OOP. We make classes so that we can derive new classes out of the existing classes. This is called inheritance which is an integral part of the Object-Oriented Programming Paradigm. Chapter 13 introduces inheritance and discusses the types of inheritance and the problems with Multiple Inheritance. The chapter also discusses the concept of Bound methods and derivation's type vis-a-vis methods.

Python provides the user with the power of operator overloading. It basically means using an existing operator for user-defined data types. However, Operator Overloading is a bit different in Python as compared to C++ or C#. In Python, we have specialized functions which must be defined to overload an operator. Chapter 14 discusses operator overloading. This section ends with the most important of all: exception handling.

INTRODUCTION TO OBJECT-ORIENTED PARADIGM

Objectives

After reading this chapter, the reader should be able to

- Understand the elements of object-oriented paradigm
- Understand the concept of a class and define an object
- Define encapsulation, inheritance, and polymorphism

11.1 INTRODUCTION

In the preceding chapters, the control structures of Python were discussed. The chapters discussed loops, conditional statements, etc. However, these constructs are an integral part of other procedural languages also. In a procedure, each instruction tells the computer what needs to be done, and these procedures constitute a program in such languages. Python not only supports procedural programming but also supports object-oriented programming (OOP). This chapter introduces the principles of OOP and explains the need for and importance of classes and objects. The chapter also discusses the difference between OOP and procedural programming.

It may be stated here that the topics discussed in this chapter will be discussed in detail in the following chapters. Some of the readers, not familiar with C++ (or for that matter C# or JAVA) may find the discussion abstract, but things will become clear as we proceed.

As stated earlier, in procedural programming, each statement tells the program what needs to be done. For example, the following code asks the user for the input, calculates the square root of the number entered by the user, and displays the result.

Code:

```
a=float(input("Enter a number\t:"))
b=math.sqrt(a)
b
```

Output:

```
Enter a number :67
8.18535277187245
```

This strategy is good if the program is very small. Often, telling the computer what to do, step by step, works if the task to be accomplished is not very complex. In such cases, no other paradigm is needed.

In the case of a moderately large program, division into functions makes the task easier. The division of a larger program into modules makes the program manageable and helps achieve code's reusability. The functions generally accomplish a clearly defined task and become handy whenever that particular task is to be accomplished. The reader is advised to go through the chapter on functions in order to understand the advantages of functions. The clubbing of functions, on some basis, gives rise to what is commonly referred to as modules. This programming paradigm is called modular programming.

The problem with the above paradigm is that the accidental clubbing together of unrelated functions is far from real-world situations and hence becomes a source of problems at some point in time. Moreover, the approach does not restrict the access of data in any module and jeopardize the sanctity of the data.

It may be noted that the data should not be accessible to all the modules. The accessibility of data must be managed with utmost care; otherwise, a module, which should not have alerted the data as per the program logic, might change the data.

In order to understand the gravity of the problem, let us take the example of C. In C, a variable can be global or local. If it is global, then any module can change it. But, on the other hand, if it is local, then other modules would not be able to access it. So, there is nothing in between.

The solution to the above problem is to model the software in such a way that the design is conceptually as close to the real world as possible. This modeling of real-world situations requires the creation of entities having both attributes and behavior. The clubbing together of data and the functions that manipulate the data would be helpful in crafting the above entities. These entities would henceforth be referred to as classes. The instances of classes are objects, and the paradigm is called object-oriented paradigm. Various programming paradigms and their disadvantages have been summarized in Figure 11.1.

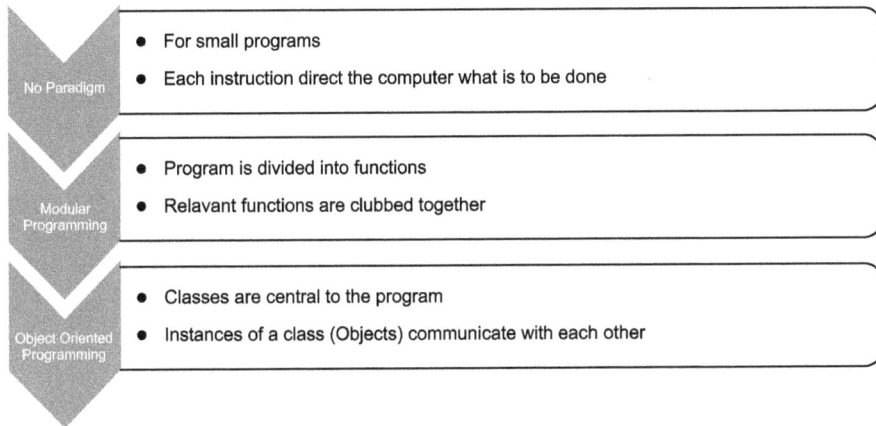

No Paradigm
- For small programs
- Each instruction direct the computer what is to be done

Modular Programming
- Program is divided into functions
- Relevant functions are clubbed together

Object Oriented Programming
- Classes are central to the program
- Instances of a class (Objects) communicate with each other

FIGURE 11.1 Programming paradigms.

11.2 CREATING NEW TYPES

Though types are not explicitly declared in Python, they were important in other languages (well most of them). For example, when one says that a "number" is of integer type, one states the type of information and its maximum and minimum value. Assume that an integer takes two bytes, the maximum value of "number" would be 32, 767, and the minimum value would be −32, 768. Moreover, saying that "number" is of integer type also restricts the operations that can be performed on the number.

The integer is a predefined type. Most of the languages also allow users to create custom types and hence extend the power of built-in types. This is essential as the ability to create new data types would help us to create programs which are near to the real world. For example, if one has to design an inventory management system, a type called "item" would make the matters uncomplicated. This "item" can have variables which are of predefined types like integers and strings.

A new type can be created by declaring a class. A class may have many components, the most important of which are the attributes and its functions. This clubbing together of functions and data forms the basis of OOP. The functions, as we will see later, generally, manipulate the data members of a class. Before proceeding any further let us have an overview of attributes and functions.

11.3 ATTRIBUTES AND FUNCTIONS

One can perceive a class as a prototype and an object as an instance of a class. For example, "movie" is a class and "The Fault in Our Stars," "Love Actually," and "Sarat" are objects (Figure 11.2). A class has attributes and behavior. The attributes, generally, store data and the behavior is implemented using functions. A class can be depicted using a class diagram. A class diagram has, generally, three parts, the first part contains the name, the second part has attributes, and the third part shows the functions of a class. The basics of attributes and behavior have been discussed in the following section. In Figure 11.2, the class diagram (Movie) has only the name.

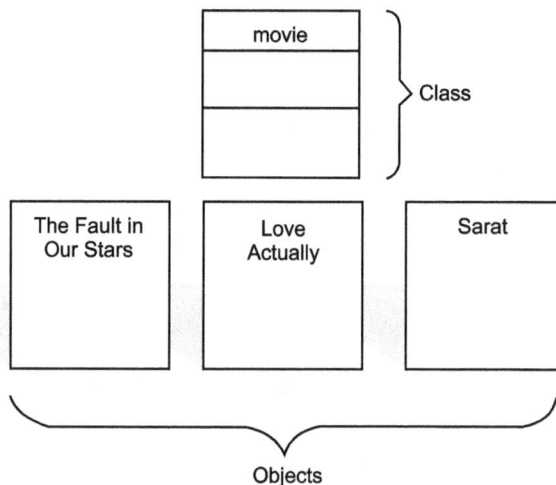

FIGURE 11.2 Example of a class and objects.

11.3.1 Attributes

The attributes, here, depict the characteristics of the entity, which we are concerned with. For example, in creating a website that gives the details of movies, a class "movie" would be needed. Say, after detailed deliberations, it was decided that this class would have attributes like name, year, genre, director, producer, actors, music_director, and story_writer.

Note that for the said purpose, only the above details are needed. Storing unnecessary details would not only make data management difficult but would also violate one of the core principles, that of including only the details pertaining to the problem at hand. These attributes are generally shown in the second section of the class diagram. In Figure 11.3, the attributes of "movie" class have been shown.

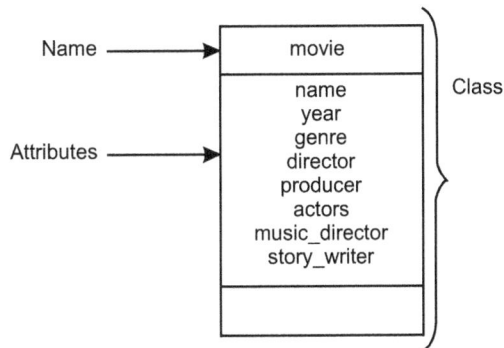

FIGURE 11.3 Name and attributes of a movie class.

11.3.2 Functions

The next step would be to include functions in the above class. In our example, there are two functions getdata() and putdata(). The getdata() function would ask for the values of the variables from the user, and the putdata() function would display the data. Functions implement the behavior of a class. The functions, as stated earlier, accomplish a particular task. In a class, there can be any number of functions, each accomplishing a particular task. We have special functions for initializing the data members of a class as well. The functions of a class would henceforth be referred to as member functions. The functions (or behavior) are shown in the third section of a class diagram. In Figure 11.4, the functions of the "Movie" class (getdata() and putdata()) have been shown in the third box.

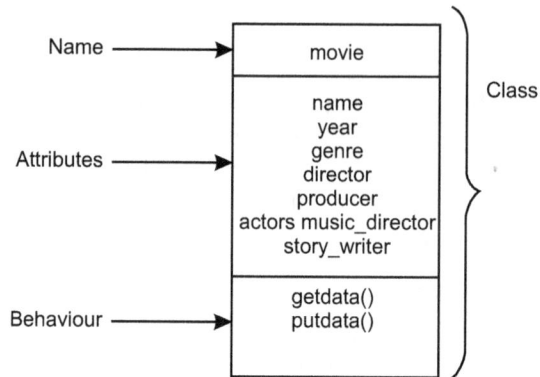

FIGURE 11.4 Name, attributes, and functions of a movie class.

The following example shows a class called movie. This class has the following data members:

- Name
- Year
- Genre
- Director
- Producer
- Actors
- Music_Director, and
- Story_writer

The class has two functions getdata(), which asks the user to enter the values of the data members and putdata(), which displays the values of the variables. In order to call the functions getdata() and putdata(), an instance of the employee class is created ("m"). As we will see later, the functions are called using the dot operator. The details regarding the syntax will be explained in the following chapter.

The following code implements the above class. Though, the syntax, etc., has not been discussed as of yet. The code has been given to give an idea of how things actually work.

Code:

```
class movie:
    def getdata(self):
        self.name=input('Enter name\t: ')
```

```
        self.year=int(input('Enter year\t:'))
        self.genre=input('Enter genre\t:')
        self.director=input('Enter the name of the director\t:')
        self.producer=input('Enter the producer\t:')
        L=[]
        item=input('Enter the name of the actor\t:')
        L.append(item)
        choice=input('Press \'y\' for more \'n\' to quit')
        while(choice == "y"):
          item=input('Enter the name of the actor\t:')
          L.append(item)
          choice=input('Enter \'y\' for more \'n\' to quit')

    self.actors=L
        self.music_director=input('Enter the name of the music
  director\t:')

def putdata(self):
        print('Name\t:',self.name)
        print('Year\t',self.year)
        print('Genre\t:',self.genre)
        print('Director\t:',self.director)
        print('Producer\t:',self.producer)
        print('Music_director\t:',self.music_director)
        print('Actors\t:',self.actors)

m=movie()
m.getdata()
m.putdata()
```

Output:

```
Enter name     :Kapoor
Enter year     :2016
Enter genre    :Drama
Enter the name of the director :ABC
Enter the producer      :Karan
Enter the name of the actor      :Siddarth
```

```
Press 'y' for more 'n' to quity
Enter the name of the actor     :Fawad
Enter 'y' for more 'n' to quitn
Enter the name of the music director     :XYZ
Name :   Kapoor
Year     2016
Genre:   Drama
Director : ABC
Producer : Karan
Music_director    : XYZ
Actors    : ['Siddarth', 'Fawad']
```

In object-oriented languages, a special function initializes the value of the data members. In languages like C++, this function, generally, has the same name as that of the class. The function is called **constructor**.

One can create a **default constructor** in a class, which does not take any parameter. The **parameterized constructor**, on the other hand, takes arguments and initializes the data members using those arguments. The implementation of constructors and their use will be dealt with in the next chapter.

When the lifetime of an object ends, a destructor is called. A **destructor** can be called using **del** in Python. The concept is explained in the next chapter of this book.

TIP!   *A Constructor is invoked when an object is created and a destructor is called when the lifetime an object ends.*

11.4 ELEMENTS OF OBJECT-ORIENTED PROGRAMMING

The following discussion briefly outlines the elements of OOP. The concepts like encapsulation, data hiding, and polymorphism have been discussed in this section. Even if things appear a bit abstract at this stage, the reader is advised not to skip the section.

11.4.1 Class

A class is a real or a virtual entity, having importance to a problem at hand and having sharp physical boundaries. A class can be a real entity. For example, when one develops software for a car wash company, then "Car," is central to the software, and hence, there would be a class called "Car." A class can also be a virtual entity; for example, in developing a student management system, a "student" class is crafted, which is a virtual entity. In both examples, the entity is crafted as it is important to the problem at hand.

The example of the "student" class can be taken further. The class would have attributes which are needed in the program. The selection of attributes would decide the physical boundaries of the class. Note that we would not need unnecessary details like the number of cars a student has or where he went last evening. This is because there is no point in storing those details for an educational institute for which we are making the student management system.

Examples of some of the classes that are central to the stated software are as follows (Table 11.1).

TABLE 11.1 Examples of classes central to various systems.

System	Class central to the software
Student management system	Student
Employee management system	Employee
Inventory control	Item
Library management	Book
Movie review	Movie
Airline management	Flight
Examination	Test

11.4.2 Object

Consider a student management system, which stores the data of each student of a school. Note that the operator would deal with an individual student, not a student's idea, while entering the data. Thus, class depicts the idea of a student, and the objects denote the individual students.

An object is an instance of a class. The objects interact with each other and get the work done. Generally, a class can have any number of objects. One can even form an array of Objects. The example "movie" had "m" as an object. As a matter of fact, we make an object and call the methods of a class (those which can be called).

In object-oriented paradigm, the program revolves around an object and therefore the type of programming is termed as object-oriented program. Calling a method of an object is equivalent to sending message to an object.

11.4.3 Encapsulation

The class is an entity, which has both data and functions. The clubbing together of the data and the functions that operate on the data is called encapsulation. Encapsulation, as a matter of fact, is one of the core principles of Object-Oriented Paradigm. Encapsulation not only makes it easier to handle objects but also improves the manageability of the Software.

Moreover, the functions in a class can be used in a variety of ways. For example, the accessibility of data members and member functions can also be managed, using access specifiers, as explained in the following subsection.

11.4.4 Data Hiding

The data hiding, is another important principle of object-oriented programming. As stated in the above discussion, the accessibility of data can be governed in a class. The data that is accessible all throughout the program is referred to as global data. The data private to a class is one that can be accessed only by the class members. There are other access specifiers as well, explained in the following sections.

In C++, for example, the data in a class is generally kept private. That is, only the member functions of the class can access the data. This ensures that the data is not accidentally changed. The functions, on the other hand, are public, in C++. The public functions can be accessed anywhere in the program (Figure 11.5). In C++, JAVA, C#, etc., there is another access specifier, which is protected. If a member is to be accessed in the class and its derived class, then protected specifier is used. C# and JAVA also have some other specifiers like internal.

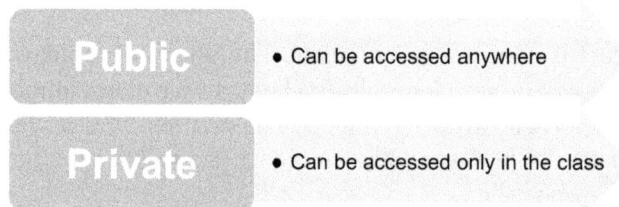

Public — • Can be accessed anywhere

Private — • Can be accessed only in the class

FIGURE 11.5 Access specifiers public and private.

Having stated the above convention, it must be clarified that deciding what is private and what is public is the discretion of the design and development team of the project. There is no hard and fast rule so as to what should be private and what should be public. The designers must decide on the accessibility of a member based on their needs.

This protection of data is not related to the security of data but to accidental change. This is needed so that the data can be changed only via the functions which have the authority to change data.

11.4.5 Inheritance

Classes are made so that they can be subclassed. This art of creating subclass(es) is called inheritance. For example, the movie class can be subclassed into various classes like art_movie, commercial_movie etc. Likewise, the student class can be subclassed into "regular student" and "part_time_student." In both examples, the subclass has many things in common in the base class (class from which the class has been derived). In addition, each subclass can have functions and data which belong to the subclass only.

For example, the student class can have attributes, namely name, date_of_birth, address, etc. The regular subclass student will use all the above data members and can also have an attribute like attendance associated with it. The class from which classes are subclassed would be called the base class, and the subclasses would be called derived classes.

For example, in Figure 11.6, the movie is the base class, and commercial_movie and art_movie are the derived classes.

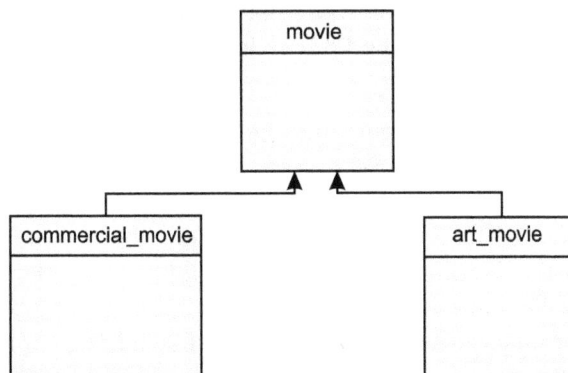

FIGURE 11.6 Deriving classes from other classes is inheritance. There are many types of inheritance. This figure shows hierarchical inheritance.

11.4.6 Polymorphism

Poly means many and morphism means forms, so polymorphism means many forms. Polymorphism can be implemented in many ways. One of the simplest examples of polymorphism is operator overloading. Operator overloading, in general, means using the same operator in more than one way. For example, "+" is used between integers to add, with strings for concatenation, and even can be used with user-defined data types, as explained in Chapter 17 of this book.

Likewise, function overloading means having more than one function with the same name, in a class, with different arguments. Various forms of polymorphism are explained in Chapter 16 of this book.

11.4.7 Reusability

The procedural programming came with almost no reusability. Modular programming allowed reusability but only to a certain extent. The functions could be used on as it is basis of modular programming. In OOP, the concept of reusability could be used in its full force. The concept of inheritance, introduced above and explained in Chapter 13 of this book, helps the programmer to reuse a code as per the requirement and that to the relevant part. As a matter of fact, reusability is one of the USPs of the object-oriented paradigm.

11.5 CONCLUSION

While designing software, one must keep in mind the entities he is going to work on. The nitty-gritty can be decided at a later stage. As a matter of fact, the popular literature does not consider the details of the operation as a matter of concern for the OOP. Hiding unnecessary details is, therefore, an important part of OOP.

For example, in developing the website for movies, the entity central to the problem is "Movie." So, one starts with an empty class called "movie." The designer must then decide on the attributes needed to implement the functions. The attributes constitute the data members of the said class. This is followed by the decision regarding the implementation details of the behavior of the class. The functions are then designed for this purpose. Finally, the things like inheritance and polymorphism, discussed later, come into play.

The journey of the formation of this class has been depicted in the following Figure (Figure 11.7).

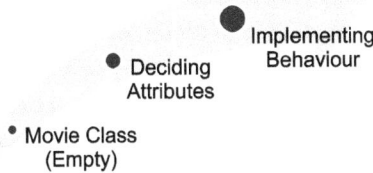

FIGURE 11.7 The design of a movie class.

Programming is an art. A good programmer should be versed in the syntax of the language, the data structures, and the concepts of algorithm analysis. In addition to the above, a programmer needs to decide the programming paradigm he will use. The chapter briefly introduces various programming paradigms and their advantages and disadvantages. Next, the chapter introduces the concept of OOP. The definitions of class, object, etc., have been discussed in the chapter. The chapter also introduces the features of OOP. The concepts introduced in this chapter will form the foundation of this section. As already stated, some of the concepts may appear abstract at this stage, but the following chapters will revisit the concepts and would demonstrate the implementation of the ideas dealt with in this chapter. In order to be able to make a program that uses OOP, one must get out of the mindset of doing things in a procedural way and start thinking about the program as centered on real-world entities having attributes and behavior. It may also be stated that designing an object-oriented program is generally preceded by designing class diagrams, sequence diagrams, etc. These are part of the Unified Modeling Language. The reader is advised to explore the references given at the end of this book to understand UML.

GLOSSARY

- **Class:** A class is a real or a virtual entity, having importance to problem at hand and having sharp physical boundaries.
- **Object:** An object is an instance of a class.
- **Encapsulation:** The clubbing together of the data and the functions that operate on the data is called encapsulation.
- **Inheritance:** The art of dividing the class into subclass(es) is inheritance.

- **Operator overloading:** Operator overloading generally means using the same operator in more than one way.
- **Function overloading:** It means having more than one function with the same name, in a class, with different arguments.

POINTS TO REMEMBER

- Telling the computer what to do, step by step, works if the task to be accomplished is not very complex.
- In the case of a moderately large program, division into functions makes the task easier.
- The division of a larger program into modules makes the program manageable and helps to achieve reusability.
- The clubbing together of functions, on some basis, gives rise to modules. This programming paradigm is called modular programming.
- A class has two important components: Attributes and Behavior.
- A constructor initializes the members of a class.
- The destructor frees the memory occupied by an object.

EXERCISES

Multiple Choice Questions

1. Which of the following is not Object-Oriented Language?

   (a) C
   (b) C++
   (c) Python
   (d) C#

2. Which of the following is Object-Oriented Language?

   (a) Python
   (b) C#
   (c) JAVA
   (d) All of the above

3. A student is a conceptual entity which acts as a blueprint for each student. The mapping is similar to which of the following?

   (a) Class and object

   (b) Method and modular programming

   (c) Both

   (d) None of the above

4. Which of the following are the two most important components of a class?

   (a) Methods and attributes   **(b)** List and tuple

   (c) Arrays and functions   **(d)** None of the following

5. In object-oriented paradigm, a variable of a class is called

   (a) Data member   **(b)** Member function

   (c) global data   **(d)** None of the above

6. In object-oriented paradigm, the functions of a class are

   (a) Member functions   **(b)** Data members

   (c) Global functions   **(d)** None of the above

7. An instance of a class is called

   (a) Object   **(b)** Subject

   (c) Inject   **(d)** None of the above

8. The clubbing together of data and the functions that operate on the data is called

   (a) Abstraction   **(b)** Encapsulation

   (c) Overloading   **(d)** None of the above

9. Allowing the selective access of data members in a class is the same as

   (a) Data Hiding   **(b)** Encapsulation

   (c) Abstraction   **(d)** None of the above

10. Having more than one function of the same name in a class is called

   (a) Function overloading   **(b)** Overriding

   (c) Encapsulation   **(d)** None of the above

11. "+" can be used for adding two number types. However, a programmer can use "+" for the addition of two user-defined data types (e.g., complex numbers). This is

   (a) Method Overloading   **(b)** Operator Overloading

   (c) Encapsulation   **(d)** None of the above

12. Inheritance is helpful in
    (a) Reusability
    (b) Redundancy
    (c) Overhead
    (d) None of the above

13. If a function in the base class is extended in the derived class, then it is
    (a) Overloading
    (b) Abstraction
    (c) Encapsulation
    (d) None of the above

14. Which of the following is not a type of inheritance?
    (a) Simple
    (b) Multiple
    (c) Hierarchical
    (d) All of them are types of inheritance

15. Which of the following initializes the members of a class?
    (a) Constructor
    (b) Destructor
    (c) Both
    (d) None of the above

16. Which of the following is true for a well-defined class?
    (a) It has importance to problem at hand
    (b) It has sharp physical boundaries
    (c) It is a real or a physical entity
    (d) All of the above

17. A language in which one can define a new data type is
    (a) Comprehensive
    (b) Extensible
    (c) Both
    (d) None of the above

18. In object-oriented paradigm, the focus is on
    (a) Data
    (b) Way a work is done
    (c) Data Types
    (d) None of the above

19. UML is
    (a) Ultra Modern Language   **(b)** Unified Modeling Language
    (c) United Model League   **(d)** None of the above

20. Which of the following is not a principle of object-oriented paradigm?
    (a) Inheritance   **(b)** Data Hiding
    (c) Encapsulation   **(d)** Divide and Conquer

Theory

1. Briefly explain the various paradigms of programming.

2. What is the difference between object-oriented paradigm and procedural programming?

3. What is a class? What are the essential components of a class? Define the attributes and functions of a class.

4. What is the relation between an object and a class?

5. What is a class diagram? Give an example of a class diagram.

6. Explain the importance of encapsulation.

7. Explain the importance of data hiding. Is it related to the security of the data?

8. What is Polymorphism? Explain the concept of operator overloading and function overloading.

9. What is the advantage of reusability? Explain the concept of reusability vis-a-vis object-oriented paradigm.

10. Explore some of the problems in Object-Oriented Programming?

Explore and Design

The reader is expected to go through the material on the database management system. The chapters on entity relationship diagrams have details of entities involved therein. Based on your research, create class diagrams of the classes mentioned in Table 14.1.

12

CLASSES AND OBJECTS

Objectives

After reading this chapter, the reader should be able to

- Understand how to create a class in Python
- Instantiate a class
- Differentiate between instance and class variables
- Use constructors and destructors
- Understand the types of Constructors

12.1 INTRODUCTION TO CLASSES

A class is a real or a virtual entity, having importance to the problem at hand and sharp physical boundaries. The concept of classes has been discussed in the previous chapter. This chapter takes the discussion forward and explores the issues involved in the implementation. It is easier to make a class in Python as compared to other programming languages. A class, in Python, can hold any kind and any amount of data. Those with C++ background might find the syntax and use of variables odd. The mechanism of classes in Python is inspired not just by C++ but also by Modula-3.

A class in Python can be **subclassed**. All types of inheritance including **multiple inheritance** are supported in Python. **Method overriding** is also allowed in Python. The dynamic nature of classes makes Python stand apart from other languages. Classes can be created at runtime.

In a class, all data members are **public** in nature. That is, they can be accessed anywhere in the program. The member functions in a class are all **virtual**. In a class, all the member functions must have the first argument as the object representing that class, henceforth referred to as **self**. Interestingly, all the build types are themselves classes, and they can be extended by the programmer.

Note that multiple names can be associated with the same object. This gives the programmer the same power as that in languages supporting pointers. Using pointers, for example, an object can be passed to a function using just one argument, and the change done by the function is visible in the calling function. In the case of Python, **aliasing** (having multiple names for the same object) can be used to accomplish the above task.

This chapter has been organized as follows. Section 12.2 discusses the definition of a class, and Section 12.3 explains the creation of an object. Section 12.4 discusses the scope of data members, and Section 12.5 presents the concept of nesting. Section 12.6 discusses constructors, and Sections 12.7 and 12.8 present a brief discussion on overloading and destructors. The last section concludes.

12.2 DEFINING A CLASS

In Python, a class can be defined using the **class** keyword. The **class** keyword is followed by the name of the class. This is followed by the body of the class (at proper indentation).

Syntax

```
class <name of the class>:
    def <function name>(<arguments>):


        . . .

    <members>
```

Consider, for example, the **employee** class, having data members **name** and **age** and member functions **getdata()** and **putdata()**. It was stated earlier that every function in the class must have at least one argument, which is **self**. The functions of a class are defined in the traditional way. The **getdata()** function asks for the values of **name** and **age** of the user. The data members are accessed via **self**, as they belong to the class and not just the function. Likewise, the **putdata()** function displays the values of the data members. Note that the members of a class are accessed via **self**.

TIP!   ■   *A class definition has functions but can also have other members.*

12.3 CREATING AN OBJECT

An object is created by associating a name with an instance of the class, initialized using the default constructor. For example, in creating an Object of the `employee` class, the following statements are used.

```
e1=employee()
```

Here, **e1** is the name of the object and **employee()** is the constructor of the class. An Object can also be created using a parameterized constructor, as explained in the following sections. The creation of an object is referred to as **instantiation**.

The function of a class can be called using the dot operator. For example, to call the **getdata()** function of the employee class, the following statements are used.

```
e1.getdata()
```

Likewise, the other methods of a class can be called using the dot operator.

Code:

```
class employee:
    def getdata(self):
        self.name=input('Enter name\t:')
        self.age=input('Enter age\t:')
    def putdata(self):
        print('Name\t:',self.name)
        print('Age\t:',self.age)
e1= employee()
e1.getdata()
e1.putdata()


>>>
=============== RUN C:/Python/Class/employee.py ================
Enter name    :Harsh
Enter age:28
Name : Harsh
Age : 28
>>>
```

| TIP! | *An object supports the following operations:* |

* *Instantiation*
* *Attribute references*

12.4 SCOPE OF DATA MEMBERS

The **scope** of a namespace is the region where it is directly accessible. In fact, in Python, scopes are used dynamically. In determining the scope of a namespace, the following rules are followed.

* First of all, the innermost scope is searched
* Then the scope of enclosing functions is searched
* This is followed by searching the global namespaces
* Finally, the build in names are seen

The nonlocal statements rebind the variables in the global scope. In order to understand this concept, consider the following code. The following points, concerning the code, are worth noting.

* The value of **a** for all instances of the class is 5, until a function, that changes the value of **a**, is called.
* In **putdata()**, **a** does not exists, **a** is local to **getdata()**
* **b** can be accessed in both the functions as **b** is a data member of the class (note that every time **b** is called, **self.b** is used)

On the basis of the above discussion, the user is expected to find out why the following code produces the following output.

Code:

```
class demo_class:
    a=5
    def getdata(self,b):
     a=7;
     self.b=b
    def putdata(self):
     print('The value of \'a\' is',a,'and that of \'b\' is',self.b)

d=demo_class()
d.getdata(9)
```

```
d.putdata()
d.putdata()
  File "C:/Python/Class/variable_visibility.py", line 7, in
    putdata
print('The value of \'a\' is',a,'and that of \'b\' is',self.b)
NameError: name 'a' is not defined
```

In the following code **b** is a member of the class. Here, **self.b=b** means the data member **b** of the class (**self.b**) is assigned value **b**, which is the second argument of the function **getdata()**. **c** is local to **getdata()**, so **c** of **getdata()** is not same as that of **putdata()**.

Definition: Instance Variable and Class Variable

An instance variable is unique to each instance, and all instances share a class variable. In the following code, **b** can be assigned a different value for each instance, but **c** remains the same.

Code:

```
class demo_class:
    a=5
    def getdata(self,b):
        c=7;
      self.b=b
      print('\'c\' is ',c,' and \'b\' is ',self.b)
    def other_function(self):
        c=3
        print('Value',c)
    def putdata(self):
      print('\'b\' is',self.b)

d=demo_class()
d.getdata(9)
print(d.a)
d.other_function()
d.putdata()
e=demo_class()
print(e.a)
>>>
```

```
========= RUN C:/Python/Class/variable_visibility2.py ==========
'c' is  7   and 'b' is   9
5
Value 3
'b' is 9
5
```

In addition to the above, a global data member can be made outside the class, which is accessible to all the methods (until the scope of the data member is changed). In the following code, **a** is common for all instances of the class, **b** is a data member of the class, **f** is global, and **c** is a local variable.

Code:

```
global f
f=7
class demo_class:
    a=5
    def getdata(self,b):
        c=7;
      self.b=b
      print('\'c\' is ',c,' and \'b\' is ',self.b,'\'f\'',f)
    def other_function(self):
        c=3
        print('Value',c)
    def putdata(self):
      print('\'b\' is',self.b)

d=demo_class()
d.getdata(9)
print(d.a)
d.other_function()
d.putdata()
e=demo_class()
print(e.a)
>>>
========= RUN C:/Python/Class/variable_visibility2.py ==========
'c' is  7   and 'b' is   9 'f' 7
5
```

```
Value 3
'b' is 9
5
```

12.5 NESTING

The designing of a class requires conceptualization of an entity, which has attributes and behavior. The object can be used in another class also. That is, a class can also have the objects of another class as its members. This is called nesting. Note that the attributes of a class can themselves be entities. For example, in the following code, an instance of the **date** class is created in the **student** class. This makes sense, as **student** is an entity having attributes that are themselves objects (like date).

Code:

```
class date:
    def getdata(self):
        self.dd=input('Enter date (dd)\t:')
        self.mm=input('Enter month (mm)\t:')
      self.yy=input('Enter year (yy)\t:')
    def display(self):
        print(self.dd,':',self.mm,':',self.yy)
class student:
    def getdata(self):
        self.name=input('Enter name\t:')
        self.dob= date()
      self.dob.getdata()
    def putdata(self):
      print('Name \t:',self.name)
      self.dob.display()
s= student()
s.getdata()
s.putdata()
============ RUN C:/Python/OOP/Nesting of classes.py ===========
Enter name        :Harsh
Enter date (dd)   :03
Enter month (mm)  :12
```

```
Enter year (yy)    :1981
Name      :Harsh
03 : 12 : 1981
>>>
```

12.6 CONSTRUCTOR

Note that each time a class is instantiated, a constructor (e.g., **e1 = employee()**) is used. In C++ terminology, a constructor is a function with the same name as the class and initializes the data members. The above examples used default constructors, which the programmer did not make. One can initialize the objects as per the need by crafting constructors. The following discussion focuses on two types of constructors: **default** and **parameterized**. A **default constructor** does not take any argument (e.g., the **employee()** constructor). In Python, the constructors are called using functions with the same name as the class's. However, they are implemented by making the **__init__()** function inside the class.

In the following code, the object **e1** behaves as expected. The values entered by the user in the **getdata()** function are displayed when **putdata()** is called. In case of **e2**, the function **getdata()** is not called; therefore, the values assigned in **__init__()** are displayed.

Code:

```
class employee:
    def getdata(self):
        self.name=input('Enter name\t:')
self.age=input('Enter age\t:')
    def putdata(self):
        print('Name\t:',self.name)
        print('Age\t:',self.age)
    def __init__(self):
        self.name='ABC'
self.age=20
e1= employee()
e1.getdata()
e1.putdata()
```

```
e2=employee()
e2.putdata()

>>>
============ RUN C:/Python/Class/Constructor1.py ===============
Enter name    :Harsh
Enter age:28
Name : Harsh
Age : 28
Name: ABC
Age:   20
>>>
```

A **parameterized constructor** is one that takes arguments, for example, in the following code, the parameterized constructor, which takes two parameters **name** and **age** has been created. In order to assign the values to the Object, the instantiation must be of the form:

```
e2=employee('Naved', 32)
```

Note that while defining the parameterized **__init__**, the first parameter is always "**self**," the rest of the parameters are the values to be assigned to different data members of the class. In the case of employee class, three parameters, "**self**," "**name**," and "**age**" are given.

Code:

```
>>>
class employee:
    def getdata(self):
        self.name=input('Enter name\t:')
        self.age=input('Enter age\t:')
    def putdata(self):
        print('Name\t:',self.name)
        print('Age\t:',self.age)
    def __init__(self, name, age):
      self.name=name
      self.age=age
    def __del__():
        print('Done')
```

```
#e1=employee()
#e1.getdata()
#e1.putdata()
e2=employee('Naved', 32)
e2.putdata()
```

```
============ RUN C:/Python/Class/Constructor2.py ===============
Name :Naved
Age :32
>>>
```

12.7 MULTIPLE __INIT__(S)

Having the same name function in a class, with different number of parameters, or different types of parameters, is called **function overloading**. In C++, JAVA, C#, etc., the constructors can also be overloaded; one can have more than one constructor, each having different parameters. In Python, however, we cannot have more than one __**init**__ in a class. For example, an error crops up if we try executing the following code.

Note, that if one makes a parameterized __**init**__, Python looks for the rest of the parameters in the instantiation.

Code:

```
class employee:
    def getdata(self):
        self.name=input('Enter name\t:')
        self.age=input('Enter age\t:')
    def putdata(self):
        print('Name\t:',self.name)
        print('Age\t:',self.age)
    def __init__(self, name, age):
        self.name=name
self.age=age
e1= employee()
e1.getdata()
e1.putdata()
```

```
e2=employee('Naved', 32)
e2.putdata()
>>>
============== RUN C:/Python/Class/Constructor2.py =============
Traceback (most recent call last):
  File "C:/Python/Class/Constructor2.py", line 11, in <module>
    e1=employee()
TypeError: __init__() missing 2 required positional arguments:
    'name' and 'age'
>>>
```

Having studied the importance and implementation of constructors, let us now implement a constructor; let us revisit the "**movie**" class, created above. The following code has a movie class, which contains a **getdata()** and **putdata()** function and **__init__(self)** for initializing the variables. Note that the object "**m**" does not call the **getdata()** function but just **putdata()**. The values assigned in the constructor are displayed.

Code:

```
class movie:
    def getdata(self):
      self.name=input('Enter name\t:')
      self.year=int(input('Enter year\t:'))
      self.genre=input('Enter genre\t:')
      self.director=input('Enter the name of the director\t:')
      self.producer=input('Enter the producer\t:')
       L=[]
       item=input('Enter the name of the actor\t:')
      L.append(item)
       choice=input('Press \'y\' for more \'n\' to quit')
      while(choice == "y"):
            item=input('Enter the name of the actor\t:')
          L.append(item)
            choice=input('Enter \'y\' for more \'n\' to quit')

self.actors=L
self.music_director=input('Enter the name of the music
    director\t:')
```

```
    def putdata(self):
        print('Name\t:',self.name)
        print('Year\t',self.year)
        print('Genre\t:',self.genre)
        print('Director\t:',self.director)
        print('Producer\t:',self.producer)
        print('Music_director\t:',self.music_director)
        print('Actors\t:',self.actors)

    def __init__(self):
        self.name='Fault'
        self.year=2015
        self.genre='Drama'
        self.director='XYZ'
        self.producer='ABC'
        self.music_director='LMN'
        self.actors=['A1', 'A2', 'A3', 'A4']

m=movie()
#m.getdata()
m.putdata()


============= RUN C:\Python\Class\class_basic2.py ==============
Name           : Fault
Year             2015
Genre          : Drama
Director       : XYZ
Producer       : ABC
Music_director    :LMN
Actors    :['A1', 'A2', 'A3', 'A4']
```

12.8 DESTRUCTORS

A constructor initializes the data members of a class, and a destructor frees the memory. The destructor is created using **__del__** and called by writing the keyword **del** and the name of the Object. The following code exemplifies a destructor in the employee class described in the previous sections.

Code:

```
class employee:
    def getdata(self):
        self.name=input('Enter name\t:')
      self.age=input('Enter age\t:')
    def putdata(self):
        print('Name\t:',self.name)
        print('Age\t:',self.age)
    def __init__(self, name, age):
        self.name=name
      self.age=age
    def __del__(self):
        print('Done')

#e1=employee()
#e1.getdata()
#e1.putdata()
e2=employee('Naved', 32)
e2.putdata()
del e2
============== RUN C:/Python/Class/Constructor2.py =============
Name : Naved
Age : 32
Done
```

The next example is the same as that of the previous one. However, the following code also demonstrates the use of **__class__.__name__**, which displays the name of the object that calls the function. This is useful as the name of the object whose destructor (or any method) is being called can be displayed while debugging.

Code:

```
class employee:
    def getdata(self):
        self.name=input('Enter name\t:')
self.age=input('Enter age\t:')
    def putdata(self):
```

```
        print('Name\t:',self.name)
        print('Age\t:',self.age)
    def __init__(self, name, age):
        self.name=name
self.age=age
    def __del__(self):
        print(__class__.__name__,'Done')

#e1=employee()
#e1.getdata()
#e1.putdata()
e2=employee('Naved', 32)
e2.putdata()
del e2
>>>
============= RUN C:/Python/Class/Constructor2.py ==============
Name :Naved
Age :32
employee Done
>>>
```

The doctsring associated with the class can be mentioned in the definition of the class within three double quotes (""" ..."""). The docstring associated with the class can be accessed through **__doc__**, as shown in the following example.

Code:

```
class employee:
    """The employee class"""
    def getdata(self):
        self.name=input('Enter name\t:')
self.age=input('Enter age\t:')
    def putdata(self):
  print('Name\t:',self.name)
  print('Age\t:',self.age)
    def __init__(self):
        self.name='ABC'
self.age=20
```

```
e1= employee()
e1.getdata()
e1.putdata()
print(e1.__doc__)
>>>
=========== RUN C:/Python/Class/employeedocstring.py ===========
Enter name    :Sakib
Enter age:17
Name :Sakib
Age :17
The employee class
>>>
```

The above chapter discusses what is referred to as an instance method. However, another type of method can be created in a class, which is referred to as a class method.

12.9 CONCLUSION

The last chapter introduced the concepts of object-oriented programming. This chapter takes the discussion forward. This chapter introduces the syntax of a class and the creation of objects. The concept of constructors, their creation, types, and implementation have also been discussed in this chapter. The chapter also introduces the idea of destructors. Ample examples have been given in the chapter explaining the implementation of the concepts introduced earlier. The following chapter presents the idea of inheritance and polymorphism, which are essential ingredients of object-oriented programming. However, to be able to inherit a class or implement operator overloading, one must be versed with the creation of a class and its use.

GLOSSARY

- The attribute of an object is **data attribute,** and function that belongs to an object is **method**.
- **Instance variable and class variable:** An instance variable is unique to each instance, and a class variable is shared by all instances.
- **Constructor:** A constructor initializes the data members.
- A **parameterized constructor** is one which takes arguments.

POINTS TO REMEMBER

▪ The classes in Python can be subclassed.

▪ All types of inheritance including multiple inheritance are supported in Python.

▪ A class can be defined using the **class** keyword, in Python. The **class** keyword is followed by the name of the class.

▪ An object is created by associating a name with an instance of the class, initialized using a constructor.

▪ The function of a class can be called using the dot operator.

▪ An object supports the following operations
   – Instantiation
   – Attribute references

▪ A constructor initializes the data members of a class, and a destructor frees the memory.

▪ The destructor is created using **__del__** and called by writing the keyword del.

▪ **__class__.__name__** displays the name of the object that calls the function.

▪ The docstring associated with the class can be accessed through **__doc__**.

EXERCISES

Multiple Choice Questions

1. A class generally has

   (a) Function and data members   **(b)** Function and Lists

   (c) Lists and Tuples   **(d)** None of the above

2. A class can have

   (a) Any number of functions   **(b)** Any type of data members

   (c) A variable local to a function **(d)** All of the above

3. **self** is

   (a) Object of the same class   **(b)** Object of the base class

   (c) Object of predefined class   **(d)** None of the above

4. Each function in Python must have at least one parameter, which is
   - **(a)** Data
   - **(b)** List
   - **(c)** Self
   - **(d)** None of the above

5. The __init__ function
   - **(a)** Initializes the data members
   - **(b)** Is compulsory
   - **(c)** Must be overloaded
   - **(d)** None of the above

6. The __init__ function in a class
   - **(a)** Must be overloaded
   - **(b)** Can be overloaded
   - **(c)** Cannot be overloaded
   - **(d)** None of the above

7. The doctring of a class can be accessed using
   - **(a)** __init__
   - **(b)** __doc__
   - **(c)** __class__
   - **(d)** None of the above

8. A global variable
   - **(a)** Can be accessed anywhere
   - **(b)** Can be accessed only in __init__
   - **(c)** Both of the above
   - **(d)** None of the above

9. The nonlocal variable
   - **(a)** Is generally associated and then used
   - **(b)** Must not be associated
   - **(c)** Does not exist
   - **(d)** None of the above

10. A variable shared by all the instances of a class is
    - **(a)** Class variable
    - **(b)** Instance variable
    - **(c)** Both
    - **(d)** None of the above

11. A variable unique to an instance is

(a) Instance variable   (b) Class variable

(c) Both   (d) None of the above

12. Which of the following keyword is used to define a class?

(a) class   (b) def

(c) del   (d) None of the above

13. Which of the following is used to define a function that acts as a destructor?

(a) del   (b) init

(c) Both   (d) None of the above

14. Which of the following operations are supported by an Object?

(a) Instantiation   (b) Attribute reference

(c) Both   (d) None of the above

15. Suppose e1 is an object, which of the following code is used to call __ del__?

(a) del e1   (b) e1.__del__

(c) Both   (d) None of the above

16. If the name of the object is to be displayed in a function of a class, then which of the following can be used?

(a) __class__.__name__   (b) __object__.__name__

(c) Both   (d) None of the above

17. In a class all variables are _ by default?

(a) Public   (b) Private

(c) Cannot say   (d) Depends on the type of variables

18. In Python, which of the following operator is used to access methods?

(a) Dot   (b) Plus

(c) []   (d) None of the above

19. Can a list of Objects be created?

(a) Yes, if the type of variables is public

(b) Yes, in all cases

(c) No, in all cases

(d) Yes, if the type of variables is private

20. Data members of a class

(a) Must be private   **(b)** Can be private

(c) Must be public   **(d)** None of the above

Theory

1. What is an Object? How is an Object created in Python?

2. Explain the scope of variables in a class? Give an example of data members of a class that are shared by all the objects and of those which are unique to an Object.

3. What is a constructor? What are the different types of constructors in Python?

4. Can we overload a constructor in Python?

5. How can one access the name of the docstring in Python?

6. How can one access the name of an Object in Python?

7. What is a destructor? How is a destructor created in Python?

8. Give an example of the use of a destructor in Python.

9. Give an example of the instantiation of a class.

10. Explain the concept of aliasing in Python.

Programming Exercises

A start-up employee interns. The following details of interns are stored by the company.

  ▪ first_name
  ▪ last_name

▪ address
▪ mobile_number
▪ e_mail

1. Create a class called **Intern**, which stores the above details. Craft two functions **getdata()**, which asks the user to enter data and **putdata()** to display the data.

2. In the above program create **__init__** which takes only one parameter (**self**).

3. In question number 1, create **__init__**, which takes six parameters, first being "**self**" and the rest the values of variables, stated above.

4. In the above question, craft a destructor and call it.

5. A library management system is to be created, in which the following details of a "**Book**" are to be stored.

▪ Name
▪ Publisher
▪ Year
▪ ISBN
▪ Authors

The **authors**, above, is a list consisting of all the authors of that book.

Create a **class** called Book, which stores the above details. Craft two functions **getdata()**, which asks the user to enter data, and **putdata()** to display the data.

6. In the above program create **__init__** which takes only one parameter (**self**).

7. In question number 6, create **__init__**, which takes 6 parameters, first being "**self**" and the rest the values of variables, stated in question number 1.

8. In the above question craft a destructor and call it.

9. Create a class called complex, having **real_part** and **ima_part** as its two data members and **getdata()** and **putdata()** and its member functions.

10. In the above question, craft **__init__** and **__del__**.

11. Create a function called **add**, which takes two complex numbers as its parameters and returns the sum of the two complex numbers.

12. Create a function called **sub**, which takes two complex numbers as its parameters and returns the difference between the two complex numbers.

13. Create a function called **multiply**, which takes two complex numbers as its parameters and returns the product of the two complex numbers.

14. Create a function called **div**, which takes two complex numbers as its parameters and returns the result of the division of the two complex numbers.

15. Create a class called **date** having day, month and year as its data members and **getdata()** and **putdata()** as its member functions.

13

INHERITANCE

Objectives

After reading this chapter, the reader should be able to

- Understand the concept and importance of Inheritance
- Differentiate between Inheritance and Composition
- Understand the types of Inheritance
- Appreciate the role of "self" in methods
- Understand the concept and importance of super
- Appreciate the need of an abstract class

13.1 INTRODUCTION TO INHERITANCE AND COMPOSITION

Readers versed with C++ must have studied the importance of inheritance and composition. Inheritance was projected as a path breaking concept, which promised to solve all the problems and bring about a change in the way programming is done. Inheritance may also create problems, much more than you can imagine.

Many programmers believe that inheritance is a black hole which somehow attracts programmers, who fall in the trap of tall claims and end up landing themselves in a situation which tempts them to use multiple inheritance. Multiple Inheritance is like Voldemort, and the Object-Oriented Programming environment is Hogwards. Therefore, it is better to avoid Multiple Inheritance, as much as possible.

Object-oriented programming has its charm, but also comes with its own problems. So, use inheritance, only if required. In which case, remember never ever to use Multiple Inheritance. Also remember that anything that can be done using inheritance can be done otherwise. Composition, introduced later in the chapter, can be easily used to accomplish most of the tasks that can be done using inheritance.

In hindsight, inheritance means a class would get features (all or some) from the parent class. So, when one writes:

```
class SoftwareDeveloper(Employee):
    ...
```

it implies that the class **SoftwareDeveloper** is a subclass of the class Employee. This relationship falls into the category of **is a** type relationship. That is, **Software Developer is-a Employee**.

The class from which class(es) are derived is called **base class** and those that inherit features from the base class are **derived classes**. In the above example, Employee is the base class and **SoftwareEmployee** is the derived class. Note that inheritance does not affect the base class. The derived class can use the modules of the base class in a variety of ways, discussed as follows.

13.1.1 Inheritance and Methods

As far as modules are concerned, Inheritance can help the programmer to derive the features by one of the following ways.

The method is not present in the child class, but only in the parent class: In such case, if an instance of the child class calls the said method, the parent class's method is called. For instance, in the following snippet, the derived class does not have method called **show()**, so calling **show** using an instance of the derived class invokes the method of the parent class.

Code:

```
class ABC:
    def show(self):
        print("show of ABC")


class XYZ(ABC):
    def show1(self):
        print("show of XYZ")
```

```
A = ABC()
A.show()
B= XYZ()
B.show()
B.show1()
```

The method is present in both the parent class and in the derived class: In such cases, if this method is invoked using an instance of the derived class, the method of the derived class is called. If the method is called using an instance of the base class, the method of the base class is called. Note that in such cases, the derived class redefines the method. This **overriding** ensures that the search of the method in the inheritance tree ends up invoking this method only. For example, in the following snippet, **B.show**() calls the **show**() method of the derived class, whereas **A.show**() calls the method of the base class.

Code:

```
class ABC:
    def show(self):
        print("show of ABC")

class XYZ(ABC):
    def show(self):
        print("show of XYZ")

A = ABC()
A.show()
B= XYZ()
B.show()
```

The inherited class modifies the method of the base class and in this process invokes the method of the base class inside the method of the derived class also. Note that, in the following snippet, the **show** method of the derived class prints a message, then calls the method of the base class and finally prints another message. Note that in this case, the method of the base class can be called by qualifying the name of the method with the name of the base class. For example, in the following snippet, the **show** method of the base class can be called using **ABC.show(self)**. The importance of the **self** argument has been explained in section 13.3.

Code:

```
class ABC:
    def show(self):
        print("show of ABC")


class XYZ(ABC):
    def show(self):
        print("Something before calling the base class function")
ABC.show(self)
print("Something after calling the base class function")
A = ABC()
A.show()
B= XYZ()
B.show()
```

The first type of inheritance would henceforth be referred to as **implicit inheritance**. In this type, the method of the base class can be called using an instance of the derived class.

The second type of inheritance would henceforth be referred to as **explicit overriding**. As stated earlier, the derived class would redefine the method of the base class and calling this method using an instance of the derived class would invoke the method of the derived class.

The third type of inheritance is the most important and practical form of overriding methods. This type of inheritance leaves the room of not making an instance of the base class, if not required, still using the function.

The following illustration combines the three types of inheritance:

Illustration 13.1:

Create a class called **Student** having **__init__** and **show** methods. The **Student** class should have a data member called **name**. The **__init__** should assign value to **name** and **show** should display the value. Create another class called **RegularStudent**, which would be the derived class of the **Student** class. The class should have two methods **__init__** and **show**. The **__init__** should assign values to **age** and should call the **__init__** of the base class and pass the value of **name** to the base class. The **show** method must display the data of the **Regular Student**. In addition to the above, both classes should have methods called **random**, both of which should be independent of each other (Figure 13.1). Find what happens when the methods of the base class and the derived classes are called using the instances of the base and the derived classes.

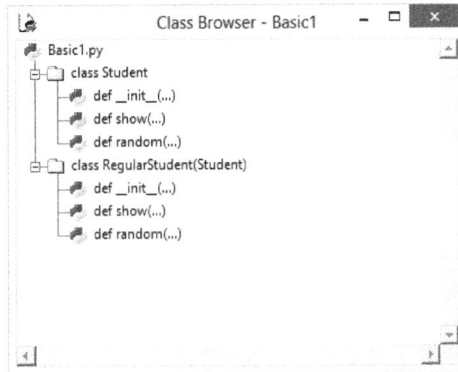

FIGURE 13.1 Class hierarchy for Illustration 13.1.

Solution:

Code:

```python
class Student:
    def __init__(self,name):
        self.name=name
    def show(self):
        print("Name\t:"+self.name)
    def random(self):
        print("A random method in the base class")


class RegularStudent(Student):
    def __init__(self,name):##overrides the base class method and
    calls the base class method
        self.age=22
        Student.__init__(self,name)
    def show(self):##redefines the base class method
        print("Name (derived class)\t:"+self.name+" Age\
            t:"+str(self.age))
    def random(self):##nothing to do with the base class method
        print("Random method in the derived class")
```

```
naks = Student("Nakul")
hari = RegularStudent("Harsh")
naks.show()
hari.show()
##The variables can be seen outside the class also
print(naks.name)
print(hari.name)
```

Output:

```
Name    :Nakul
Name (derived class)    :Harsh Age    :22
Nakul
Harsh
```

13.1.2 Composition

Making an instance of another class inside a class makes things easy and helps the programmer to accomplish many tasks. In order to understand the concept, let us consider an example. Consider that a **Student** and his **PhDguide** are subclasses of the **person** class. Also, the data of the **PhD guide** includes the list of students guided by them. Hence, composition comes into play. The instantiation of the **students** in the **PhDGuide** class can be done as explained in the following illustration.

Illustration 13.2:

Create a class called **Student**, having **name** and **email** as its data members and **__init__(self, name, email)** and **putdata(self)** as bound methods. The **__init__** function should assign the values passed as parameters to the requisite variables. The **putdata** function should display the data of the student. Create another class called **PhDGuide** having **name, email**, and **students** as its data members. Here, the **students** variable is the list of students under the guide. The **PhDGuide** class should have four bound methods: **__init__, putdata, add**, and **remove**. The **__init__** method should initialize the variables, the **putdata** should show the data of the guide, include the list of students, the **add** method should add a student to the list of students of the guide and the **remove** function should remove the student (if the student exist in the list of students of that guide) from the list of students.

Solution:

The details of the classes have been shown in Figure 13.2. It may be noted that since **students** is a list therefore a **for** loop is needed to display the list of **students**. Also, while adding the student to the list, the data of the passed parameter has been stored in **s** (an instance of Student) and **s** has been added to the list of the students. Same procedure has been adopted to remove a student. The code is as follows:

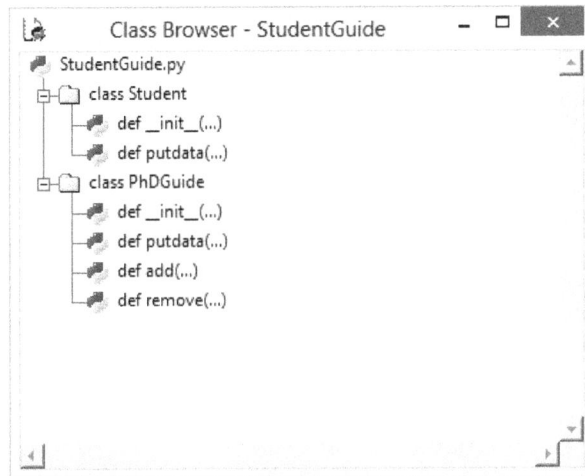

FIGURE 13.2 Details of classes for Illustration 13.2.

Code:

```python
class Student:
    def __init__(self,name,email):
        self.name=name
        self.email=email

    def putdata(self):
        print("\nStudent's details\nName\t:",self.name,"\
            nE-mail\t:",self.email)

class PhDGuide:
    def __init__(self, name, email,students):
        self.name=name
        self.email=email
        self.students=students
```

```python
    def putdata(self):
        print("\nGuide Data\nName\t:",self.name,"\nE-mail\t:",
            self.email)
        print("\nList of students\n")
        for s in self.students:
            print("\t",s.name,"\t",s.email)
    def add(self, student):
        s=Student(student.name,student.email)
        if s not in self.students:
            self.students.append(s)
    def remove(self, student):
        s=Student(student.name,student.email)
        flag=0
        for s1 in self.students:
            if(s1.email==s.email):
                print(s, " removed")
                self.students.remove(s1)
                flag=1
        if flag==0:
            print("Not found")


Harsh=Student("Harsh","i_harsh_bhasin@yahoo.com")
Nav=Student("Nav","i_nav@yahoo.com")
Naks=Student("Nakul","nakul@yahoo.com")
print("\nDetails of students\n")
Harsh.putdata()
Nav.putdata()
Naks.putdata()
KKA=PhDGuide("KKA","kka@gmail.com",[])
MU=PhDGuide("Moin Uddin","prof.moin@yahoo.com",[])
print("Details of Guides")
KKA.putdata()
MU.putdata()
MU.add(Harsh)
MU.add(Nav)
KKA.add(Naks)
print("Details of Guides (after addition of students")
```

```
KKA.putdata()
MU.putdata()
MU.remove(Harsh)
KKA.add(Harsh)
print("Details of Guides")
KKA.putdata()
MU.putdata()
```

Output:

```
Details of students

Student's details
Name   : Harsh
E-mail : i_harsh_bhasin@yahoo.com

Student's details
Name   : Nav
E-mail : i_nav@yahoo.com

Student's details
Name   : Nakul
E-mail : nakul@yahoo.com
Details of Guides

Guide Data
Name   : KKA
E-mail : kka@gmail.com

List of students

Guide Data
Name   : Moin Uddin
E-mail : prof.moin@yahoo.com

List of students

Details of Guides (after addition of students
```

```
Guide Data
Name   : KKA
E-mail : kka@gmail.com

List of students

    Nakul     nakul@yahoo.com

Guide Data
Name   : Moin Uddin
E-mail : prof.moin@yahoo.com

List of students

    Harsh     i_harsh_bhasin@yahoo.com
    Nav       i_nav@yahoo.com
<__main__.Student object at 0x03A49650>   removed
Details of Guides

Guide Data
Name   : KKA
E-mail : kka@gmail.com

List of students

    Nakul     nakul@yahoo.com
    Harsh     i_harsh_bhasin@yahoo.com

Guide Data
Name   : Moin Uddin
E-mail : prof.moin@yahoo.com

List of students

    Nav   i_nav@yahoo.com
```

13.2 INHERITANCE: IMPORTANCE AND TYPES

The concept of classes was introduced in the previous chapter. It was mentioned that classes are real or conceptual entities which have sharp physical boundaries and importance to problems at hand. A class has attributes (data members) and behavior (class methods). However, at times these classes must be extended to be able to solve some specific problem without having to meddle with the original class. To be able to do so, the language should support inheritance. As a matter of fact, the presence of classes in the language is primarily because it can be inherited. Inheritance is, as per most of the authors, one of the most essential features of Object-Oriented Language.

Using inheritance one can create new classes (derived classes) from existing class (Base Class(es)). Note that a derived class can even have more than one base class, referred to as **Multiple Inheritance,** which is one of the most undesirable forms of inheritance. Also, a base class can itself be a derived class of some other class. The derived class will have all the allowed features of the base class plus some futures of its own.

A class can be depicted using a class diagram. A class diagram is the diagrammatic representation of a class generally having three sections. In the representation used henceforth, the first section has the name of the class. The second section has the attributes, and the third section has the class methods. Figure 13.3 shows the class diagram in which the **Book** class is the base class and the **Text_Book** class is the derived class. Note that the arrow is from the derived class to the base class. The arrow indicates **"is derived from"** or **"is inherited from"**. Figure 13.4 gives the details of the two classes. Note that the **Book** class has the following attributes.

- name
- authors
- publisher
- ISBN
- year

The class methods of this class are **getdata()** and **putdata()**. The **Text_Book** class has another attribute, **course**. Figure 13.5 shows the class browser showing the two classes and the relation between them. The corresponding program is presented in Illustration 13.3.

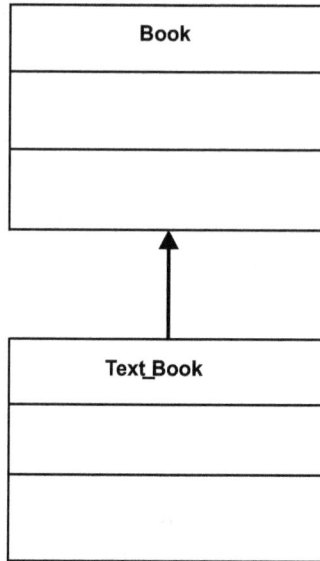

FIGURE 13.3 Text_book is the derived class of the Book class.

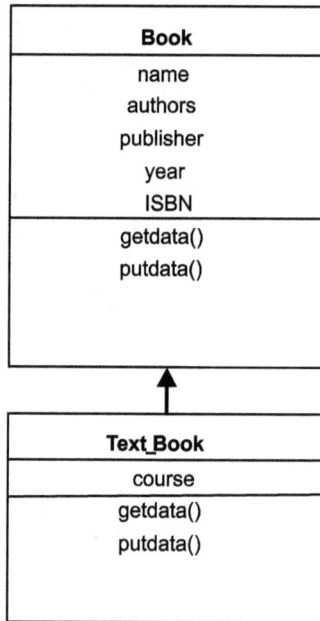

FIGURE 13.4 A class diagram, generally, has three components: the name of the class, the data members, and the methods of the class.

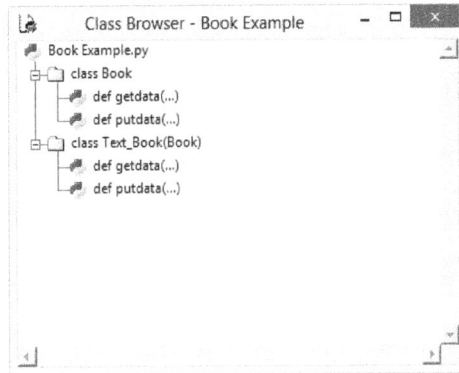

FIGURE 13.5 The book examples' class hierarchy in the class browser of Python.

13.2.1 Need for Inheritance

In very large programs, it is difficult to code and debug a class. Once the programmer has crafted a class, there is little need to meddle with it. If one needs to craft classes having the same features as the class that has been developed (and add some more features to it), then it makes sense to derive classes from the existing class. So, inheritance helps to reuse a code. Reusing the code has its own advantages. It not only saves time but also money. The reliability of the program also increases by reusing a code. One can also develop his class by extending classes developed by others. That is, inheritance helps in distributing libraries. Inheritance also helps to implement a design that is more intuitive, better, and practical. Inheritance also has some disadvantages.

Inheritance is important because of the following factors:

- *Reusability*
- *Increased Reliability*
- *Distributing libraries*
- *Intuitive, better programs*

13.2.2 Types of Inheritance

This section presents various types of inheritance and corresponding examples. Note that the reader is expected to execute the problem given in the illustrations and analyze the output. As explained earlier, Inheritance means deriving new classes from the existing classes. The class from which features have been derived is called the **base class** and the class which derives features is called the **derived class**. There are five types of inheritance: Simple, Hierarchical, Multilevel, Multiple, and Hybrid.

13.2.2.1 Simple inheritance

The simple inheritance has a single base class and a single derived class. Illustration 13.3 exemplifies this type. The following illustration has two classes: **Book** and **Text_Book**. The Book class has two methods: **getdata** and **putdata**. The **getdata** method asks the user to enter the **name of the book, number of authors, the list of authors, publisher, ISBN**, and **year**. The derived class **Text_Book** has another attribute called **course**. The **getdata** and the **putdata** methods extend the base class methods (refer to the previous section).

Illustration 13.3:

Implement the following hierarchy (Figure 13.6). The **Book** function has name, **n** (number of authors), **authors** (list of authors), **publisher, ISBN**, and **year** as its data members and the derived class has **course** as its data member. The derived class methods override (extends) the methods of the base class.

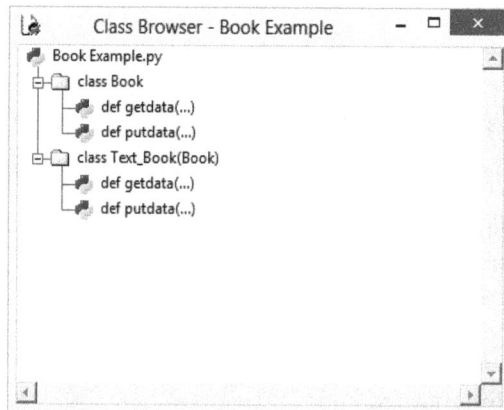

FIGURE 13.6 The class hierarchy for Illustration 13.3.

Solution:

The following code implements the above hierarchy. The output of the program follows.

Code:

```
class Book:
    def getdata(self):
        self.name=input("\nEnter the name of the book\t:")
            self.n=int(input("\nEnter the number of authors\t:"))
        self.authors=[]
        i=0
        while i<self.n:
            author=input(str("\nEnter the name of the "+str(i)+"th
                    author\t:"))
            self.authors.append(author)
            i+=1
        self.publisher=input("\nEnter the name of the publisher\t:")
        self.ISBN=input("\nEnter the ISBN\t:")
        self.year=input("\nEnter year of publication\t:")

    def putdata(self):
        print("\nName\t:",self.name,"\nAuthor(s)\t:",self.
            authors,"\nPublisher\t:",self.publisher,"\
            nYear\t:",self.year,"\nISBN\t:",self.ISBN)

class Text_Book(Book):
    def getdata(self):
        self.course=input("\nEnter the course\t:")
        Book.getdata(self)
    def putdata(self):
        Book.putdata(self)
        print("\nCourse\t:",self.course)

Book1=Book()
Book1.getdata()
Book1.putdata()
TextBook1=Text_Book()
TextBook1.getdata()
TextBook1.putdata()
```

Output:

```
Enter the name of the book        :Programming in C#

Enter the number of authors       :1

Enter the name of the 0th author  :Harsh Bhasin

Enter the name of the publisher   :Oxford

Enter the ISBN                    :0-19-809740-9

Enter year of publication         :Oxford


Name                              :Programming in C#
Author(s)                         :['Harsh Bhasin']
Publisher                         :Oxford
Year                              :Oxford
ISBN                              :0-19-809740-9

Enter the course                  :Algorithms

Enter the name of the book        :Algorithms Analysis and Design

Enter the number of authors       :1

Enter the name of the 0th author  :Harsh Bhasin

Enter the name of the publisher   :Oxford

Enter the ISBN                    :0-19-945666-6
Enter year of publication         :Oxford
Name                              :Algorithms Analysis and Design
Author(s)                         :['Harsh Bhasin']
Publisher                         :Oxford
Year                              :Oxford
ISBN                              :0-19-945666-6

Course                            :Algorithms
```

13.2.2.2 Hierarchical inheritance

In the Hierarchical inheritance, a single base class has at least two derived classes. Illustration 13.4 exemplifies this type. The following illustration has three classes: **Staff, Teaching**, and **NonTeaching**. Both **Teaching** and **NonTeaching** are the derived classes of the **Staff** class. The **Staff** class has two methods: **getdata** and **putdata**. The **getdata** method asks the user to enter the **name** and the **salary** of the member of the **staff**. The derived class **Teaching** has another attribute called **subject**. The **getdata** and the **putdata** methods extend the base class methods. Similarly, the derived class **NonTeaching** has an attribute called **department**. The **getdata** and the **putdata** methods extend the base class methods.

Illustration 13.4:

*Implement the following hierarchy (Figure 13.7). The **Staff** class has **name** and **salary** as its data members, the derived class **Teaching** has **subject** as its data member and the class **NonTeaching** has **department** as its data member. The derived class methods override (extends) the methods of the base class.*

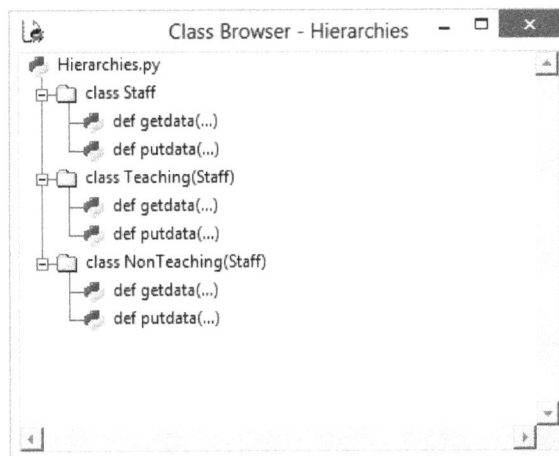

FIGURE 13.7 The class hierarchy for Illustration 13.4.

Solution:

The following code implements the above hierarchy. The output of the program follows.

Code:

```
##Hierarchies
class Staff:
    def getdata(self):
        self.name=input("\nEnter the name\t:")
        self.salary=float(input("\nEnter salary\t:"))
    def putdata(self):
        print("\nName\t:",self.name,"\nSalary\t:",self.salary)

class Teaching(Staff):
    def getdata(self):
        self.subject=input("\nEnter subject\t:")
        Staff.getdata(self)
    def putdata(self):
        Staff.putdata(self)
        print("\nSubject\t:",self.subject)

class NonTeaching(Staff):
    def getdata(self):
        self.department=input("\nEnter department\t:")
        Staff.getdata(self)
    def putdata(self):
        Staff.putdata(self)
        print("\nDepartment\t:",self.department)

X=Staff()
X.getdata()
X.putdata()
##Teacher
Y=Teaching()
Y.getdata()
Y.putdata()
##Non Teaching Staff
Z=NonTeaching()
Z.getdata()
Z.putdata()
```

Output:

```
========== RUN C:/Python/Inheritance/Hierarchies.py ============

Enter the name          :Hari

Enter salary            :50000

Name                    :Hari
Salary                  :50000.0

Enter subject           :Algorithms

Enter the name          :Harsh

Enter salary            :70000

Name                    :Harsh
Salary                  :70000.0

Subject                 :Algorithms

Enter department        :HR

Enter the name          :Prasad

Enter salary            :52000

Name                    :Prasad
Salary                  :52000.0

Department              :HR
```

13.2.2.3 Multilevel inheritance

In the Multilevel inheritance, a base class has a derived class, which itself becomes a derived class for some other class. Illustration 13.5 exemplifies this type. The following illustration has three classes: **Person, Employee**, and **Programmer**. The **Person** class is the base class. The **Employee** class has been derived from the **Person** class. The **programmer** class has been derived from the **Employee** class. The **Person** class has two attributes **name** and **age** and two methods **getdata** and **putdata**. The **getdata** method asks the user to enter the **name** and the **age** of the member of the **staff**. The derived class **Employee** has another attribute called **emp_code**. The **getdata** and the **putdata** methods extend the base class methods. Similarly, the class **Programmer** has another attribute called **language**. The **getdata** and the **putdata** methods extend its base class methods (**Employee** class).

Illustration 13.5:

Implement the following hierarchy (Figure 13.8). The **Person** class has **name** and **age** as its data members, the derived class **Employee** has **emp_code** as its data member and the class **Programmer** has **language** as its data member. The derived class methods override (extends) the methods of the base class.

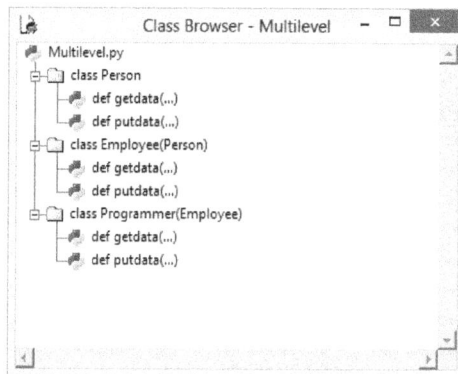

FIGURE 13.8 The class hierarchy for Illustration 13.5.

Solution:

The following code implements the above hierarchy. The output of the program follows.

Code:

```
class Person:
    def getdata(self):
        self.name=input("\nEnter Name\t:")
        self.age=int(input("\nEnter age\t:"))
```

```
    def putdata(self):
        print("\nName\t:",self.name,"\nAge\t:",str(self.age))


class Employee(Person):
    def getdata(self):
        Person.getdata(self)
        self.emp_code=input("\nEnter employee code\t:")
    def putdata(self):
        Person.putdata(self)
        print("\nEmployee Code\t:",self.emp_code)


class Programmer(Employee):
    def getdata(self):
        Employee.getdata(self)
        self.language=input("\nEnter Language\t:")
    def putdata(self):
        Employee.putdata(self)
        print("\nLanguage\t:",self.language)


A=Person()
print("\nA is a person\nEnter data\n")
A.getdata()
A.putdata()
B=Employee()
print("\nB is an Empoyee and hence a person\nEnter data\n")
B.getdata()
B.putdata()
C=Programmer()
print("\nC is a programmer hence an employee and employee is a
      person\nEnter data\n")
C.getdata()
C.putdata()
```

Output:

```
A is a person
Enter data
Enter Name              :Har
```

```
Enter age            :28
Name                 :Har
Age                  :28
B is an Empoyee and hence a person
Enter data
Enter Name           :Hari
Enter age            :29
Enter employee code  :E001
Name                 :Hari
Age                  :29
Employee Code        :E001
C is a programmer hence an employee and employee is a person
Enter data
Enter Name           :Harsh
Enter age            :30
Enter employee code  :E002
Enter Language       :Python
Name                 :Harsh
Age                  :30
Employee Code        :E002
Language             :Python
>>>
```

13.2.2.4 *Multiple inheritance and hybrid inheritance*

In Multiple Inheritance, a class has more than one base class. This type of inheritance can be problematic as it can lead to ambiguity. It is therefore advised to avoid this kind of inheritance as far as possible. However, the following sections throw some light on this type and the problems associated with this type.

A design may have a combination of more than one type of inheritance. Two classes **B** and **C** have been derived from class **A** (see Figures 13.9 and 13.10). However, for class **D**, the classes **B** and **C** are the base classes. This is an example of combining hierarchical and multiple inheritance. Such type is referred to as **Hybrid Inheritance**.

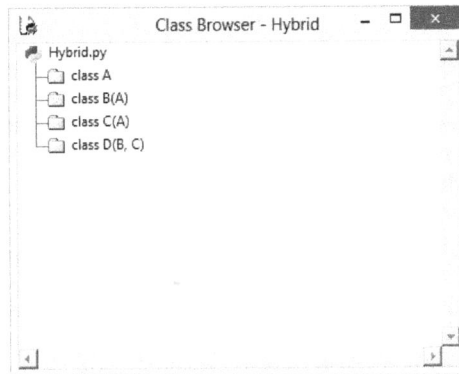

FIGURE 13.9 Classes B and C have been derived from A (Hierarchical Inheritance) and D is derived from B and C (Multiple Inheritance).

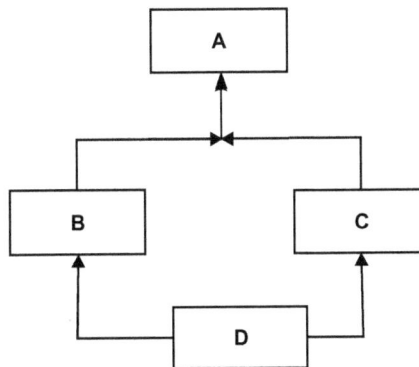

FIGURE 13.10 Classes B and C have been derived from A (Hierarchical Inheritance) and D is derived from B and C (Multiple Inheritance).

13.3 METHODS

The importance of functions and methods has already been stated in the first section of this book. Methods are, as stated earlier, just functions with a special positional parameter, within a class. Methods, in fact, help the programmer to accomplish many tasks. Methods can be bound or unbound. The unbound methods do not have **self** as a parameter. It is worth mentioning here that, in Python 3.X, the unbound methods are the same as functions, whereas in Python 2.X, it is a different type. The bound methods, on the other hand, have **self** as the first positional parameter, when a method is accessed through qualifying an instance of a class. Here, the instance of the class need not to be passed.

In spite of the above differences, the following similarities between the two types may not be missed.

- A method in Python is also an Object. Both bound and unbound methods are objects.
- Same method can be invoked as a bound method and an unbound method.

The discussion and illustrations that follow would clarify the second point.

13.3.1 Bound Methods

A method can be invoked in a variety of ways. If the first positional parameter of the method is **self**, it is bound. In such cases, the instance of the class can call the method by passing the requisite parameters.

A variable which hold **<Object name>.<method name>(Hari.display)**, in the following example, can also be used to invoke the method. Those of you, from C# background may find the concept similar to that of delegates.

A method can also be invoked by creating an unmanned instance of the class. The third call of the display method depicts this way of calling method.

Illustration 13.6:

Calling a bound method

This illustration has a class called **Student**. The **Student** class has a **display** method, which takes two arguments. The first being the positional parameter and the second being a string that is printed. Note that the **display** method is a bound method and hence is called through an instance of the class.

Code:

```
class Student:
    def display(self, something):
        print("\n"+something)

##Invoking a bound method
Hari = Student()
Hari.display("Hi I am Hari")
```

```
##display() can also be invoked through an instance of the
   method
X= Hari.display
X("Hi I am through X")
##display called again
Student().display("Calling display again")
```

Output:

```
>>>
========== RUN C:\Python\Inheritance\BoundUnbound.py ===========
Hi I am Hari
Hi I am through X
Calling display again
>>>
```

13.3.2 Unbound Method

An unbound method does not have **self**. Hence, the positional parameter need not to be passed in method. In such methods, the variables should not be qualified by **self**. Calling such methods in the same way, as before, would result in an error, as shown in the output of Listing 1 of Illustration 13.7. The second listing calls the unbound method in an appropriate way. Such methods must be called by the name of the class and not the object. In Python 3.X, as stated earlier, such methods work in the same manner as functions. Also note that normal functions can be called using the class, of which they are member, as shown in the previous illustration.

Illustration 13.7:

Calling an unbound method

This illustration extends the previous illustration and adds the **getdata** method, which does not take **self** as a parameter and hence is called by the class itself.

Code:

```
class Student:
    def display(self, something):
        print("\n"+something)
```

```python
    def getdata(name,age):
        name=name
        age=age
        print("\nName\t:",name,"\nAge\t:",age)


##Creating a new student
Naved=Student()
name=input("\nEnter the name of the student\t:")
age=int(input("\nEnter the age of the student\t:"))
Naved.getdata(name,age)
```

Output:

```
>>>
========== RUN C:/Python/Inheritance/BoundUnbound.py ===========
Enter the name of the student :Naved

Enter the age of the student  :22
Traceback (most recent call last):
File "C:/Python/Inheritance/BoundUnbound.py", line 21, in
    <module>
Naved.getdata(name,age)
>>>
```

Snippet 2:

Code:

```python
class Student:
    def display(self, something):
        print("\n"+something)
    def getdata(name,age):
        print("\nName\t:",name,"\nAge\t:",str(age))
##Creating a new student
Naved=Student()
name=input("\nEnter the name of the student\t:")
age=int(input("\nEnter the age of the student\t:"))
Student.getdata(name,age)
```

Output:

```
>>>
========= RUN C:/Python/Inheritance/BoundUnbound1.py ===========
Enter the name of the student :Naved
Enter the age of the student  :22
Name          :Naved
Age           :22
>>>
```

13.3.3 Methods are Callable Objects

Methods, like any other object in Python, can be stored in a list and called as per the requirement. In the illustration that follows, the class operations has a constructor **__init__(self, number)**, which assigns the value of the second parameter to the data member called **number**. The class has two methods **square** and **cube**. The first method calculates (and returns) the square of the number and the second calculates (and returns) the cube of the number. Two instances of the class operations have been created: **X** and **Y**. **X** is initialized to 5 and **Y** to 4. The list **List**, stores the objects **X.square, X.cube, Y.square** and **Y.cube**. The elements of the list are then called one by one and invoked.

Illustration 13.8:

Methods as callable objects

Code:

```
class operations:
    def __init__(self, number):
        self.number=number
    def square(self):
        return (self.number*self.number)
    def cube(self):
        return(self.number*self.number*self.number)

Num1=operations(5)
Num2=operations(4)
List= [Num1.square, Num1.cube, Num2.square, Num2.cube]
for callable_object in List:
print(callable_object())
```

Output:

```
>>>
======== RUN C:/Python/Inheritance/CallableObjects.py ========
25
125
16
64
>>>
```

13.3.4 The Importance and Usage of Super

A class may have data members and member functions (method). A method is just a function in a class, defined using the keyword **def**. As discussed in the earlier chapters, the methods depict the behavior of a class. Generally, the methods' first argument is an instance of the class itself. The first argument generally referred to as **self**, is similar to **this** of C++. Using **self** with the variable name indicates that the reference is to the instance variable, not that in the global scope. For example, in the following snippet, the **__init__** method has two arguments: first being **self** and second being the **name**. Assigning name to **self.name** implies that the local variable name is assigned **name** (the second argument of **__init__**). Similarly, the **putdata** method has a positional parameter indicating the data for the instance which invokes **putdata** must be shown. Note that the output reinforces the fact that **self** binds the method call with the instances.

Code:

```
class Student:
    def __init__(self,name):
        self.name=name
    def putdata(self):
        print("name\t:",self.name)
Hari=Student("Hari")
Hari.putdata()
Naks=Student("Nakul")
Naks.putdata()
>>>
```

Output:

```
============== RUN C:/Python/Inheritance/Basic.py ===============
name         : Hari
name         : Nakul
>>>
```

However, methods can also be crafted without the **self** argument. These are unbound methods. The concept has been discussed in Section 13.3.2 of this chapter. The method of a class is an instance method by default. So, generally, the method of a class can be called by creating an instance of the class and using the dot operator to call the method. Note that this was the case in languages like C#, JAVA, etc.

However, there are other types of methods as well. For example, static methods do not require the instance of a class as their first argument.

13.3.5 Calling the Base Class Function Using Super

The functions of the base class can be called **super**. In fact, **super** can be used to call any function of the base class and it clearly depicts the calling of the base class's function. In order to understand the usage of **super**, let us consider the following example. In the following example, **BaseClass** has two methods: **__init__** and **printData**. **__init__** has one positional parameter and one parameter that initializes **data** (the data member of the **BaseClass**). The **DerivedClass** is the derived class of the **BaseClass**. This class has **__init__**, which takes a positional parameter (**self**), and two other parameters. The first initializes the data member of **DerivedClass** and the second is passed onto the **__init__** of the base class (**BaseClass**) using **super**. The **super** takes the name of the class (**DerivedClass**), the positional parameter (**self**) and calls the **__init__** of the base class by passing all parameters except the positional parameter. Note that the second function also uses **super** in the same manner.

Code:

```
class BaseClass:
    def __init__(self, data):
        self.data=data
    def printData(self):
        print("Data of the base class\t:",self.data)
```

```
class DerivedClass(BaseClass):
    def __init__(self,data1, data2):
        self.data1=data1
        super(DerivedClass, self).__init__(data2)
    def printData(self):
        super(DerivedClass,self).printData()
        print("Data of the derived class\t:",self.data1)
```

Output:

```
>>>
Data of the base class    : 4
Data of the base class    : 5
Data of the derived class : 6
>>>
```

13.4 SEARCH IN INHERITANCE TREE

An object is searched in the inheritance tree, in a bottom-up fashion. First of all, the class searches for the given object. If it is found, then the found object is used to accomplish the given task. If not, then its super class (Base Class) is searched for the object. In the case of more than one base classes, ambiguity can be there.

For example, in the following illustration, the **Derived1** class has been derived from **BaseClass**. The **show()** method of this class displays

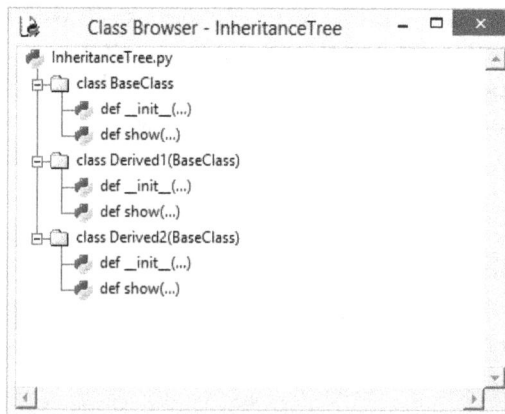

FIGURE 13.11 The class hierarchy for given illustration.

the values of **"data1"** and **"data."** The former is in the class and hence its value is displayed. However, if the former is not in the class, the inheritance tree is searched for the object. Note that **"data"** exists in the base class (**BaseClass**) and hence its value would be displayed. This is true for methods also. Even if the derived class does not have a particular method, it can be invoked if it exists in the parent class or in any other class, up in the inheritance tree.

Code:

```
class BaseClass:
    def __init__(self,data):
        self.data=data
    def show(self):
        print("\nData\t:",self.data)

class Derived1(BaseClass):
    def __init__(self,data,data1):
        self.data1=data1
        BaseClass.__init__(self,data)
    def show(self):
        print("\nData\t:",self.data1,"\nBase class data\t:",self.
    data)
class Derived2(BaseClass):
    def __init__(self,data,data2):
        self.data2=data2
        BaseClass.__init__(self,data)
    def show(self):
        print("\nData\t:",self.data2,"\nBase class data\t:",self.
            data)

X=BaseClass(1)
X.show()
print(X.data)
Y=Derived1(2,3)
Y.show()
Z=Derived2(4,5)
Z.show()
#Inheritance tree
```

Output:

```
======== RUN C:/Python/Inheritance/InheritanceTree.py ========
Data    : 1
1
Data    : 3
Base class data  : 2
Data    : 5
Base class data  : 4
```

13.5 CLASS INTERFACE AND ABSTRACT CLASSES

At times the classes are crafted so that they can be subclassed. While designing, there is no intention of instantiating these classes. That is, these classes would not be instantiated but would only be used to create derived classes that are called abstract classes. In order to understand the concept, let us consider an example. The following example has four classes: **BaseClass, Derived1, Derived2**, and **Derived3**.

The **BaseClass** has two methods: **method1** and **method2**. The first method has some tasks associated with it, whereas the second wants the derived class to implement it. The derived class should, to be able to call this method, must have a method called **action**. The first derived class (**Derived1**) replaces **method1**. So, if an instance of **Derived1** calls **method1**, the version defined in **Derived1** would be called. The second method extends **method1**, it adds something to **method1** and also calls the **BaseClass** version of **method1**. When **method1** is called from **Derived2**, the **BaseClass** version is called, as the search in the inheritance tree invokes the base class version of the method. The third derived class (**Derived3**) also implements the **action** method defined in the base class. Note that when **method2** is called through an instance of **Derived3**, the base class version of **method2** is invoked. This version calls **action** and a new search is initiated, thus resulting in the invocation of action of **Derived3**. Illustration 13.9 presents the code.

Note that the above concept can be extended, and a class may have methods that would be implemented by the derived classes. Interestingly, Python has a provision that such classes would not be instantiated until all such methods are not defined. Such base classes are called abstract classes.

Illustration 13.9:

Implement the following hierarchy. **"method1"** of **Derived1** should replace **method1** of the base class, **method1** of **Derived2** should extend **method1** of the base class and **action** of **Derived3** should implement **method2** of the **BaseClass**.

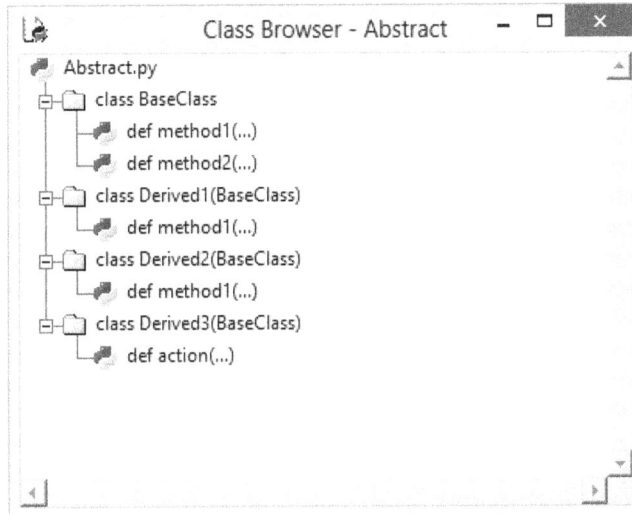

FIGURE 13.12 Class hierarchy for Illustration 13.9.

Solution:

```python
class BaseClass:
    def method1(self):
        print("In BaseClass from method1")
    def method2(self):
        self.action()
class Derived1(BaseClass):
    def method1(self):
        print("A new method, has got nothing to do with that of
            the base class")
class Derived2(BaseClass):
    def method1(self):
        print("A method that extends the base class method")
        BaseClass.method1(self)
```

```
class Derived3(BaseClass):
    def action(self):
        print("\nImplementing the base class method")

for className in (Derived1, Derived2, Derived3):
        print("\nClass\t:",className)
        className().method1()
X=Derived3()
X.method2()
```

Output:

```
Class  : <class '__main__.Derived1'>
A new method, has got nothing to do with that of the base class
Class  : <class '__main__.Derived2'>
A method that extends the base class method
In BaseClass from method1
Class  : <class '__main__.Derived3'>
In BaseClass from method1
Implementing the base class method
```

13.6 CONCLUSION

This chapter introduced the concept of Inheritance, which is one of the most important ingredients of Object-Oriented Programming. Inheritance, as explained in the chapter helps the program in reusing the code and making the program more structured. However, it should be used wisely, as in many cases, it leads to problems like ambiguity. The reader must also appreciate that it is not always necessary to use inheritance. Most of the tasks can be accomplished using composition. However, even if using inheritance becomes a necessity, be clear about the type of inheritance required, the type of methods calls required and the use of bound methods. The discussion on Object-Oriented Programming Paradigms continues in the next chapter also, wherein the concept of Operator Overloading has been introduced. The last chapter, this chapter, and the next one would help the reader to successfully develop software using OOP.

GLOSSARY

▨ **Inheritance:** Inheritance is the process of creating subclasses from existing classes.

▨ **Base class and derived class:** The class from which other classes are derived is the base class and the classes that inherit from the base class are the derived classes.

▨ **Implicit inheritance:** In this type, the method of the base class can be called using an instance of the derived class.

▨ **Explicit overriding:** In this type of inheritance, the derived class redefines the method of the base class and calling this method using an instance of the derived class invokes the method of the derived class.

POINTS TO REMEMBER

▨ Inheritance provides reusability and increased reliability.

▨ Types of inheritance are Simple Inheritance, Multiple Inheritance, Multilevel Inheritance, Hierarchical Inheritance, and Hybrid Inheritance.

▨ Multiple inheritance may lead to ambiguity.

▨ A bound method has a **"self"** parameter whereas an unbound method does not have **"self"** parameter.

▨ A class can also be instantiated in another class.

▨ **super** can be used to access the base class methods.

▨ The inheritance tree is searched to find the version of the method to be invoked.

EXERCISES

Multiple Choice Questions

1. A class that cannot be instantiated until all its methods have been defined · by its subclass(es) is called

   (a) Abstract class   (b) Meta class

   (c) Base class   (d) None of the above

   A class called **operation** has an __init__, that takes a positional parameter and an integer as an argument. Two instances of operations **Num1** and **Num2** have been defined, as follows. The class has two functions, the first calculates the square of a number and the second calculates the cube. A list called **List1** is created which contains the names of the four methods (two of **Num1** and two of **Num2**). A **for** loop is then used to call the methods as shown in the following snippet.

```
Num1=operations(5)
Num2=operations(4)
List= [Num1.square, Num1.cube, Num2.square, Num2.cube]
for callable_object in List:
    print(callable_object())
```

2. The program containing the above code (Assume that the rest of the code is correct)

(a) Has no syntax error but does not execute

(b) Has syntax error

(c) Has no syntax error and executes

(d) Insufficient information

3. In the question above, what would be the output (if the code is correct)?

(a) The code is not correct   **(b)** 25   125   16   64

(c) 125 25 64 16   **(d)** None of the above

4. The names of the methods in the list (question number 2) are similar to (in C#)

(a) Meta classes   **(b)** Delegates

(c) Both   **(d)** None of the above

5. "self" in Python is similar to

(a) "this" is C#   **(b)** "me" in C#

(c) delegate in C#   **(d)** None of the above

6.
```
class Student:
    def display(self, something):
        print("\n"+something)
    def getdata(name,age):
        name=name
        age=age
        print("\nName\t:",name,"\nAge\t:",age)
```

In the above snippet, how would you invoke **getdata** (assume that **Hari** is an instance of **Student**).

(a) Student.getdata("Harsh", 22)   **(b)** Hari.getdata("Harsh", 24)

(c) Both are correct   **(d)** None of the above

7. Can a method also be invoked by creating an unnamed instance of a class?

(a) Yes

(b) No

(c) Insufficient data

(d) There in nothing called unnamed instance of a class in Python

8. Which of the following is used in searching for an inheritance tree?

(a) Breadth First search   **(b)** Depth First search

(c) Both   **(d)** None of the above

9. In an inheritance tree, at the same level, which policy is used to search for an object?

(a) Left to right   **(b)** Right to left

(c) Any   **(d)** None of the above

10. "super" can be used to call

(a) The __init__ of the base class

(b) Any method of the base class

(c) Cannot be used to call methods of the base class

(d) None of the above

11. Which type of inheritance leads to ambiguity?

(a) Multiple   **(b)** Multilevel

(c) Both   **(d)** None

12. Which type of inheritance has just one base class and a single derived class?

   (a) Simple   (b) Hierarchical

   (c) Multiple   (d) None of the above

13. Which type of inheritance has more than one base class(es) and a single derived class?

   (a) Simple   (b) Hierarchical

   (c) Multiple   (d) None of the above

14. Which type of inheritance has more than one derived class(es) and a single derived class?

   (a) Simple   (b) Hierarchical

   (c) Multiple   (d) None of the above

15. Can a derived class be a base class of some other class?

   (a) Yes   (b) No

   (c) Insufficient data   (d) None of the above

Theory

1. What is inheritance? Explain the importance of inheritance.

2. What are the disadvantages of inheritance? Explain with reference to multiple inheritance.

3. What are the various types of inheritance? Give examples.

4. What are the problems in implementing Multiple Inheritance? How are they resolved?

5. What is composition? Is it a type of inheritance?

6. Is inherence mandatory in object-oriented programming? Justify.

7. What is the difference between **is a** and **has a** relationship. Explain with the help of examples.

8. Which is better: inheritance or composition? Can all that can be achieved using inheritance be done using composition?

9. Explain the use of **super**. How can it be used to call methods of the base class?

10. Are methods in Python, objects? Justify your answer. What is meant by callable object?

11. What is an abstract class? How does an abstract class helps in achieving the goals of OOPs?

12. What are bound methods? What are the various ways of invoking a bound method?

13. Differentiate between a bound and an unbound method. Give examples.

14. What is the importance of **self** in Python?

15. Explain the mechanism of search in an inheritance tree.

Programming Exercises

1. A class called **Base1** has two methods: **method1**(self, message) and **method2**(self). The first method prints the message passed as an argument to the method. The second invokes another method called **action1**(self), which would be defined by the subclass (**Derived2**) of **Base1**. **Derived1**, another derived class of **Base1**, redefines **method1**, and does nothing with **method2**. **Derived2**, on the other hand, does nothing with **method1**. Implement the hierarchy and find what happens in the following cases.

   (a) An instance of **Base1** calls **method1**

   (b) An instance of **Derived1** calls **method1**

   (c) An instance of **Derived2** calls **method1**

   (d) An instance of **Base1** calls **method2**

   (e) An instance of **Derived2** calls **method2**

   (f) An instance of **Derived1** calls **method2**

   (g) An instance of **Derived2** calls **action**

2. A class called **operation** has an **__init__**, that takes a positional parameter and an integer as an argument. The class has two functions, the first calculates the square root of a number and the second calculates the cube root. Two instances of operations: **Num1** and **Num2** are to be created. A list called **List1** is to be created which contains the names of the four methods (two of **Num1** and two of **Num2**). Implement the above and use a **for** loop to call all the callable objects from the list.

3. A class **employee** has two methods **getdata**(name, age) and **getdata1**(self, name, age). The **getdata** method stores the value in the local variables. Another method called **putdata** shows the data. Write a program to call the methods (the first is not bound but the second is) and display the data.

4. Create the following hierarchy and explain the search process of method called **show**.

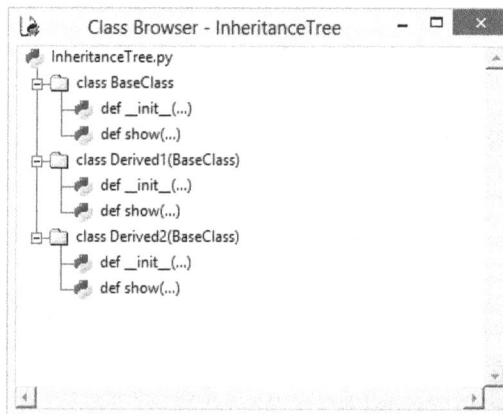

5. Create a class called **Employee**, having **name**, **email**, **age** and **phone_number** as data members. The methods **getdata** assign values to the variables. The **putdata** method should display the data. The class must have an **__init__**. Programmer is a derived class of **Employee**, which has another data member called **language**. Override the **getdata** and the **putdata** function. **Manager** is another derived class of the **Employee** class. The **Manager** class should also have a list of **Employees** that work under him. The class has the usual **getdata** and **putdata** methods and also has special methods to add or remove employees.

6. Implement the above hierarchy and carry out the following tasks:

(a) Create two managers: **manager1** and **manager2** and ask the user to enter the data (including the **employees** that work under the two managers)

(b) Create a menu driven program to add or remove employees under a manager

(c) Find out how many employees working under **manager1** are programmers?

(d) Find how many employees working under **manager2** are not programmers?

(e) Find the maximum age of the employees working under **manager1.**

(f) Find the mean age of the employees working under **manager2.**

(g) Find the phone numbers of all the employees having **age>35**.

(h) Find the manager who has more employees under him.

(i) Which manager has more programmers under him?

(j) Is **manager1** a programmer (Design classes accordingly and choose the type of inheritance).

14

OPERATOR OVERLOADING

Objectives

After reading this chapter, the reader should be able to

- Understand the importance of Operator Overloading
- Understand the issues in constructor overloading
- Use various methods used for overloading operators
- Implement Operator Overloading for Complex Numbers and Fractions

14.1 INTRODUCTION

Operators operate on operands to give some results. At times, an operator performs more than one task. For example, in Python, the + operator adds two numbers or two floats or concatenates two strings. That is, the + operator operates on both integers and strings. However, for user defined data types, the programmer cannot use these operators directly. Operator Overloading helps the programmer to define existing operator for user defined objects. This makes the language powerful and the work simple. This simplicity and intuitiveness in turn makes programming fun and increases the readability of the code.

We will also define methods to implement operator overloading for user defined data types. Operator Overloading can be used to intercept Python Operators by classes and even to overload built-in operations. These particular methods, which help in Operator Overloading, have been specially named and one can call these methods when instances of classes use the associated operator.

This chapter discusses various aspects of Operator Overloading. It has been organized as follows:

- Section 14.2 revisits __init__ and discusses the issues in overloading __init__.
- Section 14.3 presents some of the common Operator Overloading methods.
- Section 14.4 presents an example of overloading the binary operators.
- Section 14.5 discusses the __iadd__ method.
- Section 14.6 discusses the comparison operators.
- Section 14.7 discusses **bool** and **len**. The last section concludes.

14.2 __INIT__ REVISITED

The __init__ function has already been explained. This function initializes the members of a class. Earlier it was stated that __init__ could not be overloaded, which is partly true. Though one cannot have two __init__'s in the same class, there is a way to implement constructor overloading, explained in the following discussion.

Let us revisit __init__. The purpose of __init__ is to initialize the members of the class. In the following example (Illustration 14.1), a class called **Complex** has two members: **real** and **imaginary**, which are initialized by the parameters of the __init__ function. Note that, the members of the class are denoted by **self.real** and **self.imaginary** and the parameters of the functions are **real** and **imaginary**. The example has a function called **putData** to display the values of the members. In the __main__() function, **c1** is an instance of the class **Complex**. The object **c1** is initialized by 5 and 3 and the **putData()** of the class has been called to display the **"Complex Number."**

Illustration 14.1:

Create a class called **Complex**, having two members **real** and **imaginary**. The class should have __init__, which takes two parameters to initialize the values of **real** and **imaginary** respectively and a function called **putData**, to display the Complex number. Create an instance of the Complex number in the __main__() function, initialize it by (5, 3) and display the number by invoking the **putData** function.

Code:

```
class Complex:
    def __init__(self, real, imaginary):
        self.real = real
        self.imaginary = imaginary
```

```
def putData(self):
    print(self.real," +i ",self.imaginary)

def __main__():
    c1=Complex(5, 3)
    c1.putData()
__main__()
```

Output:

```
5 +i 3
```

Let us consider another example (Illustration 14.2) which deals with the implementation of vectors. In the example, a class called **Vector** has two data members namely **args** and **length**. Since **args** can contain any number of items, the **__init__** has ***args** as the parameter. The **putData** function displays the **Vector** and the **__len__** function calculates the length of the **Vector** (the number of arguments).

Illustration 14.2:

Create a class called **Vector**, which can be instantiated with a vector of any length. Design the requisite **__init__** function and a function to overload the **len** operator. The class should also have a **putData** function to display the vector. Instantiate the class with a vector having:

- no element
- one element
- two elements
- three elements

Display each vector and display the length.

Code:

```
class Vector:
    def __init__(self, * args):
        self.args=args
    def putData(self):
        print(self.args)
        print('Length ',len(self))
    def __len__(self):
        self.length = len(self.args)
        return(self.length)
```

```
def __main__():
    v0= Vector()
    v0.putData()
    v1 = Vector(2)
    v1.putData()
    v2 = Vector(3, 4)
    v2.putData()
    v3 = Vector(7, 8, 9)
    v3.putData()


__main__()
```

Output:

```
()
Length  0
(2,)
Length  1
(3, 4)
Length  2
(7, 8, 9)
Length  3
>>>
```

Note that the above example, **__init__** has the same effect as having many constructors with different parameters. Though **__init__** has not been overloaded in the literal sense, the program has same effect as that of one having overloaded constructors.

14.2.1 Overloading __init__(Sort of)

Constructors can be overloaded (part of) by assigning **None** to the arguments (some or all, except for the positional argument). To understand the point, consider a class called **Complex**. The class must have two constructors, one which takes two arguments and one where no argument is given. In the first case the **real** and **imaginary** part of **Complex** should be initialized with the arguments of **__init__** and in the second the **real** and **imaginary** should become zero. One of the simplest solutions is to check if the two arguments are **NONE** or not. If both of them are **NONE**, the data members should be zero. In the second case, they should contain the arguments, passed in **__init__**. Though, the following program (Illustration 14.3) does not have two **__init__'s**, the above task has been accomplished.

Illustration 14.3:

*Construct a class called **Complex** having **real** and **ima** as its data members. The class should have an **__init__**, for initializing the data remembers and **putData** for displaying the complex number.*

Code:

```
class Complex:
    def __init__(self, real=None, ima=None):
        if ((real == None)&(ima==None)):
            self.real=0
            self.ima=0
        else:
            self.real=real
            self.ima=ima
    def putData(self):
        print(str(self.real)," +i ",str(self.ima))
c1=Complex(5,3)
c1.putData()
c2=Complex()
c2.putData()
```

Output:

```
5  +i  3
0  +i  0
```

14.3 METHODS FOR OVERLOADING BINARY OPERATORS

The following methods (Table 14.1) help in overloading the binary operators like +, −, *, and /. The operators operate on two operands: **self** and another instance of the requisite class. When an operator is used between objects, the corresponding methods are invoked. For example, for objects **c1** and **c2** of a class called **Complex, c1+c2** invokes the **__add__** method. Similarly, the − operator would invoke the **__sub__** method, the * operator would invoke the

__mul__ method and so on. Table 14.1 shows the methods and the operator due to which the method is invoked.

TABLE 14.1 Methods for overloading binary operators.

Task	Method	Explanation
Addition	__add__	Helps in overloading the + operator. Generally, takes two arguments: The positional parameter and the instance to be added.
Subtraction	__sub__	Helps in overloading the – operator. Generally, takes two arguments: The positional parameter and the instance to be subtracted.
Multiplication	__mul__	Helps in overloading the * operator. Generally, takes two arguments: The positional parameter and the instance to be multiplied.
Division	__truediv__	Helps in overloading the / operator. Generally, takes two arguments: The positional parameter and the instance to be divided.

The use of the above operators has been explained in the following section.

14.4 OVERLOADING BINARY OPERATORS: THE FRACTION EXAMPLE

The overloading of the operators shown in the table above can be easily understood by the example that follows. The following example overloads the addition (+), subtraction (–), multiplication (*), and division (/) operator for a class **fraction**. The **fraction** class depicts the standard fraction, having a numerator and a denominator.

1. **__init__**

    The **__init__** initializes the class members by setting the numerator to 0 and the denominator to 1. The statement

    ```
    x=fraction()
    ```

    therefore, creates a fraction 0/1.

2. **__add__**

    The **__add__** method helps in overloading the + operator. The statement

    ```
    z=x+y
    ```

calls the **__add__** method of **x** and takes **y** as the "other" argument. Therefore, it must have two arguments: a **positional** parameter (**self**) and a **fraction** (**other**). The addition of two fractions $\dfrac{a_1}{b_1}$ and $\dfrac{a_2}{b_2}$ is done as follows. The LCM of b_1 and b_2 becomes the denominator of the resultant fraction. The numerator is calculated using the following formula.

$$\text{numerator} = \left(\frac{LCM}{b_1}\right) \times a_1 + \left(\frac{LCM}{b_2}\right) \times a_2$$

Note that the resultant fraction is stored in another fraction (s). The method **__add__** returns the sum.

3. **__sub__**

The **__sub__** method helps in overloading the – operator. The statement

```
t=x-y
```

calls the **__sub__** method of **x** and takes **y** as the **"other"** argument. Therefore, it must have two arguments: a positional parameter (**self**) and a **fraction** (**other**). The difference of two fractions $\dfrac{a_1}{b_1}$ and $\dfrac{a_2}{b_2}$ is done as follows. The LCM of b_1 and b_2 becomes the denominator of the resultant fraction. The numerator is calculated by using the following formula.

$$\text{numerator} = \left(\frac{LCM}{b_1}\right) \times a_1 - \left(\frac{LCM}{b_2}\right) \times a_2$$

Note that the resultant fraction is stored in another fraction (d). The method **__sub__** returns **d**.

4. **__mul__**

The **__mul__** method helps in overloading the * operator. The statement

```
prod=x*y
```

calls the **__mul__** method of **x** and takes **y** as the **"other"** argument. Therefore, it must have two arguments: a positional parameter (**self**) and a fraction (**other**). The product of two fractions $\dfrac{a_1}{b_1}$ and $\dfrac{a_2}{b_2}$ is calculated as follows. The numerator is calculated using the following formula.

```
numerator = a₁ × a₂
```

and the denominator is calculated as follows.

$$\text{denominator} = b_1 \times b_2$$

Note that the resultant **fraction** is stored in another **fraction (m)**. The method **__mul__** returns **m**.

5. **__truediv__**

The **__truediv__** method helps in overloading the / operator (which returns an integer). The statement

```
div=x/y
```

calls the **__truediv__** method of **x** and takes **y** as the **"other"** argument. Therefore, it must have two arguments: a positional parameter (**self**) and a fraction (**other**). The division of two fractions $\dfrac{a_1}{b_1}$ and $\dfrac{a_2}{b_2}$ is done as follows. The numerator is calculated using the following formula.

$$\text{numerator} = a_1 \times b_2$$

and the denominator is calculated as follows.

$$\text{denominator} = b_1 \times a_2$$

Note that the resultant **fraction** is stored in another **fraction** (answer). The method **__truediv__** returns the answer.

Illustration 14.4:

Create a class **fraction** having numerator and denominator as its members. Overload the following operators for the class:

- +
- −
- *
- /

Create LCM and GCD methods in order to accomplish the above tasks. The LCM method should find the LCM of two numbers and the GCD method should find the GCD of the two numbers. Note that $LCM(x, y) \times GCD(x, y) = x \times y$.

Solution:

The implementation has already been discussed. The following code performs the requisite task, and the output follows.

Code:

```
##fractions
class fraction:
    def __init__(self):
        self.num=0;
        self.den=1;
    def getdata(self):
        self.num=input("Enter the numerator\t:")
        self.den = input("Enter the denominator\t:")
    def display(self):
        print(str(int(self.num)),"/",str(int(self.den)))
    def gcd(first, second):
        if(first<second):
            temp=first
            first=second
            second=temp
        if(first%second==0):
            return second
        else:
            return(fraction.gcd(second, first%second))
    def lcm(first, second):
        ##print("GCD is",str(fraction.gcd(first,second)))
        return((first*second)/fraction.gcd(first,second))
    def __add__(self,other):
        s=fraction()
        lc=fraction.lcm(int(self.den), int(other.den))
        s.num=((lc/int(self.den))*int(self.num))+((lc/int(other.
            den))*int(other.num))
        s.den=lc
        return(s)
    def __sub__(self,other):
        lc=fraction.lcm(int(self.den), int(other.den))
```

```
        d=fraction()
        d.num=((lc/int(self.den))*int(self.num))-((lc/int(other.
            den))*int(other.num))
        d.den=lc
        return(d)
    def __mul__(self,other):
        m=fraction()
        m.num=int(self.num)*int(other.num)
        m.den=int(self.den)*int(other.den)
        return(m)
    def __truediv__(self,other):
        answer=fraction()
        answer.num=int(self.num)*int(other.den)
        answer.den=int(self.den)*int(other.num)
        return(answer)
x =fraction()
x.getdata()
print("First fraction\t:")
x.display()
y=fraction()
y.getdata()
print("Second fraction\t:")
y.display()
z=(x+y)
print("Sum\t:")
z.display()
t=x-y
print("Difference\t:")
t.display()
prod=x*y
print("Product")
prod.display()
div=x/y
print("Division")
div.display()
```

Output:

```
Enter the numerator        :2
Enter the denominator      :3
First fraction             :2 / 3
Enter the numerator        :4
Enter the denominator      :5
Second fraction            :4 / 5
Sum                        :22 / 15
Difference                 :-2 / 15
Product
8 / 15
Division
10 / 12
>>>
```

Was it really needed?

Note that the above illustration has been included in the chapter to explain the overloading of binary operators. Python, as such provides addition, subtraction, multiplication, and division for fractions. The same task could be done without overloading the operators as follows:

```
from fractions import Fraction
X=Fraction(20,4)
X
Y=Fraction(3,5)
Y
X+Y
```

Output:

```
Fraction(28, 5)
```

```
X-Y
```

Output:

```
Fraction(22, 5)
```

```
X*Y
```

Output:

```
Fraction(3, 1)
```

```
X/Y
```

Output:

```
Fraction(25, 3)
```

14.5 OVERLOADING THE += OPERATOR

The += operator adds a quantity to the given object. For example, if the value of "a" is 5, then a+=4 would make it 9. However, the operator works for integer, reals, and strings. The use of += for integer, real, and string has been shown as follows.

Code:

```
a=5
a+=4
a
```

Output:

```
9
```

Code:

```
a=2.3
a+=1.3
a
```

Output:

```
3.599999999999996
```

Code:

```
a="Hi"
a+=" There"
a
```

Output:

```
'Hi There'
>>>
```

However, to make it work for a user defined data type (or objects), it needs to be overloaded. The **__idd__** function helps in accomplishing this task. The following illustration (Illustration 14.5) depicts the use of **__idd__** for an object of the **complex** class. A complex number has a **real** part and an **imaginary** part. Adding another complex number to a given complex number adds their respective real parts and imaginary parts. The program follows. Note that, **__iadd__** takes two arguments. The first being the positional parameter and the second is another object called **"other."** The real part of **"other"** is added to the real part of the object and the imaginary part of **"other"** is added to the imaginary part. The **__iadd__** returns **"self."** Likewise, the reader may overload the **__iadd__** operator for his class, as per the requirements.

Illustration 14.5:

Overloading += for Complex Class (Illustration 14.1 and Illustration 14.3)

Code:

```
##overloading += for Complex class
class Complex:
    def __init__(self, real, imaginary):
        self.real=real
        self.imaginary=imaginary
    def __iadd__(self, other):
        self.real+=other.real
        self.imaginary+=other.imaginary
        return self
    def display(self):
        print("Real part\t:",str(self.real)," Imaginary
        part\t:",str(self.imaginary))
X=Complex(2,3)
Y=Complex(4,5)
X.display()
Y.display()
X+=Y
X.display()
X+=Y
X.display()
```

Output:

```
Real part : 2    Imaginary part   : 3
Real part : 4    Imaginary part   : 5
Real part : 6    Imaginary part   : 8
Real part : 10   Imaginary part   : 13
>>>
```

14.6 OVERLOADING THE > AND < OPERATORS

The greater than (>) and less than (<) operators work in the usual manner for the integers, fractions, and some other predefined types. However, to be able to use these operators for user defined classes, the programmer must overload the operators. In Python, greater than (>) and less than(<) can be overloaded using the **__gt__** and **__lt__**. The **__gt__** returns a **true** or a **false** depending upon whether the first object is greater than the second or not. Similarly, the **__lt__** returns a **true** or a **false** depending upon whether the first object is less than the second or not.

The following example overloads the **__gt__** and **__lt__** for a class called **Data**. The **Data** class has a data member called **"value."** The **__gt__** compares the value of the instance (**self**) and that of another instance (**other**). If the value of the instance is greater than that of the other instance then a **true** is returned, otherwise a **false** is returned. Similarly, the **__lt__** compares the value of the instance (**self**) and that of another instance(**other**). If the value of the instance is smaller than that of the other instance then a **true** is returned, otherwise a **false** is returned.

Illustration 14.6:

Write a program to create a class called **Data** having **"value"** as its data member. Overload the (>) and the (<) operator for the class. Instantiate the class and compare the objects using **__lt__** and **__gt__**.

Solution:

The mechanism of the **__gt__** and **__lt__** has already been discussed. The program follows.

Code:

```
class Data:
    def __init__(self, value):
self.value=value
    def display(self):
        print("data is ",str(value))
    def __lt__(self,other):
        return(self.value<other.value)
    def __gt__(self,other):
        return(self.value>other.value)
X= Data(5)
Y= Data(4)
print(X>Y)
print(X<Y)
```

Output:

```
True
False
>>>
```

14.7 OVERLOADING THE __BOOL__ OPERATOR: PRECEDENCE OF __BOOL__ OVER __LEN__

In using **"if"** and **"while,"** the programmer checks the condition passed in **"if"** or **"while."** If the condition is true, the block following **"if"** or **"while"** is executed, otherwise it is not executed. We can also define the Boolean operators for a user defined object. In order to accomplish this task, the programmer would require some method that helps in the overloading. Python provides two Boolean operators **__bool__** and **__len__**. The **__bool__** method returns a **true** if the requisite condition is met, otherwise it returns a **false**. The **__len__** method finds the length of the data member and returns false if it is null. The Boolean condition can be checked using the **__len__** method, only if the **__bool__** for that class is not defined. In case both **__len__** and **__bool__** are defined in a class, **__bool__** takes precedence over **__len__**.

For example, in the following illustration, writing if(**X**), where **X** is an instance of the class, returns a false if no argument is passed while instantiating the class. Note that, the first listing uses **__len__**. The next illustration (Illustration 14.7) checks the length of the data member **"value"** to return a true if **"value"** is not null and **false** otherwise.

Illustration 14.7:

The following illustration creates a class called **Data**. If no argument is passed while instantiating the class, a **false** is returned, otherwise a **true** is returned.

Program:

```
class Data:
    def __len__(self): return 0
X= Data()
if X:
    print("True")
else:
    print("False")
```

Output:

```
False
```

Illustration 14.8:

Another variant of the above example has **value** as its data member. If the **value** is null, a **false** is returned, otherwise a **true** is returned.

Program:

```
class Data:
    def __init__(self, value):
        self.value=value
    def __len__(self):
        if len(self.value)==0:
            return 0
        else:
            return 1
Y= Data("hi")
if Y:
    print("True")
```

```
else:
    print("False")
X = Data("")
if X:
    print("Ture")
else:
    print("False")
```

Output:

```
True
False
```

Also, note that if **__bool__** is also defined in the class, then it takes precedence over the **__len__** method. The **__bool__** returns a true or a false as per the given condition. Although overloading **__bool__** may not be of much use as every object is either **true** or **false** in Python. Illustration 14.9 presents an example in which both **__bool__** and **__len__** are defined.

Illustration 14.9:

An example in which both **__bool__** and **__len__** are defined.

Program:

```
class Data:
    def __init__(self, value):
        self.value=value
    def __len__(self):
        if len(self.value)==0:
            return 0
        else:
            return 1
    def __bool__(self):
        if len(self.value)==0:
            print("From Bool")
            return False
        else:
            print("From Bool")
            return True
```

```
Y= Data("hi")
if Y:
    print("True")
else:
    print("False")
X= Data("")
if X:
    print("True")
else:
    print("False")
```

Output:

```
From Bool
True
From Bool
False
```

14.8 CONCLUSION

This chapter presented a brief overview of operator overloading. The overloading of binary operators, +=, len, and bool have been discussed in this chapter. The reader is expected to attempt the exercises to get hold of the concept. The next chapter introduces Exception Handling and discusses some of the methods to deal with run-time errors.

GLOSSARY

- **Operator Overloading:** It is the mechanism of assigning a new meaning to an existing operator.

POINTS TO REMEMBER

- Operator Overloading helps the programmer to define existing operator for user defined objects.
- In Python, all expression operators can be overloaded.
- Operator Overloading can be implemented using special methods.
- __bool__ has higher priority over __len__.

EXERCISES

Multiple Choice Questions

1. Using Operator Overloading, the programmer can

    (a) Define an existing operator for user defined data type

    (b) Create new operators

    (c) Both

    (d) None of the above

2. In Python, Operator Overloading can be implemented by

    (a) Defining corresponding methods in the class for which user defined objects would be made

    (b) Operators are redefined in the same way as C++

    (c) Python has predefined methods for defining operators

    (d) None of the above

3. Can **__init__** be overloaded?

    (a) Yes

    (b) No

    (c) It can be overloaded only for specific classes

    (d) None of the above

4. Same **__init__** is to be designed to accept varying number of arguments, which of the following correctly is the correct representation?

    (a) def __init__(self)

    (b) def __init__(self, *args)

    (c) def __init__(self, args)

    (d) Both b and c

5. Can the above task be accomplished in some other way?

    (a) By not giving any arguments in __init__

    (b) By equating some of the arguments to NONE

    (c) Both

    (d) None of the above

6. Which of the following methods is used to overload the + operator?

    (a) __add__   (b) __iadd__

    (c) __sum__   (d) None of the above

7. Which of the following is used to overload the – operator?

    (a) __diff__   (b) __sub__

    (c) __minus__   (d) None of the above

8. Which of the following is used to overload the * operator?

    (a) __prod__   (b) __mul__

    (c) Both   (d) None of the above

9. For which of the following, Operator Overloading is really needed in Python?

    (a) Complex   (b) Fraction

    (c) Polar coordinates   (d) None of the above

10. Which of the following is overloaded using __iadd__?

    (a) +   (b) +=

    (c) ++   (d) None of the above

11. Can > and < operators be overloaded in Python?

    (a) Yes   (b) No

    (c) Only for specific classes   (d) None of the above

12. Which has more priority __bool__ or __len__?

    (a) __bool__   (b) __len__

    (c) Both   (d) None of the above

Theory

1. What is Operator Overloading? Explain its importance.

2. Explain the mechanism of Overloading Operators in Python.

3. Can all Python Operators be overloaded?

4. The membership can be tested using the "in" operator. The __contains__ method can be used for testing the membership, in Python. Create a class having three lists and overload the membership operator for the class.

5. Explain the following methods and explain Operator Overloading using the operators.

(a) __add__   (b) __iadd__

(c) __sub__   (d) __mul__

(e) __div__   (f) __len__

(g) __bool__   (h) __gt__

(i) __lt__   (j) __del__

6. The following methods have not been discussed in the chapter. Explore the following (refer to the Bibliography for details).

(a) __getIitem__   (b) __setIitem__

(c) __iter__   (d) __next__

Programming Exercises

1. Create a class called `Distance` having `meter` and `centimetre` as its data members. The member functions of the class would be `putData()`, which takes the values of `meter` and `centimetre` from the user; `putData()`, which displays the data members and add, which adds the two distances.

   The addition of two instances of distances (say d1 and d2), would require addition of corresponding centimeters" (d1.centimeter +s2.centimeter), if the sum is less than 100, otherwise it would be (d1.centimeter +s2.centimeter)%100. The "meter" of the sum would be the sum of meters of the two instances (d1.meter +d2.meter), if (d1.centimeter +d2.centimeter)<100, otherwise it would be (d1.meter+d2.meter+1).

2. Overload the + operator for the above class. The + operator should carry out the same task as is done by the add function.

3. The subtraction of two instances of distances (say d1 and d2), would require subtraction of corresponding centimeters' (d1.centimeter -s2.centimeter). The "meter" of the difference would be the sum of meters of the two instances (d1.meter - d2.meter).

Overload the – operator for the `distance` class. Assume that d1-d2, would always mean d1>d2.

4. Overload the += operator for `Distance` class. The += operator (that is d1+=d2) would require the addition of d1 and d2 (as explained earlier) and updating d1 with (d1+d2). Note that the value of d2 is not altered.

5. Overload the * operator for the `Distance` class.

The government of a developing country intends to do away with the present currency and intends to introduce a barter system, in which 12 bottles of "Tanjali" would be equivalent to a unit of currency. This in turn would increase the sales of the company also. Hari and Aslam have 37 and 92 bottles of "Tanjali" and would like to exchange the bottles to buy tickets of a movie. If each ticket is 60 Units, would they be able to watch the movie.

6. Now, help the people of the country by developing a program having a class called `nat_currency` and overload the + operator, which adds two instances of `nat_currency`.

7. For the above question, overload the – operator.

8. For the `nat_currency` class of question 8, overload the += operator.

9. For the `nat_currency` class of question _, overload the * operator.

10. Create a class called `date` having members `dd`, `mm` and `yyyy` (date, month, and year). Overload the + operator, which adds the two instances of the `date` class.

A hypothetical number called `irr`, of the form a $+c\sqrt{b}$, has b constant. Two instances of `irr` can be added as follows. If the first `irr` number is r_1 = $a_1 + c_1\sqrt{d}$ and the second is $r_2 = a_2 + 2\sqrt{d}$, the addition of r_1 and r_2 can be defined as follows.

$$r = r_1 + r_2 = (a_1 + a_2) + (c_1 + c_2)\sqrt{d}.$$

The difference of r_1 and r_2 would be.

$$r = r_1 - r_2 = (a_1 - a_2) + (c_1 - c_2)\sqrt{d}.$$

The product of r_1 and r_2 would be.

$$r = r_1 \times r_2 = (a_1 a_2 + c_1 c_2 d) + (a_1 c_2 + a_2 c_1)\sqrt{d}.$$

11. Create a class called `irr` and overload the + operator.

12. For the `irr` class, overload the − operator.

13. For the `irr` class, overload the + = operator

14. For the `irr` class, overload the * operator.

    A vector is written as .. , where $\hat{i}$ is a unit vector in the x-axis, $\hat{j}$ is the unit vector in the y-axis and $\hat{k}$ is the unit vector is the z-axis. The addition of two `vectors`, requires the addition of the corresponding $\hat{i}$ components, addition of the corresponding $\hat{j}$ components and the addition of the corresponding $\hat{k}$ components. That is, for two `vectors` $v_1 = a_1\hat{i} + b_1\hat{j} + c_1\hat{k}$ and $v_2 = a_2\hat{i} + b_2\hat{j} + c_2\hat{k}$, the sum would be $v = v_1 + v_2 = (a_1 + a_2)\hat{i} + (b_1 + b_2)\hat{j} + (c_1 + c_2)\hat{k}$. Likewise, the difference of two vectors requires the subtraction of the corresponding $\hat{i}$ components, subtraction of the corresponding $\hat{j}$ components and the subtraction of the corresponding $\hat{k}$ components. That is, for two vectors $v_1 = a_1\hat{i} + b1\hat{j} + c_1\hat{k}$ and $v_2 = a_2\hat{i} + b_2\hat{j} + c_2\hat{k}$, the difference would be $v = v_1 - v_2 = (a_1 - a_2)\hat{i} + (b_1 - b_2)\hat{j} + (c_1 - c_2)\hat{k}$.

15. Create a class called vector having three data members a, b, and c. The class must have the `getData()` function to ask the user to enter the values of a, b, and c; the `putData()` function to display the `vector`.

16. Overload the + operator for the `vector` class.

17. Overload the − operator for the `vector` class.

18. Overload the += operator for the `vector` class.

EXCEPTION HANDLING

Objectives

After reading this chapter, the reader should be able to

- Understand the concept of Exception Handling
- Appreciate the importance of Exception Handling
- Use **try/except**
- Manually throw exceptions
- Craft program that raises user defined exceptions

15.1 INTRODUCTION

Writing a program is an involved task. It requires due deliberation, command over the syntax, and problem-solving capabilities. Despite all efforts, there is a possibility of some error cropping up, or an unexpected output. These errors can be classified as compile time errors and run time errors. The compile time errors can be intercepted by the compiler. The programmer would have to remove these, to be able to execute the program. While compiling a program, if some errors exist then, some standard message appears. These errors can be handled by learning the syntax or by changing the code as per the requirement of the problem at hand. The following is an example of a code having syntax error. Note that the closing parenthesis is missing in the statement `fun1('Harsh'`.

If we try to execute the following code, a pop-up message displaying syntax error will appear.

Code:

```
def fun1(a):
    print('\nArgument\t:',a)
    print('\nType\t:',type(a))
fun1(34)
fun1(34.67)
fun1('Harsh'
```

Output:

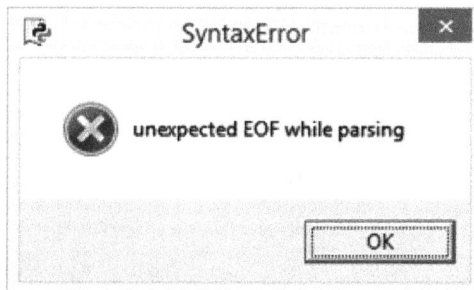

The second type is more complex in which the program stops working or behaves in an undesirable way while executing. This may be due to:

- incorrect user input,
- inability to open a file,
- accessing something which the program does not have authority to and so on.

These are referred to as exceptions. Exceptions are "Events that modify the flow of the program"[1]. Python invokes exception handling mechanism in case of such errors. The exception handling can also be invoked by the programmer.

Exceptions handling mechanisms are used to handle some undesirable situations. So, if something undeniable comes up, the control must have a place to go (part of the code), where that situation can be handled. To understand the point, consider the following example.

Suppose you intend to design a machine learning technique to identify whether a given EEG shows an epileptic spike. You must decide the algorithm, the language, the tool etc. However, you are not able to get the data. What will you do? Simply abandon the project and go to the exception handling part.

That is, an exception will occur when situations like the above crop up. Let us consider one of the most common examples of exception handling. If one is crafting a program to divide two numbers entered by the user, an exception should be raised if the denominator entered is zero.

One of the most common ways of handling exception is to craft a block, where one expects exception to occur. If somewhere in that block, an exception is raised, the control goes to the part which handles the exception. The block where you expect exception to come is the **try** block and the part where it will be handled is in the **except** block. The chapter discusses some more ways to handle exceptions. However, the readers from C++ or C# background would be familiar with the above technique and hence would find it easy and intuitive. Though Python has a mechanism to handle exceptions, the reader should also learn how to create his own classes to handle exceptions. Therefore, the reader must revisit the chapter on classes and objects.

Exception handling, in Python, can be done using either of the following.

- try/except
- try/except/finally
- raise and
- assert

The chapter, though, concentrates on the first three. The chapter has been organized as follows. Section 15.2 discusses the importance and mechanism of Exception Handling, Section 15.3 presents some of the most common Exceptions in Python, Section 15.4 summarizes the process by taking an example, Section 15.5 presents another example of Exception Handling and the last section concludes.

15.2 IMPORTANCE AND MECHANISM

Exception handling mechanism can help the programmer to notify something. For example, consider the problem discussed in the previous section. You have the EEG of the patients, and you want to find the epileptic spike in the EEG. If you are not able to find the spike, you can simply raise an exception. This technique is better compared to the conventional method of returning an integer code on being able to find something (or for that matter, not find something). Likewise, on detecting some special case or an unusual condition, an exception can be raised.

At the runtime, if an error crops up, an exception is raised. This exception can be handled by the corresponding **except** or can be simply ignored. Moreover, if there is no provision to handle the exception, in the code, the Pythons' default error handling mechanism comes into play. As stated earlier, on encountering an error condition, the execution is restored after the **try** statement.

Python also has the try /finally statements to handle the termination condition. Those of you from JAVA background must be familiar with **finally**. It is for handling the termination condition, whether an exception has occurred. For example, in designing software the concluding screen must appear, whether the exception has occurred or for that matter the memory of objects must be reclaimed at the end. For such type of situations **except–finally** is immensely helpful.

15.2.1 An Example of Try/Except

In order to understand how exception handling works, consider the following example. A list contains an ordered set of students. The first location contains the name of the students who got the highest marks. Likewise, the student who got the second highest marks has his name at the second position and so on.

```
>>>L=['Harsh', 'Naved', 'Snigdha', 'Gaurav']
```

In order to access an element at a given location, the user is asked to enter the index

```
>>>Index=input('Enter the index')
```

Now, the element at that position is accessed using the following statement

```
>>> print(L[int(index)])
```

So, if the user enters 1, "Naved" would be the output, if he enters 2, "Snigdha" would be the output. However, the following message appears, if he enters anything above 3.

```
Traceback (most recent call last):
  File "<pyshell#5>", line 1, in <module>
    print(L[int(index)])
IndexError: list index out of range
>>>
```

This condition can be handled by using **try/except** as shown.

Code:

```
L= ['Harsh', 'Naved' 'Snigdha', 'Gaurav']
try:
    index=input('Enter index\t:')
    print(L[index])
except IndexError:
    print('List index out of bound')
print('This statement always executes')
```

Note that the try block contains the part of the code where the exception may appear. If a runtime error is there, the **except** part handles it. Also note, that the **except** may have the name of the predefined exception. The statement after the **except** always executes, whether or not exception has been raised. The reader can take note of the fact that the control does not go back to the point where the expectation really occurred. It can only handle the exception in the requisite block. After which the normal execution continues. The syntax of the exception handling mechanism is as follows.

Syntax

```
try:
    ##code where exception is expected
except<Exception>:
    ##code to handle the exception
## rest of the program
```

15.2.2 Manually Raising Exceptions

The discussion so far has concentrated on the situations wherein exceptions were raised and caught by Python itself. In Python, one can manually raise the exceptions. The keyword **"raise"** is used to explicitly trigger an exception. The keyword is followed by the <exception name> (same as that which is caught). The mechanism of handling such exceptions is the same as that described above. That is, the corresponding **except** would handle the thrown exception. The syntax is as follows.

Syntax

```
try:
    raise <something>
except <something>:
    ##code which handles the exception
##rest of the code
```

If such exceptions do not occur, they can be handled in the same fashion as in the above section. The examples in Section 15.4 presents codes where the exceptions have been raised and caught.

15.3 BUILD-IN EXCEPTIONS IN PYTHON

If the programmer can raise the specific exception, the program would be more effective. To be able to do so one must know the predefined Exceptions in Python and then use these in appropriate situations. This section presents some of the most common exceptions in Python. The following sections represent the use of these exceptions.

- **AssertionError**
  When an assert statement fails, the **AssertionError** is raised.

- **AttributeError**
  When an assignment fails, the **AttributeError** is raised

- **EOFError**
  When the last word of the file is reached and the program attempts to read any further, the **EOFError** is raised.

- **FloatingPointError**
  This exception is raised when floating point operations fail.

- **ImportError**
  If the import statement, written in the code, cannot load the said module, this exception is raised. This is same as the **ModuleNotFoundError** in the later versions of Python.

- **IndexError**
  When the sequence is out of range, the exception is raised.

- **KeyError**
  If, in a dictionary, the key is not found, then this exception is raised.

▪ **OverflowError**
Note that each data type can hold some value and there is always a maximum limit to what it can hold. When this limit is reached, the **OverflowError** is raised.

▪ **RecursionError**
While executing a code that uses recursion, at times maximum iteration depth is reached. At this point in time, the **recursionError** is raised

▪ **RuntimeError**
If an error occurs and it does not fall in any of the said categories, then this exception is raised.

▪ **StopIteration**
If one is using the __**next**__() and there are no more methods that can be processed, then this exception is raised.

▪ **SyntaxError**
When the syntax of the code is incorrect, this exception is raised.

▪ **IndentationError**
When incorrect use of indentation is done, then this exception comes up.

▪ **TabError**
An inconsistent use of spaces or tabs lead to this type of error.

▪ **SystemError**
If some internal error is found, then this exception is raised. The exception displays the problem that was encountered due to which the exception is raised.

▪ **NotImplementedError**
If an object is not supported or the part that provides support has not been implemented, then the **NotImplementedError** is raised.

▪ **TypeError**
If an argument is passed and is not expected, the **TypeError** is raised. For example, in a program that divides two numbers, entered by the user, a character is passed, then **TypeError** is raised.

▪ **ValueError**
When an incorrect value is passed in a function (or an attempt is made to enter it in a variable), the **ValueError** is raised. For example, if a value which is outside the bounds of an integer is passed, this exception is raised.

- **UnboundLocalError**
  This exception is raised when a reference is made to a variable which does not have any value in that scope.

- **UnicodeError**
  It is raised when errors related to Unicode encoding or decoding come up.

- **ZeroDivisionError**
  The division and the modulo operation have two arguments. If the second argument is zero, this exception is raised.

15.4 THE PROCESS

This section revisits the division of two numbers and summarizes how to apply the concepts studied so far.

15.4.1 Example

Consider a function that takes two numbers as input and divides the two numbers. If the function is called and two integers are passed as arguments (say, 3 and 2), an expected output is produced, if the second number is not zero. However, if the second number is zero, a runtime error occurs, and an error message (shown as follows) is produced. That is, Python handles exceptions automatically.

The program can be made user-friendly, by printing a comprehendible, easy to understand message. This can be done by using exception handling.

Code:

```
def divide(a,b):
    result =a/b
print('Result is\t:',result)
divide(3,2)
divide(3,0)
>>>
```

Output:

```
Result is   : 1.5
Traceback (most recent call last):
```

```
File "C:/Python/Exception handling/No Exception.py", line 5, in
    <module>
divide(3,0)
File "C:/Python/Exception handling/No Exception.py", line 2, in
    divide
  result =a/b
ZeroDivisionError: division by zero
```

15.4.2 Exception Handling: Try/Except

The above problem can be handled by using the **try/except** construct to handle the run time error. The part of the code where the exception is likely to be raised is put in the **try** block. If an exception is raised, it would be handled in the **except** block. The **except** block can have the user-friendly error message or the code which would handle the exception. The following code shows the use of the **try** block and displays how a run time error can be handled in the **except** block. Note that the statement which divides the two numbers is in the **try** block. If the second number is zero an exception would be raised and the statements in the **except** block would be executed.

Code:

```
def divide(a, b):
    try:
        d=a/b
print('Result is\t:',str(d))
    except:
print('Exception caught')

divide(2,3)
divide(2,0)
```

Output:

```
Result is   : 0.6666666666666666
Exception caught
```

15.4.3 Raising Exceptions

One can also raise specific exceptions. For example, the following code raises the **ZeroDivisionError**, if the second number is zero. Note that the corresponding **except** block catches this exception. This can be done if the user is sure which exception to raise in each situation. Moreover, there is a chance that the programmer fails to raise the correct exception, thus leading to the invocation of the automatic exception handling mechanism of Python.

Some of the common exceptions and their meaning have already been presented in Section 15.3. However, there are many more. The list of such exceptions can be found at the link provided in the References at the end of this book.

Code:

```
def divide(a, b):
    try:
        if b==0:
            raise ZeroDivisionError
        d=a/b
        print('Result is\t:',str(d))
    except ZeroDivisionError :
        print('Exception caught:ZeroDivisionError ')

divide(2,3)
divide(2,0)
```

Output:

```
Result is   : 0.6666666666666666
Exception caught:ZeroDivisionError
>>>
```

15.5 CRAFTING USER DEFINED EXCEPTIONS

So far, we have seen the automatic exception handling capabilities of Python. That is, even if there is no **try/except**, Python handles exceptions. The use of **try/except** has also been discussed. The use of **raise** makes the exception handling more meaningful, as one can raise specific exceptions as per the needs. However, so far, we have not seen how to deal with the situation, which requires us to raise a user defined exception. This section discusses the crafting and use of user defined exceptions.

Suppose there is a situation, where a specific exception (as per the need of the program), is to be raised. However, there is no predefined exception to handle that situation. In such cases, a class which would handle the exception needs to be created. The particular class should be a subclass of the **Exception** class, so that we can use it for raising exceptions. When the situation arises, the exception could be raised, as shown in the following illustration. In the illustration that follows, a class called **MyError**, which derives **Exception** has been created. The **__init__** of this class may contain the message, which would be printed, when the exception occurs. While raising the exception, the keyword **raise**, followed by the name of the class has written. The reader should observe the output and understand that it will print the message written in **__init__** followed by the message in the **except** block. Though, this is just a dummy example, it gives an idea of how to craft classes that handle exceptions.

Code:

```
class MyError (Exception):
    def __init__(self):
       print('My Error type error')
def divide(a, b):
    try:
       if b==0:
          raise MyError
       d=a/b
       print('Result is\t:',str(d))
    except MyError:
       print('Exception caught : MyError ')
divide(2,3)
divide(2,0)
>>>
```

Output:

```
Result is   : 0.6666666666666666
My Error type error
Exception caught: My Error
>>>
```

15.6 AN EXAMPLE OF EXCEPTION HANDLING

The following program finds the maximum number from a given list. The idea is simple. Initially, we assigned first item to **max**. Then we traverse the items of the given list. While traversing, if any element is greater than that stored in **max**, then assign that number to the variable **max**. At the end, print the value of **max**. The following program and the corresponding output will illustrate this.

Code:

```
def findMax(L):
    max =L[0]
    for item in L:
        if item>max:
            max =item
    print('Maximum\t:',str(max))


L=[2, 10, 5, 89, 9]
findMax(L)
>>>
```

Output:

```
Maximum:89
>>>
```

Note that if the contents of the list L are strings (e.g. L=['Harsh', 'Nakul', 'Naved', 'Sahil']), the strings would be compared as per the rules and the largest ("Sahil") would be printed. That is, the program works for integers, strings, or floats. However, for the following list, an exception would be raised.

```
L= [2, 'Harsh', 3.67]
```

Output:

```
Traceback (most recent call last):
File "C:/Python/Exception handling/Example/findMax.py", line 15,
    in <module>
    findMax(L)
File "C:/Python/Exception handling/Example/findMax.py", line 4,
    in findMax
    if item>max:
```

```
TypeError: unorderable types: str() > int()
>>>
```

The problem can be handled by putting the part of the code where the problem is likely to come, in the **try** block. Moreover, if all the items of the list were entered by the user, the possibility of a runtime error cropping up would be higher. In such cases, the programmer must make sure that everything, including the input of items and calling the function should be in the try block. The following code presents the version of the program, where items are entered by the user and exception handling should be implemented. Note that the first run produces an expected result, whereas the second run results in runtime error and hence, it will invoke exception handling mechanism.

Code:

```
def findMax(L):
    max =L[0]
    for item in L:
        if item>max:
            max =item
    print('Maximum\t:',str(max))


L=[]
item=input('Enter items (press 0 to end)\n')
try:
    while int(item) !=0:
        L.append(item)
        item=input('Enter item (press 0 to end)\n')
        #print('\nItem entered \t:',str(item))
    print('\nList \n')
    print(L)
    findMax(L)
except:
    print('Run time error')
```

Output (First run):

```
>>>
Enter items (press 0 to end)
3
```

```
Enter item (press 0 to end)
2
Enter item (press 0 to end)
5
Enter item (press 0 to end)
12
Enter item (press 0 to end)
8
Enter item (press 0 to end)
98
Enter item (press 0 to end)
1
Enter item (press 0 to end)
0

List

['3', '2', '5', '12', '8', '98', '1']
Maximum  : 98
>>>
```

Output (Second run):

```
Enter items (press 0 to end)
2
Enter item (press 0 to end)
8
Enter item (press 0 to end)
Harsh
Run time error
>>>
```

Also note that if a **finally** is added to the code. The statements in finally will always be executed, whether the exception occurs. The code that contains both **finally** and **except** is represented as follows. Note that the first output produces the expected result, and it also prints the statements given in **finally**. The second output results in a runtime error and invokes exception handling mechanism and it also prints the message in **finally**. The reader should appreciate that there was no need, whatsoever, of the **except** as **finally** is already there. The code would have run correctly, as it will handle the runtime error with a **finally**, however, both have been included to bring home the point that **except** does its intended job with a **finally** and **finally** can be used for cleanup actions or for de-allocating memory and so on.

Code:

```
def findMax(L):
    max =L[0]
    for item in L:
        if item>max:
            max =item
    print('Maximum\t:',str(max))
L=[]
item=input('Enter items (press 0 to end)\n')
try:
    while int(item)!=0:
        L.append(item)
        item=input('Enter item (press 0 to end)\n')
        #print('\nItem entered \t:',str(item))
print('\nList \n')
    print(L)
    findMax(L)
except:
    print('Run time error')
finally:
    print('This is always executed')
```

Output (first run):

```
Enter items (press 0 to end)
1
Enter item (press 0 to end)
4
Enter item (press 0 to end)
2
Enter item (press 0 to end)
89
```

```
Enter item (press 0 to end)
3
Enter item (press 0 to end)
0

List

['1', '4', '2', '89', '3']
Maximum  : 89
This is always executed
>>>
```

Output (second run):

```
Enter items (press 0 to end)
3
Enter item (press 0 to end)
1
Enter item (press 0 to end)
7
Enter item (press 0 to end)
harsh
Run time error
This is always executed
>>>
```

15.7 CONCLUSION

The chapter presented a remarkable way to deal with exceptions. Though Python has an inbuilt mechanism to deal with exceptions, the knowledge of Exception Handling makes programs more effective, user-friendly, and robust. The first step would be to identify the part of the code, where exceptions are likely to come, and put the part in the **try** block. The exceptions can also be manually caught and handled in the **except** block. The **finally** block handles the unhandled exceptions and also executes even if there is no exception. The chapter also presents some of the most common Exceptions that can be caught in Python. The reader should use the concepts learned in this chapter in their programs. Happy Programming!

GLOSSARY

try/except :Syntax

```
try:
    ##code where exception is expected
except<Exception>:
    ##code to handle the exception
## rest of the program
```

Manually raising exceptions: Syntax:

```
try:
    raise <something>
except<something>:
    ##code which handles the exception
##rest of the code
```

POINTS TO REMEMBER

- At the runtime, if an error crops up, an exception is raised.
- Exception handling, Python can be done using either of the following.
  - try/catch
  - try/finally
  - raise
  - assert
- In Python, one can also manually raise the exceptions.
- The part of the code, where the exception is likely to be raised is put in the **try** block. If an exception is raised, it would be handled in the **except** block.
- The class that helps to raise user defined exceptions should be a subclass of the **Exception** class.
- The statements in **finally** will always execute, whether exception occurs or not.

EXERCISES

Multiple Choice Questions

1. Exception handling

    (a) Handles runtime errors in a program

    (b) Provides robustness

(c) Both

(d) None of the above

2. Exception handling is needed for

   (a) Syntax Errors

   (b) Run time errors

   (c) Both

   (d) None of the above

3. Which of the following is not supported in Python?

   (a) Nested try

   (b) Re-throwing an exception

   (c) Both are supported

   (d) None of the above is supported

4. Which of the following is raised in the case of division by zero?

   (a) Divide

   (b) ZeroDivide

   (c) Both

   (d) None of the above

5. Which of the following is raised in the case, when an index outside the bounds are accessed?

   (a) Array Index Out of Bound

   (b) Out of Bound

   (c) Array

   (d) None of the above

6. Which of the following is true?

   (a) For each try there is exactly one catch

   (b) Every try must include a raise

   (c) A catch can handle any type of exception

   (d) A catch can handle the exception for which it is designed, unless it catches all exceptions, (in this case it handles all the exceptions).

7. How many exceptions can a try have?

   (a) Single

   (b) Two, only in specific conditions

   (c) Any number of catch

   (d) None of the above

8. Which type of exception can occur?

   (a) Predefined

   (b) User defined

   (c) Both

   (d) None of the above

9. What is the base class of a class, of whose exception is to be raised?

(a) Exception

(b) Error

(c) Both

(d) None of the above

10. Which is the correct syntax of raise?

(a) raise <name of the exception>

(b) (raise (<name of the exception>))

(c) (raise (new<user defined exception>))

(d) All of the above

Theory

1. What is the difference between compile time and run time error?

2. What is exception handling?

3. Explain the mechanism of exception handling?

4. Explain how to create a class that derives the Exception class. How is this class used to raise exceptions?

5. Mention which following classes must be explained?

Programming Exercises

The roots of a quadratic equation $ax^2 + bx + c = 0$ are given by the formula $x = \dfrac{-b + \sqrt{b^2 - 4ac}}{2a}$. Write a program to ask the user to enter the values of a, b, and c and calculate the roots.

1. Use try/except in the above question to handle the following situations

(a) Calculating root of a negative number

(b) Division by zero

(c) Incorrect format

2. Create a class called negative_discriminant, which is a subclass of the Exception class. Now, in question number 1, raise the negative_discriminant exception when the value of $b^2 - 4ac$ is negative.

   The division of two complex numbers is defined as follows. If $c_1 = a_1 + ib_1$ is the first complex number and $c_2 = a_2 + ib_2$ is the second complex number, then the complex number

   $$c = (a_1 \times a_2 - b_1 b_2)/(a_2^2 + b_2^2) + i(a_1 \times b_2 + b_1 a_2)/(a_2^2 + b_2^2)$$

3. Create a class called Complex and implement Exception Handling in the method that carries out division.

4. For the Complex class, defined in the previous question, use Exception Handling to prevent the user from entering a nonreal number (as real or imaginary part).

5. In the complex class create a function that converts complex number to the polar form.

6. Implement Stacks using lists. Incorporate Exception Handling.

7. Implement Queues using lists. Incorporate Exception Handling.

8. Implement the operations of Linked List, throw an exception when the number entered by the user is negative. Assume that the data part of the linked list would contain numbers only.

9. Write a program that takes the ppm of chlorine in water from the user and finds whether the ppm is within a permissible limit. In the other case, the program should raise an exception.

10. Write a program that finds the inverse of a given matrix. The program should raise an exception when the determinant of the matrix is zero.

SECTION IV
NUMPY, PANDAS, AND MATPLOTLIB

You are now empowered with Python and know how to write programs. Let's now move to the more sophisticated part. This section has six chapters. The first two deal with NumPy, the next two deal with MatplotLib, and the last two deal with Pandas. This section is your door to Machine Learning and Data Science. Let's dive into the real Python!

NUMPY–I

16.1 INTRODUCTION

The **numpy** package is primarily used for scientific computing. It is a powerful package that provides us with an N-dimensional array of objects. There are numerous data types and tools in **numpy** like broadcasting which helps us to implement sophisticated algorithms easily and efficiently. It also provides functions for finding out various transformations like the Fourier Transform of a given series and sophisticated functions for generating all types of random numbers. This package also makes the integration of a project developed in Python with existing projects seamlessly easy.

To import the numpy package, the following statement is written

```
import numpy as np
```

The installation of **Anaconda** has already been explained in the previous section. On installing **Anaconda**, the **numpy** and **scipy** packages are automatically installed. This book uses **Jupyter** IDE for running the code. This and the next chapter discuss the various aspects of **numpy** and its uses to accomplish various tasks. The reader is advised to attempt the questions given at the end of this chapter to get hold of the nuances of **numpy**. The chapter has been organized as follows (Figure 16.1).

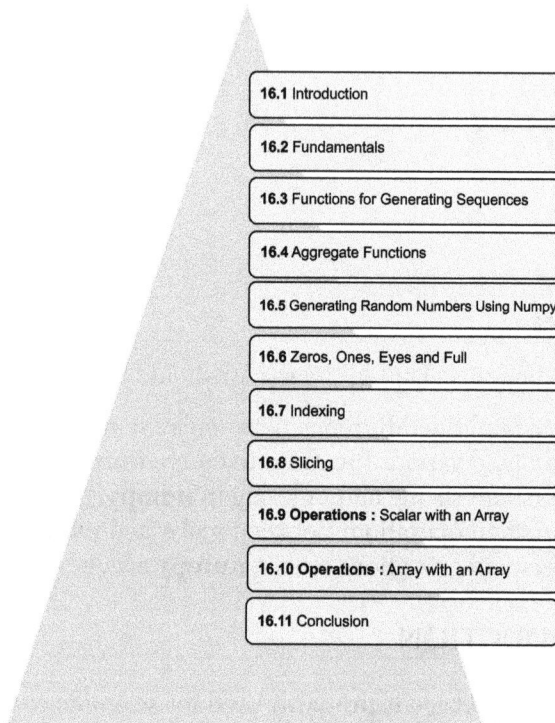

16.1 Introduction

16.2 Fundamentals

16.3 Functions for Generating Sequences

16.4 Aggregate Functions

16.5 Generating Random Numbers Using Numpy

16.6 Zeros, Ones, Eyes and Full

16.7 Indexing

16.8 Slicing

16.9 Operations : Scalar with an Array

16.10 Operations : Array with an Array

16.11 Conclusion

FIGURE 16.1 Organization of the chapter.

16.2 FUNDAMENTALS

This section briefly presents the various building blocks of a program capable of carrying out practical tasks. The explanations are precise and concise to accelerate your moment toward the more important topics.

Array: An array is a group of elements of the same type stored at consecutive memory locations. In the following example, the numbers 8, 1, 5, 89, and 45 are stored in an array called **arr**.

arr = *array*([8, 1, 5, 89, 45])

Note that the elements are of the same type (all of them are integers) and are stored at consecutive memory locations.

Zero-based indexing: The arrays are zero-based indexed, that is the first element can be accessed by writing **arr[0]**, the second by **arr[1]**, and so on.

Vectors: A 1-dimensional array is called a vector.

Multi-dimensional arrays: A multi-dimensional array may have more than one dimension. For example,

$$A = \begin{bmatrix} 1 & 2 & 3 \\ 4 & 5 & 6 \\ 7 & 8 & 9 \end{bmatrix}$$

is a 2-dimensional array containing three rows and three columns. The element at the intersection of the first row and the first column is denoted as $A[0, 0]$, that at the third row and second column is denoted as $A[2, 1]$, and so on. These arrays would henceforth be called matrices (or matrix (singular)).

Storage: The matrices can be stored in one of the following formats:

Row major: In this technique, the first row is stored in the memory, followed by the second row, and so on. For example,

$$A = \begin{bmatrix} 1 & 2 & 3 \\ 4 & 5 & 6 \\ 7 & 8 & 9 \end{bmatrix}$$

is stored as

$$1, 2, 3, 4, 5, 6, 7, 8, 9$$

in the row-major format.

Column major: In this technique, the first column is stored in the memory, followed by the second column, and so on. For example,

$$A = \begin{bmatrix} 1 & 2 & 3 \\ 4 & 5 & 6 \\ 7 & 8 & 9 \end{bmatrix}$$

is stored as

$$1, 4, 7, 2, 5, 8, 3, 6, 9$$

in the column-major format.

Axes: In **numpy**, axis = i denotes the axes. In a 1-dimensional array, the elements are stored at axis = 0. In the case of a 2-dimensional array axis = 0 denotes rows, axis = 1 denotes columns (Figure 16.2) and in the case of a

3-dimensional array axis = 2 denotes the number of matrices. Axes denote the dimensions. So, for 1 dimensional array, axis = 0; for a 2-dimensional array, axis = 0 and 1 and for a 3-dimensional array axis = 0, 1, and 2.

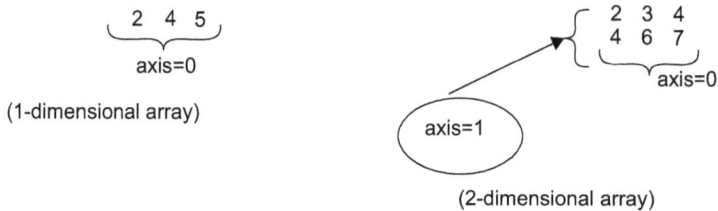

FIGURE 16.2 Axes in a **numpy** array.

Shape: The shape of a 1-dimensional array is the number of elements in the array. For example, if

$$A = array([8,1,5,89,45])$$

then $A.$ *shape* = 5.

The shape of a 2-dimensional array is (*the number of rows, the number of columns*). For example, if

$$A = \begin{bmatrix} 1 & 2 & 3 \\ 4 & 5 & 6 \\ 7 & 8 & 9 \end{bmatrix}$$

then the shape of the array is (3, 3).

Rank: The number of axes in a **numpy** array is the rank of the array. That is, the rank of a 3-D array would be 3.

List to an array: The function **numpy.array(L)**, converts the list **L** to a **numpy** array.

16.2.1 Similarity and Differences Between a List and a NumPy Array

The similarity between list and numpy array

Both lists and **numpy** array are zero-based indexed.

Differences between lists and array:

(a) numpy array is homogeneous whereas a list can contain different types of elements.

(b) The elements of a **numpy** array are stored at consecutive memory locations.

- **Numpy arrays support vectorized operations:** If an operation can be performed on each element of an array, it is vectorized. The **numpy** array supports vectorized operations, whereas a list does not.

16.3 FUNCTIONS FOR GENERATING SEQUENCES

Having gone through the basics of the **numpy** arrays, let us visit some of the remarkable functions, which help in creating useful arrays.

16.3.1 arange()

The **arange** function helps to print a sequence, having some initial value **(start)**, some final value **(stop)**, the difference between the consecutive terms **(step)**, and the data type **(dtype)**. The syntax of the function and the description of each of the parameters are as follows.

numpy.arange(start, stop, step, dtype)

- **Start:** The starting value of the sequence
- **Stop:** The value up to which the sequence is generated (not inducing the value itself)
- **Step:** The difference between the consecutive values
- **dtype:** The data type of the elements

The above function has been exemplified in the following code, which generates an Arithmetic Progression having first term 3, the last term 23 (less than 25), and the difference between the consecutive terms equals 2. The data type of the elements is specified as **"int."**

```
a=np.arange(3,25,2, int)
print(a)
```

Output:

```
array([3, 5, 7, 9, 11, 13, 15, 17, 19, 21, 23])
```

arange can also take a single argument. For example, writing **np.arange**(6) would generate a sequence having first value 0, the difference between the consecutive terms as 1, and the last term 5, that is

```
b=np.arange(6)
print(b)
```

Output:

```
array([0, 1, 2, 3, 4, 5)]
```

```
c=np.arange(6, dtype=float)
print(c)
```

Output:

```
array([0., 1., 2., 3., 4., 5.])
```

16.3.2 linspace()

The **linspace** function divides the given range into a specified number of segregations and returns the sequence so formed. The function takes the following parameters:

- **start:** The first value of the sequence
- **stop:** The last value (included by default unless the endpoint parameter is set to **False**)
- **num:** The number of items
- **endpoint:** If the endpoint is False then the **stop** value is not included in the sequence
- **retstep:** If this is True, the **stepsize** is returned
- **dtype:** The data type of elements can be specified using this parameter.

In the example that follows, the first value of the sequence is 1 and the last value is 27. The number of elements in the sequence is 11.

```
d=np.linspace(11, 27, 11)
print(d)
```

Output:

```
array([11., 12.6, 14.2, 15.8, 17.4, 19., 20.6, 22.2, 23.8, 25.4,
27.])
```

If the value of the **endpoint** argument is **False**, the last value (in this case, 27) is not included.

```
e=np.linspace(11, 27, 11, endpoint=False)
print(e)
```

Output:

```
array([11., 12.45454545, 13.90909091, 15.36363636, 16.01010102,
18.27272727, 19.72727273, 21.10101010, 22.63636364, 24.09090909,
25.54545455])
```

The value of the **step** can be viewed by assigning **True** to the **retstep** argument. For example, in the sequence generated by dividing the range 11-26 in 11 parts, the gap (the last argument of the result) is 1.45454545454546.

```
f=np.linspace(11, 27, 11, endpoint=False, retstep=True)
print(f)
```

Output:

```
(array([11., 12.45454545, 13.90909091, 15.36363636, 16.01010102,
18.27272727, 19.72727273, 21.10101010, 22.63636364, 24.09090909,
25.54545455]), 1.4545454545454546)
```

16.4 AGGREGATE FUNCTIONS

The **numpy** module contains many aggregate functions. The various functions, along with their brief explanations, are as follows:

- **numpy.sum:** It finds the sum of the elements of the argument (e.g. a list or an array).
- **numpy.prod:** It finds the product of the elements of the argument (e.g. a list or an array).
- **numpy.mean:** It finds the mean of the elements of the argument (e.g. a list or an array).
- **numpy.std:** It finds the standard deviation of the elements of the argument.
- **numpy.var:** It finds the variance of the elements of the argument.
- **numpy.max:** It finds the maximum element of the argument. In the case of a list or a 1D array, the maximum element would be displayed. However, in the case of a 2D array, the axis along which the maximum element is desired can also be mentioned. Here, **axis=0** indicates rows and **axis=1** indicates columns.
- **numpy.min:** It finds the minimum element of the argument. In the case of a list or a 1D array, the minimum element would be displayed.

However, in the case of a 2D array, the axis along which the minimum element is desired can also be mentioned. Here, **axis=0** indicates rows and **axis=1** indicates columns.

- **numpy.argmin:** It finds the position (index) of the minimum element.
- **numpy.argmax:** It finds the position of the maximum element.
- **numpy.median:** It finds the median of the elements of the argument.
- **numpy.percentile:** It finds the percentile of the elements of the argument. The percentile (25 etc.) is the second argument.
- **numpy.any:** It finds if any element of the given argument is present in the list.
- **numpy.all:** It finds if all the elements of the given argument are present in the list.

The following code exemplifies the above functions by generating a set of 50 values. The values are between 0 and 100 (the **np.random.random**(50) has been multiplied by 100). The maximum, minimum, index of the maximum, index of the minimum, average, median, standard deviation, variance, 25th percentile, and 75th percentile of the elements has been found using the above functions.

Code:

```
Values1=100*(np.random.random(50))
print(Values1)
```

Output:

```
array([ 7.89901504e+01, 3.10353272e+01, 4.55247975e+01,
2.09271021e+01, 1.28704552e+01, 8.83259317e+01, 4.34685519e+01,
6.47957990e+01, 3.94568075e+01, 8.14517974e+01, 1.30191468e+01,
8.69577211e+01, 9.94997332e+01, 5.33860103e+01, 5.67079066e+01,
9.98534029e+01, 3.22963592e+01, 4.98089020e+01, 6.68875653e+01,
9.65255635e+01, 4.94490583e+01, 7.37397326e+01, 3.40551969e+01,
4.37639703e+01, 4.48223897e+01, 3.25917428e+01, 9.59794929e+01,
5.87367182e+01, 9.87710458e+01, 4.37364340e+01, 1.97519881e+00,
6.03630476e+01, 8.92749410e-02, 9.06113729e+01, 7.97883172e+01,
8.95203320e+01, 1.69638876e+01, 8.40854179e+00, 3.45767708e+01,
3.24516258e+01, 9.71498648e+01, 1.29033485e+01, 7.12565243e+01,
3.77831919e+01, 6.59571908e+01, 6.80006473e+01, 3.69824712e+00,
9.23685114e+01, 3.10464585e+01, 3.48051930e+01])
```

Code:

```
Max=np.max(Values1)
Max_Index=np.argmax(Values1)
```

```
Min=np.min(Values1)
Min_Index=np.argmin(Values1)
Sum=np.sum(Values1)
Prod=np.prod(Values1)
Mean=np.mean(Values1)
SD=np.std(Values1)
Variance=np.var(Values1)
Med=np.median(Values1)
Per25=np.percentile(Values1,25)
Per75=np.percentile(Values1,75)
print("Max\t:",Max,"\nIndex\t:",Max_Index,"\nMin\t:",Min,"\
    nIndex\t:", Min_Index,"\nAverage\t:",Mean,"\nStdDeviation\
    t:",SD,"\nVariance t:",Variance," nMedian\t:",Med,"
    \nPercentile 25\t:",Per25,"\nPercentile 75\t:",Per75)
```

Output:

```
Max             : 99.8534028512
Index           : 15
Min             : 0.0892749410473
Index           : 32
Average         : 53.3430471879
Stad Deviation  : 29.5561768206
Variance        : 873.567588252
Median          : 49.6289801762
Percentile 25   : 32.4866550288
Percentile 75   : 79.5887755051
```

Code:

```
import numpy as np
v1 = np.array([True,False,True])
v2 = np.array([True,True,True])
v3 = np.array([False,False,False])
print(np.any(v1))
print(np.all(v1))
print(np.any(v2))
print(np.all(v2))
print(np.any(v3))
print(np.all(v3))
```

Output:

```
True
False
True
True
False
False
```

16.5 GENERATING RANDOM NUMBERS USING NUMPY

The **numpy.random.random, numpy.random.normal,** and **numpy.random.randint** are the most common tools for the generation of an array of random numbers (or a single random number for that matter). The arguments and explanations of these three functions are presented in Table 16.1. The illustrations that follow explicate their usage.

TABLE 16.1 Generating random numbers using numpy.

Function	Arguments	Explanation
numpy.random.random	**size:** The size of the N-D array. The default value of this argument is None	This function returns a random float in the interval $[0.0, 1.0)$
numpy.random.normal	**loc:** The mean of the normal distribution	This function generates a sample from the parameterized normal distribution
	scale: The standard deviation of the normal distribution	
	size: The size of the N-D array	
numpy.random.randint	**low:** This argument sets the lowest number of the generated distribution	This function generates random integers in each range
	high: This argument sets the highest number of the generated distribution	
	size: This argument sets the size of the generated distribution	
	dtype: This argument sets the datatype of the numbers of the generated distribution	

Illustration 16.1:

*Generate a random number using **numpy.random.random.***

Code:

```
import numpy as np
num=np.random.random()
print(num)
```

Output:

```
0.5829500009456373
```

Illustration 16.2:

*Generate a random number using **numpy.random.random** between 3 and 9.*

Code:

```
import numpy as np
a=3
b=9
num=(b-a)*(np.random.random())+a
print(num)
```

Output:

```
5.287550385736058
```

Illustration 16.3:

*Generate 5 random numbers using **numpy.random.normal** from a normal distribution having mean 3 and standard distribution 2. Also, print the mean and the standard deviation of the numbers generated.*

Code:

```
import numpy as np
rand_num_list=np.random.normal(3, 2, 5)
print(rand_num_list)
print(np.mean(rand_num_list))
print(np.std(rand_num_list))
```

Output:

```
[1.98930695 1.87258593 4.1491093 2.10198921 4.17106858]
2.8568119957867197 1.066613900848987
```

The reader is expected to take note of the difference in the mean, passed as an argument and that generated. Likewise, reason out the variation in the standard deviation.

TRY   *Note the difference between the generated mean and the mean passed as the argument.*

16.6 ZEROS, ONES, EYES, AND FULL

The **numpy.zeros, numpy.ones,** and **numpy.eye** are used for generating arrays containing zeros, ones, and identity matrices respectively. The arguments and explanations of the three are presented in Table 16.2. The illustrations that follow explicate their usage.

TABLE 16.2 Zeros, Ones, eyes, and full.

Function	Arguments	Explanation
numpy.zeros	**shape:** The shape of the N-D array **dtype:** This argument sets the **datatype** of the zeros.	This function generates an array of zeros with the given shape.
numpy.ones	**shape:** The shape of the N-D array **dtype:** This argument sets the datatype of the ones.	This function generates an array of ones with the given shape.
numpy.eye	**n:** The number of rows **m:** The number of columns **dtype:** This argument sets the data type of the elements.	This function generates an identity matrix of the given shape.
numpy.full	**shape:** The shape of the N-D array **fill_value:** The value with which the array would be filled **dtype:** This argument sets the data type of the elements.	This function generates an array filled with fill_values.

Illustration 16.4:

Generate an array filled with zeros, having 5 rows and 2 columns.

Solution:

```
Z=np.zeros((5,2))
print(Z)
```

Output:

```
[[0. 0.]
 [0. 0.]
 [0. 0.]
 [0. 0.]
 [0. 0.]]
```

Illustration 16.5:

Generate an array filled with ones, having 3 rows and 4 columns.

Solution:

```
O=np.ones((3,4))
print(O)
```

Output:

```
[[1. 1. 1. 1.]
 [1. 1. 1. 1.]
 [1. 1. 1. 1.]]
```

Illustration 16.6:

Generate an array filled with "11," having 2 rows and 6 columns.

Solution:

```
F=np.full((2,6),fill_value=11)
print(F)
```

Output:

```
[[11 11 11 11 11 11]
 [11 11 11 11 11 11]]
```

Illustration 16.7:

Generate an identity matrix of order 3.

Solution:

```
I=np.eye(3)
print(I)
```

Output:

```
[[1. 0. 0.]
 [0. 1. 0.]
 [0. 0. 1.]]
```

16.7 INDEXING

A **numpy** 1-D array is zero-based indexed. The first element of the array can be accessed by writing the name of the array followed by square brackets containing 0, that is

If arr=[364, 3748, 347, 849, 374], then **arr[0]** is 364.

Likewise, the second element can be accessed using **arr[1]**, the third by **arr[2]**, and so on. The last element can be accessed by writing **arr[-1]**. The second last element is accessed by writing **arr[-2]** and so on. Figure 16.3 presents examples of the indexes of a given array.

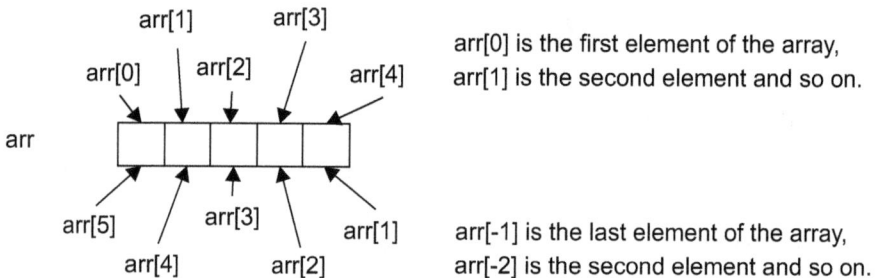

FIGURE 16.3 Indexing.

In the case of 2-D arrays, **arr[i, j]** indicates the element at the ith row and the jth column. For example, consider an array with 5 rows and 3 columns, wherein each element is between 7 and 80. The array can be generated by writing the following code.

Code:

```
arr1=np.random.randint(7,80,(5,3))
print(arr1)
```

Output:

```
[[46 69 53]
 [79 52 53]
```

```
[69 60 12]
[38 32 76]
[ 7 40 75]]
```

Now, consider the following code and observe the output.

Code:

```
print(arr1[0,1])
print(arr1[3,0])
print(arr1[3,2])
print(arr1[-1,2])
print(arr1[2,-2])
print(arr1[-3,-1])
```

Output:

```
69
38
76
75
60
12
```

Figure 16.4 explains this indexing.

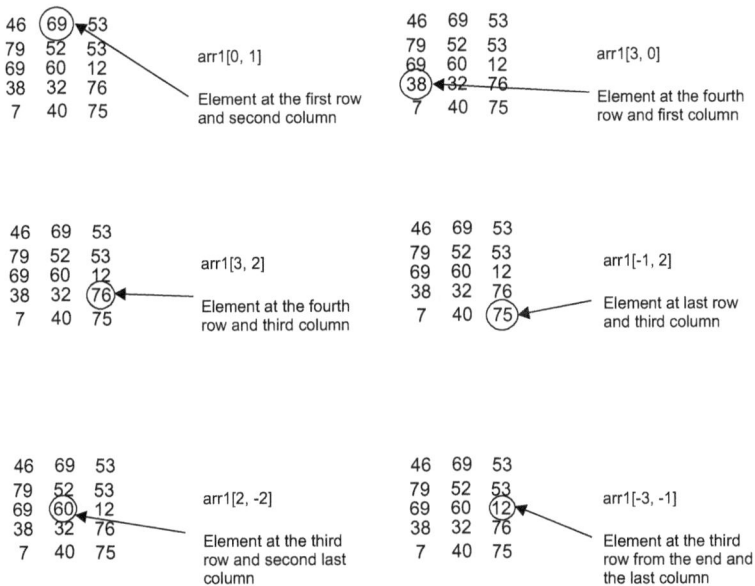

FIGURE 16.4 Indexing in a 2D-numpy array.

16.8 SLICING

Slicing can be used to find the subset of a **numpy-array**. For a 1-D array, arr

- arr[:k] generates an array consisting of elements from indices 0 to (k-1).
- arr[k:] generates an array consisting of elements from indices k to the last element.
- arr[k:m] generates an array consisting of elements from indices k to that at index (m-1).
- arr[k:-1] generates an array consisting of elements from indices k to the second last.
- arr[-k:-m] generates an array consisting of elements from the kth last to $(m-1)^{th}$ last.

Table 16.3 shows slicing in 1D arrays.

TABLE 16.3 Slicing in 1D numpy arrays.

Array	Output if the input is arr1
arr1	[36 18 44 50 27 59 17 63 72 20]
arr1[:3]	array([36, 18, 44])
arr1[3:]	array([50, 27, 59, 17, 63, 72, 20])
arr1[2:7]	array([44, 50, 27, 59, 17])
arr1[3:-1]	array([50, 27, 59, 17, 63, 72])
arr1[-5:-1]	array([59, 17, 63, 72])

Slicing in a 2D array creates a subarray. To understand the above concept, refer to Table 16.4. The topic has been revisited in the next chapter under the heading advanced indexing and slicing arrays. In the examples, given in the table, the following array is used.

TABLE 16.4 Slicing in a 2-D array.

Function	Arguments	Explanation
arr2[:2,:3]	First, the second and third columns of the first and second rows.	array([[4, 20, 9], [50, 24, 65]])
arr2[2:,3:]	Subarray consisting of elements from fourth column onwards and third row onwards.	array([[28, 75], [9, 15]])
arr2[1:3,2:3]	Subarray consisting of elements from second and the third row and third column.	array([[65], [38]])

Illustration 16.8:

Ask the user to enter the value of n (the number of elements) and generate an array of n random numbers between 0 and 1.

Solution:

```
import numpy as np
n=int(input('Enter the number of elements\t:'))
random_num=np.random.random(n)
print(random_num)
```

Output:

```
Enter the number of elements : 8
[0.68487933 0.19415796 0.19550152 0.61254812 0.22986881
 0.80080749 0.41303666 0.57186362]
```

Illustration 16.9:

Ask the user to enter the number of rows (r) and the number of columns (c) and generate a two-dimensional array of order r × c.

Solution:

```
r=int(input('Enter the number of rows\t:'))
c=int(input('Enter the number of columns\t:'))
random_num_mat=np.random.random((r,c))
print(random_num_mat)
```

Output:

```
Enter the number of rows    : 3
Enter the number of columns : 4
[[0.46332079 0.67111524 0.58163963 0.08379198]
 [0.72509431 0.48099346 0.12884776 0.40432988]
 [0.60843099 0.7945741  0.92891968 0.54176968]]
```

Illustration 16.10:

Ask the user to enter the number of rows (r), the number of columns (c), the mean of the normal distribution (mean) and its standard distribution (std) generate a two-dimensional array of order r × c.

Solution:

```
r=int(input('Enter the number of rows\t:'))
c=int(input('Enter the number of columns\t:'))
mean=float(input('Enter the mean of the distribution\t:'))
std=float(input('Enter the standard deviation of the
    distribution'))
random_nd_mat=np.random.normal(mean,std,(r,c))
print(random_num_mat)
```

Output:

```
Enter the number of rows                              : 4
Enter the number of columns                           : 5
Enter the mean of the distribution                    : 3
Enter the standard deviation of the distribution      : 4
[[0.46332079 0.67111524 0.58163963 0.08379198]
 [0.72509431 0.48099346 0.12884776 0.40432988]
 [0.60843099 0.7945741  0.92891968 0.54176968]]
```

Illustration 16.11:

Flatten the above array, sort the numbers, and plot them.

Solution:

```
numbers_nd=random_nd_mat.reshape(((random_nd_mat.shape[0]*ran-
dom_nd_mat.shape[1]),1))
print(numbers_nd.shape)
numbers_nd_sorted=np.sort(numbers_nd)
print(numbers_nd_sorted)
index=np.arange(0, numbers_nd_sorted.shape[0] )
plt.plot(index,numbers_nd_sorted)
plt.show()
```

Illustration 16.12:

Ask the user to enter the number of rows and the number of columns and generate a matrix with random numbers between the given range.

Solution:

```
r=int(input('Enter the number of rows\t:'))
c=int(input('Enter the number of columns\t:'))
a=int(input('Enter the range: From\t:'))
```

```
b=int(input('Enter the range: To\t:'))
array_random_range=np.random.randint(a,b,(r,c))
print(array_random_range)
```

Output:

```
Enter the number of rows       : 3
Enter the number of columns    : 4
Enter the range: From          : 7
Enter the range: To            : 21
[[11 13 10 17]
 [10 13 16  9]
 [17 17 16 10]]
```

Illustration 16.13:

Generate an identity matrix of the size entered by the user.

Solution:

```
n=int(input('Enter the number of rows\t:'))
I=np.eye(n)
print(I)
```

Output:

```
Enter the number of rows : 5
[[1. 0. 0. 0. 0.]
 [0. 1. 0. 0. 0.]
 [0. 0. 1. 0. 0.]
 [0. 0. 0. 1. 0.]
 [0. 0. 0. 0. 1.]]
```

Illustration 16.14:

Generate an array containing zeros having shape entered by the user.

Solution:

```
r=int(input('Enter the number of rows\t:'))
c=int(input('Enter the number of columns\t:'))
Z=np.zeros((r,c))
print(Z)
```

Output:

```
Enter the number of rows       : 4
Enter the number of columns    : 5
[[0. 0. 0. 0. 0.]
 [0. 0. 0. 0. 0.]
 [0. 0. 0. 0. 0.]
 [0. 0. 0. 0. 0.]]
```

Illustration 16.15:

*In the above question, how will you generate an array of zeros, with datatype as **integer**?*

Solution:

```
r=int(input('Enter the number of rows\t:'))
c=int(input('Enter the number of columns\t:'))
Z=np.zeros((r,c), dtype=int)
print(Z)
```

Output:

```
Enter the number of rows       : 4
Enter the number of columns    : 5
[[0 0 0 0 0]
 [0 0 0 0 0]
 [0 0 0 0 0]
 [0 0 0 0 0]]
```

Illustration 16.16:

*How will you generate an array of ones with datatype **integer**.*

Solution:

```
r=int(input('Enter the number of rows\t:'))
c=int(input('Enter the number of columns\t:'))
O=np.ones((r,c), dtype=int)
print(O)
```

Output:

```
Enter the number of rows        : 3
Enter the number of columns     : 5
[[1 1 1 1 1]
 [1 1 1 1 1]
 [1 1 1 1 1]]
```

Illustration 16.17:

*How will you generate an array of ones with datatype **float**.*

Solution:

```
r=int(input('Enter the number of rows\t:'))
c=int(input('Enter the number of columns\t:'))
O=np.ones((r,c), dtype=float)
print(O)
```

Output:

```
Enter the number of rows        : 2
Enter the number of columns     : 5
[[1. 1. 1. 1. 1.]
 [1. 1. 1. 1. 1.]]
```

Illustration 16.18:

Ask the user to enter the number of rows and the number of columns of a 2-D matrix. Generate an array filled with the same number (entered by the user).

Solution:

```
r=int(input('Enter the number of rows\t:'))
c=int(input('Enter the number of columns\t:'))
item=int(input('Enter the item\t:'))
F=np.full((r,c), item)
print(F)
```

Output:

```
Enter the number of rows        : 4
Enter the number of columns     : 5
Enter the item                  : 78
```

```
[[78 78 78 78 78]
 [78 78 78 78 78]
 [78 78 78 78 78]
 [78 78 78 78 78]]
```

Illustration 16.19:

Generate an array of n elements and find the element at the position entered by the user:

- from the beginning
- from the end

Solution:

```
n=int(input('Enter the number of elements\t:'))
random_num=np.random.random(n)
print(random_num)
pos=int(input('Enter position\t:'))
element=random_num[pos]
print('Element at ',pos,' ',element)
pos1=int(input('Enter position from end\t:'))
pos1=-1*pos1
element1=random_num[pos1]
print('Element at ',pos1,' from end ',element1)
```

Output:

```
Enter the number of elements    : 6
[0.43713108 0.33305433 0.18057763 0.76521652 0.91477333 0.59816926]
Enter position                  : 3
Element at 3 is                 : 0.7652165243368857
Enter position from end         : 2
Element at 2 from end           : 0.9147733324270373
```

Illustration 16.20:

Write a program to find the element at the location specified by the user in a 2-D array.

Solution:

```
r=int(input('Enter the number of rows\t:'))
c=int(input('Enter the number of columns\t:'))
random_num_mat=np.random.random((r,c))
print(random_num_mat)
pos1=int(input('Enter row number\t:'))
pos2=int(input('Enter col number\t:'))
element=random_num_mat[pos1, pos2]
print('Element at (', pos1,',',pos2,' ) is', element)
```

Output:

```
Enter the number of rows              : 3
Enter the number of columns           : 4
[[0.3587718  0.10041165 0.16053454 0.47189602]
 [0.5167602  0.93113063 0.3259402  0.93074249]
 [0.29976931 0.78217725 0.11342043 0.40755382]]
Enter row number :2 Enter col number : 3
Element at ( 2 , 3 ) is 0.4075538186361203
```

Illustration 16.21:

In the above question, what is the element at (pos1, -1*pos2).

Solution:

```
Element at (2 , -3) is 0.7821772493742479
```

Illustration 16.22:

In Illustration 16.20, what is the element at (-1*pos1, pos2).

Solution:

```
Element at (-2 , 3) is 0.9307424897203977
```

Illustration 16.23:

In Illustration 16.20, what is the element at (-1*pos1, -1*pos2).

Solution:

```
Element at (-2 , -3) is 0.9311306325583609
Hint:
pos1_new=-1*pos1
pos2_new=-1*pos2
element=random_num_mat[pos1, pos2_new]
print('Element at (', pos1,',',pos2_new,') is', element)
element=random_num_mat[pos1_new, pos2]
print('Element at (', pos1_new,',',pos2,') is', element)
element=random_num_mat[pos1_new, pos2_new]
print('Element at (', pos1_new,',',pos2_new,') is', element)
```

16.9 OPERATIONS: SCALAR WITH AN ARRAY

This section discusses various operations on the **numpy** arrays with a scalar. The standard operations namely: addition, subtraction, multiplication, division, modulo, and power have been presented in this section.

16.9.1 Addition

16.9.1.1 *Using the + operator*

Adding a scalar with array results in the addition of that scalar to each element of the given array. For example, if

$$A = \begin{matrix} 1 & 2 & 3 \\ 4 & 5 & 6 \\ 7 & 8 & 9 \end{matrix}$$

Then, A+ 3 becomes

$$\begin{matrix} 1 & 2 & 3 \\ 4 & 5 & 6 \\ 7 & 8 & 9 \end{matrix} + 3 =$$

$$\begin{matrix} 4 & 5 & 6 \\ 7 & 8 & 9 \\ 10 & 11 & 12 \end{matrix}$$

Illustration 16.24:

Ask the user to enter the number of rows and columns of a given matrix and generate a matrix of random numbers (between 0 and 1). Add 5 to each element of the matrix and show the result.

Solution:

The process has already been explained. The following code performs the task, and the output follows the code.

Code:

```
import numpy as np
r=int(input('Enter the number of rows\t:'))
c=int(input('Enter the number of columns\t:'))
array1=np.random.random((r,c))
print(array1)
array2=array1+5
print('The resultant matrix\t:')
print(array2)
```

Output:

```
Enter the number of rows    : 4
Enter the number of columns : 5
[[0.89144523 0.71331741 0.36089562 0.93356356 0.84708016]
 [0.55455128 0.31488978 0.95140343 0.34885856 0.76283298]
 [0.52345739 0.97636114 0.92736399 0.91055891 0.82454989]
 [0.09921104 0.26160407 0.02555853 0.05173145 0.91408363]]
The resultant matrix :
[[5.89144523 5.71331741 5.36089562 5.93356356 5.84708016]
 [5.55455128 5.31488978 5.95140343 5.34885856 5.76283298]
 [5.52345739 5.97636114 5.92736399 5.91055891 5.82454989]
 [5.09921104 5.26160407 5.02555853 5.05173145 5.91408363]]
```

16.9.1.2 Using the numpy. add function

The addition of a scalar to a matrix can also be performed using the **numpy. add** function. The signature of the function is as follows.

numpy.add(x1, x2, out, where, casting, order, dtype)

Where, **x1** and **x2** are arrays or scalars, and **out** is the location where the result is stored, and **where** is the condition which is broadcast over input.

The following illustration demonstrates the use of this function.

Illustration 16.25:

*Refer to Illustration 16.24. Perform the task using the **numpy.add** function.*

Solution:

The process has already been explained. The following code performs the task.

Code:

```
r=int(input('Enter the number of rows\t:'))
c=int(input('Enter the number of columns\t:'))
array1=np.random.random((r,c))
print(array1)
array2=np.add(array1,5)
print(array2)
```

16.9.2 Subtraction

16.9.2.1 Using the − operator

Subtracting a scalar from array results in the subtraction of that scalar from each element of the given array. For example, if

$$A = \begin{matrix} 1 & 2 & 3 \\ 4 & 5 & 6 \\ 7 & 8 & 9 \end{matrix}$$

Then, A - 1 becomes

$$\begin{matrix} 1 & 2 & 3 \\ 4 & 5 & 6 \\ 7 & 8 & 9 \end{matrix} -1 =$$

$$\begin{matrix} 0 & 1 & 2 \\ 3 & 4 & 5 \\ 6 & 7 & 8 \end{matrix}$$

Illustration 16.26:

Ask the user to enter the number of rows and columns of a given matrix and generate a matrix of random numbers (between 0 and 1). Subtract 2 from each element of the matrix and show the result.

Solution:

The process has already been explained. The following code performs the task, and the output follows the code.

Code:

```
r=int(input('Enter the number of rows\t:'))
c=int(input('Enter the number of columns\t:'))
array1=np.random.random((r,c))
print(array1)
array2=array1-2
print('The resultant matrix\t:')
print(array2)
```

Output:

```
Enter the number of rows       : 4
Enter the number of columns    : 5
[[0.39753919 0.89568943 0.56239721 0.00751823 0.51524542]
 [0.80519441 0.53579197 0.04500465 0.98211826 0.22254055]
 [0.80278989 0.84783075 0.03665448 0.64768028 0.72306593]
 [0.88770246 0.55138696 0.88508308 0.56852102 0.32854805]]
The resultant matrix :
[[-1.60246081 -1.10431057 -1.43760279 -1.99248177 -1.48475458]
 [-1.19480559 -1.46420803 -1.95499535 -1.01788174 -1.77745945]
 [-1.19721011 -1.15216925 -1.96334552 -1.35231972 -1.27693407]
 [-1.11229754 -1.44861304 -1.11491692 -1.43147898 -1.67145195]]
```

16.9.2.2 *Using the numpy.subtract function*

The subtraction of a scalar from a matrix can also be performed using the **numpy.subtract** function. The signature of the function is as follows.

numpy.subtract(x1, x2, out, where, casting, order, dtype)

Where, **x1** and **x2** are arrays or scalars, and **out** is the location where the result is stored, and **where** is the condition which is broadcast over input.

The following illustration demonstrates the use of this function.

Illustration 16.27:

*Refer to Illustration 16.26. Perform the task using the **numpy.subtract** function.*

Solution:

The process has already been explained. The following code performs the task.

Code:

```
r=int(input('Enter the number of rows\t:'))
c=int(input('Enter the number of columns\t:'))
array1=np.random.random((r,c))
print(array1)
array2=np.subtract(array1,2)
print('The resultant matrix\t:')
print(array2)
```

16.9.3 Multiplication

16.9.3.1 Using the * operator

Multiplying a scalar to array results in the multiplication of that scalar to each element of the given array. For example, if

$$A = \begin{matrix} 1 & 2 & 3 \\ 4 & 5 & 6 \\ 7 & 8 & 9 \end{matrix}$$

Then, A * 2 becomes

$$\begin{matrix} 1 & 2 & 3 \\ 4 & 5 & 6 \\ 7 & 8 & 9 \end{matrix} \times 2 =$$

$$\begin{matrix} 2 & 4 & 6 \\ 8 & 10 & 12 \\ 14 & 16 & 18 \end{matrix}$$

Illustration 16.28:

Ask the user to enter the number of rows and columns of a given matrix and generate a matrix of random numbers (between 0 and 1). Multiply 3 to each element of the matrix and show the result.

Solution:

The process has already been explained. The following code performs the task, and the output follows the code.

Code:

```
r=int(input('Enter the number of rows\t:'))
c=int(input('Enter the number of columns\t:'))
array1=np.random.random((r,c))
print(array1)
array2=array1*3
print('The resultant matrix\t:')
print(array2)
```

Output:

```
Enter the number of rows        : 4
Enter the number of columns     : 5
[[0.22004619 0.49134188 0.0220215  0.89062779 0.53269214]
 [0.78610415 0.5981393  0.82281589 0.82419645 0.1447323 ]
 [0.23443228 0.0144362  0.93418983 0.88136115 0.21863549]
 [0.30558265 0.2338263  0.7479892  0.91168111 0.91016941]]
The resultant matrix :
[[0.66013856 1.47402565 0.06606451 2.67188337 1.59807643]
 [2.35831244 1.7944179  2.46844768 2.47258936 0.4341969 ]
 [0.70329685 0.04330859 2.80256949 2.64408345 0.65590647]
 [0.91674796 0.70147889 2.2439676  2.73504333 2.73050824]]
```

16.9.3.2 Using the numpy.multiply function

The multiplication of a scalar to a matrix can also be performed using the **numpy.multiply** function. The signature of the function is as follows.

numpy.multiply(x1, x2, out, where, casting, order, dtype)

Where, **x1** and **x2** are arrays or scalars, and **out** is the location where the result is stored, and **where** is the condition which is broadcast over input.

The following illustration demonstrates the use of this function.

Illustration 16.29:

*Refer to Illustration 16.28. Perform the task using the **numpy.multiply** function.*

Solution:

The process has already been explained. The following code performs the task.

Code:

```
r=int(input('Enter the number of rows\t:'))
c=int(input('Enter the number of columns\t:'))
array1=np.random.random((r,c))
print(array1)
array2=np.multiply(array1,3)
print('The resultant matrix\t:')
print(array2)
```

16.9.4 Division

16.9.4.1 Using the / operator

Dividing a scalar to array results in the division of each element of the given array with that scalar. For example, if

$$A = \begin{matrix} 2 & 4 & 4 \\ 4 & 6 & 8 \\ 10 & 8 & 14 \end{matrix}$$

Then, A / 2 becomes

$$\begin{matrix} 2 & 4 & 4 \\ 4 & 6 & 8 \\ 10 & 8 & 14 \end{matrix} / 2 =$$

$$\begin{matrix} 1 & 2 & 2 \\ 2 & 3 & 4 \\ 5 & 4 & 7 \end{matrix}$$

Illustration 16.30:

Ask the user to enter the number of rows and columns of a given matrix and generate a matrix of random numbers (between 0 and 1). Divide each element of the matrix by 3 and show the result.

Solution:

The process has already been explained. The following code performs the task, and the output follows the code.

Code:

```
r=int(input('Enter the number of rows\t:'))
c=int(input('Enter the number of columns\t:'))
array1=np.random.random((r,c))
print(array1)
array2=array1/3
print('The resultant matrix\t:')
print(array2)
```

Output:

```
Enter the number of rows        : 4
Enter the number of columns     : 5
[[0.50390765 0.06109499 0.45480563 0.82350075 0.01273838]
 [0.09852436 0.67678827 0.41592833 0.71897566 0.29971024]
 [0.28596086 0.50947303 0.38423805 0.51570389 0.803762]
 [0.54649368 0.10232744 0.66355593 0.60196455 0.55477669]]
The resultant matrix:
[[0.16796922 0.020365   0.15160188 0.27450025 0.00424613]
 [0.03284145 0.22559609 0.13864278 0.23965855 0.09990341]
 [0.09532029 0.16982434 0.12807935 0.1719013  0.26792067]
 [0.18216456 0.03410915 0.22118531 0.20065485 0.18492556]]
```

16.9.4.2 Using the numpy.divide function

The division of a scalar to a matrix can also be performed using the **numpy. divide** function. The signature of the function is as follows.

numpy.divide(x1, x2, out, where, casting, order, dtype)

Where, **x1** and **x2** are arrays or scalars, and **out** is the location where the result is stored, and **where** is the condition which is broadcast over input.

The following illustration demonstrates the use of this function.

Illustration 16.31:

*Refer to Illustration 16.30. Perform the task using the **numpy.divide** function.*

Solution:

The process has already been explained. The following code performs the task.

Code:

```
r=int(input('Enter the number of rows\t:'))
c=int(input('Enter the number of columns\t:'))
array1=np.random.random((r,c))
print(array1)
array2=np.divide(array1,3)
print('The resultant matrix\t:')
print(array2)
```

16.9.5 Remainder

16.9.5.1 Using the % operator

Finding the remainder after dividing an array by a number then results in the creation of an array in which each element will be the remainder after the division of an element at that place with the given number.

For example, if

$$A = \begin{matrix} 2 & 4 & 3 \\ 5 & 6 & 8 \\ 10 & 7 & 14 \end{matrix}$$

Then, A % 2 becomes

$$\begin{matrix} 2 & 4 & 3 \\ 5 & 6 & 8 \\ 10 & 8 & 14 \end{matrix} \%2 =$$

$$\begin{matrix} 0 & 0 & 1 \\ 1 & 0 & 0 \\ 0 & 1 & 0 \end{matrix}$$

Illustration 16.32:

Ask the user to enter the number of rows and columns of a given matrix and generate a matrix of random numbers (integers between 10 and 100). Find modulo 3 of the matrix and show the result.

Solution:

The process has already been explained. The following code performs the task, and the output follows the code.

Code:

```
r=int(input('Enter the number of rows\t:'))
c=int(input('Enter the number of columns\t:'))
array1=np.random.randint(10,100,(r,c))
print(array1)
array2=array1%3
print('The resultant matrix\t:')
print(array2)
```

Output:

```
Enter the number of rows        : 4
'Enter the number of columns    : 5
[[29 35 51 52 74]
 [61 28 65 48 42]
 [20 55 78 92 34]
 [22 66 98 60 35]]

The resultant matrix :

[[2 2 0 1 2]
 [1 1 2 0 0]
 [2 1 0 2 1]
 [1 0 2 0 2]]
```

16.9.5.2 Using the numpy.remainder function

The above task can also be performed using the **numpy.remainder** function. The signature of the function is as follows.

numpy.remainder(x1, x2, out, where, casting, order, dtype)

Where, **x1** and **x2** are arrays or scalars, and **out** is the location where the result is stored, and **where** is the condition which is broadcast over input.

The following illustration demonstrates the use of this function.

Illustration 16.33:

*Refer to Illustration 16.32. Perform the task using the **numpy.remainder** function.*

Solution:

The process has already been explained. The following code performs the task.

```
r=int(input('Enter the number of rows\t:'))
c=int(input('Enter the number of columns\t:'))
array1=np.random.randint(10,100,(r,c))
print(array1)
array2=np.remainder(array1,3)
print('The resultant matrix\t:')
print(array2)
```

16.9.6 Power

16.9.6.1 Using the ** operator

An array to the power of scalar results in the creation of a matrix in which each element is raised to the power of that number.

For example, if

$$A = \begin{matrix} 2 & 4 & 3 \\ 5 & 6 & 8 \\ 0 & 1 & 4 \end{matrix}$$

Then, A ** 2 becomes

$$
\begin{array}{ccc}
2 & 4 & 3 \\
5 & 6 & 8 \\
0 & 1 & 4
\end{array} \; **2 =
$$

$$
\begin{array}{ccc}
4 & 16 & 9 \\
25 & 36 & 64 \\
0 & 1 & 16
\end{array}
$$

Illustration 16.34:

Ask the user to enter the number of rows and columns of a given matrix and generate a matrix of random numbers (integers between 10 and 100). Find the power 2 of the matrix and show the result.

Solution:

The process has already been explained. The following code performs the task, and the output follows the code.

Code:

```
r=int(input('Enter the number of rows\t:'))
c=int(input('Enter the number of columns\t:'))
array1=np.random.randint(10,100,(r,c))
print(array1)
array2=array1**2
print('The resultant matrix\t:')
print(array2)
```

Output:

```
Enter the number of rows:4
Enter the number of columns:5
[[30 48 22 32 16]
 [20 46 33 65 24]
 [61 40 12 80 95]
 [65 49 59 25 33]]
```

```
The resultant matrix :
[[ 900 2304 484 1024 256]
 [ 400 2116 1089 4225 576]
 [3721 1600 144 6400 9025]
 [4225 2401 3481 625 1089]]
```

16.9.6.2 Using the numpy.power function

The above task can also be performed using the **numpy.power** function. The signature of the function is as follows.

numpy.power(x1, x2, out, where, casting, order, dtype)

Where, **x1** and **x2** are arrays or scalars, and **out** is the location where the result is stored, and **where** is the condition which is broadcast over input.

The following illustration demonstrates the use of this function.

Illustration 16.35:

*Refer to Illustration 16.34. Perform the task using the **numpy.power** function.*

Solution:

The process has already been explained. The following code performs the task.

Code:

```
r=int(input('Enter the number of rows\t:'))
c=int(input('Enter the number of columns\t:'))
array1=np.random.randint(10,100,(r,c))
print(array1)
array2=np.power(array1,2)
print('The resultant matrix\t:')
print(array2)
```

16.10 OPERATIONS: ARRAY WITH AN ARRAY

This section discusses various operations on the **numpy** arrays. The standard operations namely: addition, subtraction, multiplication, division, modulo, and power have been presented in this section.

16.10.1 Addition

16.10.1.1 Using the + operator

Adding an array with another array results in the generation of an array in which each element is the sum of the corresponding elements of the two arrays. Note that two arrays can be added only if their shape/order is the same. For example, consider a 3×3 matrix:

$$A = \begin{matrix} 1 & 2 & 3 \\ 4 & 5 & 6 \\ 7 & 8 & 9 \end{matrix}$$

and another 3×3 matrix

$$B = \begin{matrix} 2 & 3 & 4 \\ 5 & 6 & 7 \\ 8 & 9 & 10 \end{matrix}$$

The addition of the matrices, A + B, results in

$$\begin{matrix} 1 & 2 & 3 \\ 4 & 5 & 6 \\ 7 & 8 & 9 \end{matrix}$$

$$+$$

$$\begin{matrix} 2 & 3 & 4 \\ 5 & 6 & 7 \\ 8 & 9 & 10 \end{matrix}$$

$$=$$

$$\begin{matrix} 3 & 5 & 7 \\ 9 & 11 & 13 \\ 15 & 17 & 19 \end{matrix}$$

Illustration 16.36:

*Ask the user to enter the number of rows and columns of a given matrix and generate a matrix (**array1**) of random numbers (between 0 and 1). Generate another matrix using the same method (**array2**). Add the two matrices and show the result.*

Solution:

The process has already been explained. The following code performs the task, and the output follows the code.

Code:

```
r=int(input('Enter the number of rows\t:'))
c=int(input('Enter the number of columns\t:'))
array1=np.random.random((r,c))
array2=np.random.random((r,c))
print('First \n',array1)
print('Second Array\n',array2)
array3=array1+array2
print('Resultant')
print(array3)
```

Output:

```
Enter the number of rows :3
Enter the number of columns :4
First
[[0.1135447 0.27965675 0.87464505 0.01639248]
 [0.81933991 0.40652706 0.84641946 0.20416069]
 [0.1686387 0.23038878 0.09283358 0.67776872]]
Second Array
[[0.32315667 0.344913 0.14198167 0.94796551]
 [0.42761314 0.33848426 0.35338747 0.25145462]
 [0.26085704 0.5791141 0.74523068 0.45423378]]
Resultant
[[0.43670137 0.62456975 1.01662671 0.96435799]
 [1.24695305 0.74501133 1.19980693 0.45561531]
 [0.42949574 0.80950288 0.83806426 1.1320025 ]]
```

16.10.1.2 Using the numpy.add function

The addition of two arrays can also be performed using the **numpy.add** function. The signature of the function is as follows:

numpy.add(x1, x2, out, where, casting, order, dtype)

Where, **x1** and **x2** are arrays or scalars, and **out** is the location where the result is stored, and **where** is the condition which is broadcast over input.

The following illustration demonstrates the use of this function.

Illustration 16.37:

*Refer to Illustration 16.36. Perform the task using the **numpy.add** function.*

Solution:

The process has already been explained. The following code performs the task.

Code:

```
r=int(input('Enter the number of rows\t:'))
c=int(input('Enter the number of columns\t:'))
array1=np.random.random((r,c))
array2=np.random.random((r,c))
print('First \n',array1)
print('Second Array\n',array2)
array3=np.add(array1,array2)
print('Resultant')
print(array3)
```

16.10.2 Subtraction

16.10.2.1 Using the — operator

Subtracting an array from another array results in the generation of an array in which each element is the difference of the corresponding elements of the two arrays. Note that the two arrays can be subtracted only if their shape/order is the same. For example, consider a 3 × 3 matrix:

$$A = \begin{matrix} 1 & 2 & 3 \\ 4 & 5 & 6 \\ 7 & 8 & 9 \end{matrix}$$

and another 3 × 3 matrix

$$B = \begin{matrix} 2 & 3 & 4 \\ 5 & 6 & 7 \\ 8 & 9 & 10 \end{matrix}$$

The subtraction of the matrices, A – B, results in

$$
\begin{array}{ccc}
1 & 2 & 3 \\
4 & 5 & 6 \\
7 & 8 & 9
\end{array}
$$

$$-$$

$$
\begin{array}{ccc}
2 & 3 & 4 \\
5 & 6 & 7 \\
8 & 9 & 10
\end{array}
$$

$$=$$

$$
\begin{array}{ccc}
-1 & -1 & -1 \\
-1 & -1 & -1 \\
-1 & -1 & -1
\end{array}
$$

Illustration 16.38

*Ask the user to enter the number of rows and columns of a given matrix and generate a matrix (**array1**) of random numbers (between 0 and 1). Generate another matrix using the same method (**array2**). Subtract the two matrices and show the result.*

Solution:

The process has already been explained. The following code performs the task, and the output follows the code.

Code:

```
r=int(input('Enter the number of rows\t:'))
c=int(input('Enter the number of columns\t:'))
array1=np.random.random((r,c))
array2=np.random.random((r,c))
print('First Array\n',array1)
print('Second Array\n',array2)
array3=array1-array2
print('Resultant')
print(array3)
```

Output:

```
Enter the number of rows        : 3
Enter the number of columns     : 4
First Array
[[0.39961261 0.3053622   0.69981721 0.48149646]
 [0.65316203 0.78056023 0.91684406 0.03676577]
 [0.79883245 0.41177364 0.0832075  0.78792757]]
Second Array
[[0.14347229 0.09637157 0.71285074 0.88376306]
 [0.34381271 0.79531302 0.54566008 0.27621104]
 [0.17137891 0.30418202 0.31075256 0.43768344]]
Resultant
[[0.25614032  0.20899063 -0.01303353 -0.4022666 ]
 [0.30934932 -0.01475279  0.37118398 -0.23944527]
 [0.62745354  0.10759161 -0.22754506  0.35024414]]
```

16.10.2.2 *Using the numpy.subtract function*

The subtraction of two matrices can also be performed using the **numpy. subtract** function. The signature of the function is as follows:

numpy.subtract(x1, x2, out, where, casting, order, dtype)

Where, **x1** and **x2** are arrays or scalars, and **out** is the location where the result is stored, and **where** is the condition which is broadcast over input.

The following illustration demonstrates the use of this function.

Illustration 16.39:

*Refer to Illustration 16.38. Perform the task using the **numpy.subtract** function.*

Solution:

The process has already been explained. The following code performs the task.

Code:

```
r=int(input('Enter the number of rows\t:'))
c=int(input('Enter the number of columns\t:'))
array1=np.random.random((r,c))
```

```
array2=np.random.random((r,c))
print('First Array\n',array1)
print('Second Array\n',array2)
array3=np.subtract(array1,array2)
print('Resultant')
print(array3)
```

16.10.3 Multiplication

16.10.3.1 Using the * operator

Multiplying two arrays results in the generation of an array in which each element of the resultant array is the product of the corresponding elements of the two arrays. Note that two arrays can be multiplied only if their shape/order is the same. Also, note that this is not the same as matrix multiplication. For example, consider a 3×3 matrix:

$$A = \begin{matrix} 2 & 4 & 3 \\ 5 & 6 & 8 \\ 10 & 7 & 14 \end{matrix}$$

and another 3×3 matrix

$$B = \begin{matrix} 2 & 2 & 2 \\ 4 & 3 & 4 \\ 3 & 2 & 5 \end{matrix}$$

The multiplication of the matrices, A * B, results in

$$\begin{matrix} 2 & 4 & 3 \\ 5 & 6 & 8 \\ 10 & 7 & 14 \end{matrix}$$

$$*$$

$$\begin{matrix} 2 & 2 & 2 \\ 4 & 3 & 4 \\ 3 & 2 & 5 \end{matrix}$$

$$=$$

$$\begin{matrix} 4 & 8 & 6 \\ 20 & 18 & 32 \\ 30 & 14 & 70 \end{matrix}$$

Illustration 16.40:

*Ask the user to enter the number of rows and columns of a given matrix and generate a matrix (**array1**) of random numbers (between 0 and 1). Generate another matrix using the same process (**array2**). Multiply the two matrices and show the result.*

Solution:

The process has already been explained. The following code performs the task, and the output follows the code.

Code:

```
r=int(input('Enter the number of rows\t:'))
c=int(input('Enter the number of columns\t:'))
array1=np.random.random((r,c))
array2=np.random.random((r,c))
print('First Array\n',array1)
print('Second Array\n',array2)
array3=array1*array2
print('Resultant')
print(array3)
```

Output:

```
Enter the number of rows        : 3
Enter the number of columns     : 4
First Array
[[0.25006456 0.8206732  0.43550708 0.23582437]
 [0.93062724 0.95464531 0.85118977 0.91312727]
 [0.29366056 0.32679254 0.59282643 0.60242423]]
Second Array
[[0.63569123 0.15618804 0.21294601 0.18053783]
 [0.18802424 0.75908975 0.64221508 0.2679345 ]
 [0.26360182 0.07651079 0.25618303 0.21757418]]
Resultant
[[0.15896385 0.12817934 0.0927395  0.04257522]
 [0.17498048 0.72466147 0.54664691 0.2446583 ]
 [0.07740946 0.02500316 0.15187207 0.13107196]]
```

16.10.3.2 Using the numpy.multiply function

The multiplication of two matrices can also be performed using the ***numpy.multiply*** function. The signature of the function is as follows:

numpy.multiply(x1, x2, out, where, casting, order, dtype)

Where, **x1** and **x2** are arrays or scalars, and **out** is the location where the result is stored, and **where** is the condition which is broadcast over input.

The following illustration demonstrates the use of this function.

Illustration 16.41:

*Refer to Illustration 16.40. Perform the task using the **numpy.multiply** function.*

Solution:

The process has already been explained. The following code performs the task.

Code:

```
r=int(input('Enter the number of rows\t:'))
c=int(input('Enter the number of columns\t:'))
array1=np.random.random((r,c))
array2=np.random.random((r,c))
print('First Array\n',array1)
print('Second Array\n',array2)
array3=np.multiply(array1,array2)
print('Resultant')
print(array3)
```

16.10.4 Division

16.10.4.1 Using the / operator

Dividing an array by another array results in the generation of an array in which each element is the result of the division of the corresponding elements of the two arrays. Note that two arrays can be divided only if their shape/order is the same. Also, note that this does not represent the mathematical division of two arrays. For example, consider a 3 × 3 matrix:

$$A = \begin{matrix} 2 & 4 & 3 \\ 5 & 6 & 8 \\ 10 & 7 & 14 \end{matrix}$$

and another 3 × 3 matrix

$$B = \begin{matrix} 2 & 2 & 2 \\ 4 & 3 & 4 \\ 3 & 2 & 5 \end{matrix}$$

The division of the matrices, A / B results in

$$
\begin{matrix} 2 & 4 & 3 \\ 5 & 6 & 8 \\ 10 & 7 & 14 \end{matrix}
$$

$$/$$

$$
\begin{matrix} 2 & 2 & 2 \\ 4 & 3 & 4 \\ 3 & 2 & 5 \end{matrix}
$$

$$=$$

$$
\begin{matrix} 1 & 2 & 1.5 \\ 1.25 & 2 & 2 \\ 3.33 & 3.5 & 2.8 \end{matrix}
$$

Illustration 16.42:

*Ask the user to enter the number of rows and columns of a given matrix and generate a matrix (**array1**) of random numbers (between 0 and 1). Generate another array using the same process (**array2**). Divide the matrices and show the result.*

Solution:

The process has already been explained. The following code performs the task, and the output follows the code.

Code:

```
r=int(input('Enter the number of rows\t:'))
c=int(input('Enter the number of columns\t:'))
```

```
array1=np.random.random((r,c))
array2=np.random.random((r,c))
print('First Array\n',array1)
print('Second Array\n',array2)
array3=array1/array2
print('Resultant')
print(array3)
```

Output:

```
Enter the number of rows :3
Enter the number of columns :4
First Array
[[0.10288708 0.05215324 0.781327 0.26069244]
 [0.73494996 0.65236699 0.73814042 0.18073924]
 [0.00529993 0.20723622 0.31061191 0.67826563]]
Second Array
[[0.07239933 0.48740549 0.52364227 0.9965622 ]
 [0.68518607 0.80388506 0.16278473 0.83231087]
 [0.45852829 0.96797794 0.39793673 0.71281227]]
Resultant
[[1.42110536 0.10700175 1.49210071 0.26159174]
 [1.07262829 0.81151775 4.5344575 0.21715352]
 [0.01155856 0.21409189 0.78055603 0.95153473]]
```

16.10.4.2 Using the numpy.divide function

The division of two matrices can also be performed using the **numpy.divide** function. The signature of the function is as follows:

numpy.divide(x1, x2, out, where, casting, order, dtype)

Where, **x1** and **x2** are arrays or scalars, and **out** is the location where the result is stored, and **where** is the condition which is broadcast over input.

The following illustration demonstrates the use of this function.

Illustration 16.43:

*Refer to Illustration 16.42. Perform the task using the **numpy.divide** function.*

Solution:

The process has already been explained. The following code performs the task.

```
r=int(input('Enter the number of rows\t:'))
c=int(input('Enter the number of columns\t:'))
array1=np.random.random((r,c))
array2=np.random.random((r,c))
print('First Array\n',array1)
print('Second Array\n',array2)
array3=np.divide(array1,array2)
print('Resultant')
print(array3)
```

16.10.5 Remainder

16.10.5.1 Using the % operator

Finding the remainder after dividing an array by another array results in the generation of an array in which each element is the remainder obtained by dividing the corresponding elements of the two arrays. Note that this operation can be performed only if their shape/order is the same. For example, consider a 3×3 matrix:

$$A = \begin{matrix} 2 & 4 & 3 \\ 5 & 6 & 8 \\ 10 & 7 & 14 \end{matrix}$$

and another 3×3 matrix

$$B = \begin{matrix} 2 & 2 & 2 \\ 4 & 3 & 4 \\ 3 & 2 & 5 \end{matrix}$$

The addition of the matrices, A + B, results in

$$\begin{matrix} 2 & 4 & 3 \\ 5 & 6 & 8 \\ 10 & 7 & 14 \end{matrix}$$

$$\%$$

$$
\begin{array}{ccc}
2 & 2 & 2 \\
4 & 3 & 4 \\
3 & 2 & 5 \\
& = & \\
0 & 0 & 1 \\
1 & 0 & 0 \\
1 & 1 & 4
\end{array}
$$

Illustration 16.44:

*Ask the user to enter the number of rows and columns of a given matrix and generate a matrix (**array1**) of random numbers (integers between 2 and 6). Generate another array in the same manner (**array2**). Find the remained obtained by dividing **array1** modulo **array2** and show the result.*

Solution:

The process has already been explained. The following code performs the task, and the output follows the code.

Code:

```
r=int(input('Enter the number of rows\t:'))
c=int(input('Enter the number of columns\t:'))
array1=np.random.randint(2,6,(r,c))
array2=np.random.randint(2,6,(r,c))
print('First Array\n',array1)
print('Second Array\n',array2)
array3=array1%array2
print('Resultant')
print(array3)
```

Output:

```
Enter the number of rows :3
Enter the number of columns :4
First Array
[[3 5 5 5]
 [5 3 5 2]
 [3 5 2 3]]
```

```
Second Array
[[5 2 2 4]
 [5 5 5 5]
 [5 3 2 4]]
Resultant
[[3 1 1 1]
 [0 3 0 2]
 [3 2 0 3]]
```

16.10.5.2 Using the numpy.mod function

The above task can also be performed using the **numpy.remainder** function. The signature of the function is as follows.

numpy.mod(x1, x2, out, where, casting, order, dtype)

Where, **x1** and **x2** are arrays or scalars, and **out** is the location where the result is stored, and **where** is the condition which is broadcast over input.

The following illustration demonstrates the use of this function.

Illustration 16.45:

*Refer to Illustration 16.44. Perform the task using the **numpy.remainder** function.*

Solution:

The process has already been explained. The following code performs the task.

Code:

```
r=int(input('Enter the number of rows\t:'))
c=int(input('Enter the number of columns\t:'))
array1=np.random.randint(2,6,(r,c))
array2=np.random.randint(2,6,(r,c))
print('First Array\n',array1)
print('Second Array\n',array2)
array3=np.mod(array1,array2)
print('Resultant')
print(array3)
```

16.10.6 Power

16.10.6.1 Using the ** operator

If two arrays are given, the first array to the power of another results in the generation of an array in which each element is **a** to the power of **b**, where **a** is an element of the first array, and **b** is the corresponding element of the second array. Note that this operation can be applied to the arrays only if their shape/order is the same. For example, consider a 3 × 3 matrix:

$$A = \begin{matrix} 2 & 4 & 3 \\ 5 & 6 & 8 \\ 0 & 1 & 4 \end{matrix}$$

and another 3 × 3 matrix

$$B = \begin{matrix} 3 & 1 & 2 \\ 2 & 2 & 1 \\ 4 & 2 & 4 \end{matrix}$$

The addition of the matrices, A + B, results in

$$\begin{matrix} 2 & 4 & 3 \\ 5 & 6 & 8 \\ 0 & 1 & 4 \end{matrix}$$

$$**$$

$$\begin{matrix} 3 & 1 & 2 \\ 2 & 2 & 1 \\ 4 & 2 & 4 \end{matrix}$$

$$=$$

$$\begin{matrix} 8 & 4 & 9 \\ 25 & 36 & 8 \\ 0 & 1 & 25 \\ & & 6 \end{matrix}$$

Illustration 16.46:

*Ask the user to enter the number of rows and columns of a given matrix and generate a matrix (**array1**) of random numbers (integers between 2 and 6).*

*Generate another array using the same process (**array2**). Find **array1** to the power of **array2**.*

Solution:

The process has already been explained. The following code performs the task, and the output follows the code.

Code:

```
r=int(input('Enter the number of rows\t:'))
c=int(input('Enter the number of columns\t:'))
array1=np.random.randint(2,6,(r,c))
array2=np.random.randint(2,6,(r,c))
print('First Array\n',array1)
print('Second Array\n',array2)
array3=array1**array2
print('Resultant')
print(array3)
```

Output:

```
Enter the number of rows    :2
Enter the number of columns :3
First Array
[[3 2 5]
 [4 5 3]]
Second Array
[[5 5 3]
 [5 4 5]]
Resultant
[[ 243 32 125]
 [1024 625 243]]
```

16.10.6.2 Using the numpy.power function

The above task can also be performed using the **numpy.power** function. The signature of the function is as follows:

numpy.power(x1, x2, out, where, casting, order, dtype)

Where, **x1** and **x2** are arrays or scalars, and **out** is the location where the result is stored, and **where** is the condition which is broadcast over input.

The following illustration demonstrates the use of this function.

Illustration 16.47:

*Refer to Illustration 16.46. Perform the task using the **numpy.power** function.*

Solution:

The process has already been explained. The following code performs the task.

Code:

```
r=int(input('Enter the number of rows\t:'))
c=int(input('Enter the number of columns\t:'))
array1=np.random.randint(2,6,(r,c))
array2=np.random.randint(2,6,(r,c))
print('First Array\n',array1)
print('Second Array\n',array2)
array3=np.power(array1,array2)
print('Resultant')
print(array3)
```

16.11 CONCLUSION

This chapter discussed the importance and features of **numpy** arrays. The function used to create **numpy** arrays have been discussed in detail. The chapter also discusses the concepts of indexing and slicing. The operations between a scalar and **numpy** array and operations between **numpy** arrays have also been discussed in detail.

This discussion continues in the next chapter, wherein the splitting and joining of arrays have been discussed. Along with this, the applications of **numpy** arrays in regression, finding the correlation, etc. have also been discussed in the next chapter. The reader may hit the exercises and float in the ocean of **numpy** to begin a journey toward becoming a data scientist.

EXERCISES

Multiple Choice Questions

1. Which of the following follows zero-based indexing?

    (a) List
    (b) Numpy array
    (c) Both
    (c) None of the above

2. In which of the following elements are stored at consecutive memory locations?

    (a) List
    (b) Numpy array
    (c) Both
    (c) None of the above

3. Which of the following support vectorized operations?

    (a) List
    (b) Numpy array
    (c) Both
    (d) None of the above

4. Which of the following is more efficient?

    (a) List
    (b) Numpy array
    (c) Both are equally efficient
    (d) None of the above

5. Which of the following are the storage mechanisms for numpy arrays?

    (a) Row major
    (b) Column major
    (c) Both
    (d) None of the above

6. In a 1-dimensional numpy array, which of the following can be the correct value of an axis?

    (a) 0
    (b) 1
    (c) 2
    (c) None of the above

7. If the value of one of the axes of a numpy array is 3, what is the minimum dimension of the array?

    (a) 4
    (b) 3
    (c) 2
    (d) 5

8. For a numpy matrix having four rows and five columns, which of the following is the correct value of shape?

    (a) 4×3 **(b)** 3×4

    (c) 12×1 **(d)** None of the above

9. In the above question, what is the rank of the array?

    (a) 2   **(b)** 3

    (c) 4   **(d)** None of the above

10. In question 8, what is the value of the axis for performing operations on rows?

    (a) 0   **(b)** 1

    (c) 2   **(c)** None of the above

11. In question 8, what is the value of the axis for performing operations on columns?

    (a) 0   **(b)** 1

    (c) 2   **(d)** None of the above

12. Which of the following is used to generate an array of zeros?

    (a) zeros   **(b)** ones

    (c) full   **(d)** None of the above

13. Which of the following is used to generate an array of ones?

    (a) zeros   **(b)** ones

    (c) full   **(d)** None of the above

14. Which of the following is used to generate an array filled with the same number?

    (a) zeros   **(b)** ones

    (c) full   **(c)** None of the above

15. Which of the following is used to generate random integers?

    (a) randint   **(b)** random.random

    (c) random.normal   **(d)** None of the above

16. Which of the following is used to generate a random sample from a normal distribution?

(a) randint

(b) random.random

(c) random.normal

(d) None of the above

17. The addition of scalar with a vector is

(a) vectorized

(b) not vectorized

(c) cannot say

(d) Data insufficient

18. Which of the following functions is used to find the remainder of a vector with a scalar?

(a) Remainder

(b) Mod

(c) Both

(d) None of the above

19. Which of the following functions is used to find the remainder of a vector with a vector?

(a) Remainder

(b) Mod

(c) Both

(d) None of the above

20. Which of the following packages can be used for finding Fourier Transform?

(a) Numpy

(b) Array

(c) Both

(c) None of the above

Theory

1. What is **numpy**? Explain its features?

2. Compare **numpy** arrays with lists.

3. Define an array.

4. By taking an example of a matrix, explain the following terms.

(a) Axes   (b) Shape   (c) Rank

5. How do you convert a List to an array? Explain the similarity between a list and a **numpy** array.

6. Discuss the storage of **numpy** arrays in memory.

7. Explain the functions for the following operations between a scalar and a matrix.

   (a) Addition **(b)** Subtraction

   (c) Multiplication **(d)** Division

   (e) Remainder **(f)** Power

8. Explain the functions for the following operations between a matrix and a matrix

   (a) Addition **(b)** Subtraction

   (c) Multiplication **(d)** Division

   (e) Remainder **(f)** Power

9. Explain the formation of a random array in **numpy**.

10. Explain the formation of different types of arrays in **numpy**.

17

NUMPY–II

Objectives

After reading this chapter, the reader should be able to

- Understand the methods to join arrays
- Understand various functions to split arrays
- Understand the concept of variance, covariance, and correlation

17.1 INTRODUCTION

The previous chapter introduced the **numpy** package and discussed the similarities and dissimilarities between **numpy** arrays and lists. The chapter also discussed various methods to generate different types of arrays in **numpy** and explained some of the most important operators.

This chapter takes the discussion forward and introduces methods to join and split arrays. Primarily, this chapter discusses three major topics: joining arrays, splitting arrays, and variance related tools.

The chapter has been organized as follows. The second section discusses the joining of arrays; the third section introduces methods to split the arrays; the fourth section discusses variance, the fifth discusses covariance, and the sixth discusses correlation and the last section concludes.

17.2 JOINING ARRAYS

This section discusses three methods used for joining **numpy** arrays. These are **hstack**, **vstack**, and **concatenate**.

17.2.1 hstack

This method concatenates two arrays horizontally and produces an array having the same number of rows. The number of columns, in the resultant array, is the sum of the number of columns of the two arrays.

Arguments: This function takes a tuple as the argument containing arrays having the same number of rows.

(array1, array2, array3, …)

The following examples demonstrate the use of **hstack**.

Illustration 17.1:

Generate two arrays containing 10 random numbers each. The numbers should be between 5 and 20. Now generate an array by horizontally stacking them.

Solution:

```
arr1=np.random.randint(5,20,10)
arr2=np.random.randint(5,20,10)
arr_result1=np.hstack((arr1, arr2))
print('Result is\n',arr_result1)
```

Output:

```
Result is [17 14 5 17 5 16 14 15 17 7 8 10 13 12 18 15 14 6 19 11]
Shape : (20,)
```

Illustration 17.2:

Ask the user to enter the number of rows and columns for creating two arrays. Generate the arrays having numbers between 5 and 20. Generate a final array by horizontally stacking them. (Note that in this example, the two input arrays have the same size)

Solution:

```
r=int(input('Enter the number of rows\t:'))
c=int(input('Enter the number of columns\t:'))
arr1=np.random.randint(5,20,(r,c))
arr2=np.random.randint(5,20,(r,c))
arr_result1=np.hstack((arr1, arr2))
```

```
print('Result is\n',arr_result1)
print('Shape\t:',arr_result1.shape)
```

Output:

```
Enter the number of rows :3
Enter the number of columns :4
Result is
[[19 11 10 12 8 14 10 17]
 [16 12 9 18 9 16 8 10]
 [15 15 5 6 7 15 18 11]]
Shape : (3, 8)
```

17.2.2 vstack

This method concatenates two arrays vertically and produces an array having the same number of columns. The number of rows, in the resultant array, is the sum of the number of rows of the two arrays.

Arguments: This function takes a tuple as the argument containing arrays having the same number of columns.

(array1, array2, array3, …)

The following examples demonstrate the use of **vstack**.

Illustration 17.3:

Generate two arrays containing 10 random numbers each. The numbers should be between 5 and 20. Now generate an array by vertically stacking them.

Solution:

```
arr1=np.random.randint(5,20,10)
arr2=np.random.randint(5,20,10)
arr_result2=np.vstack((arr1, arr2))
print('Result is\n',arr_result2)
print('Shape\t:',arr_result2.shape)
```

Output:

```
Result is
[[17 14 5 17 5 16 14 15 17 7]
 [ 8 10 13 12 18 15 14 6 19 11]]
Shape : (2,10)
```

Illustration 17.4:

Ask the user to enter the number of rows and columns of two input arrays. Now, generate two arrays, having numbers between 5 and 20. Generate an array by vertically stacking them.

Solution:

```
r=int(input('Enter the number of rows\t:'))
c=int(input('Enter the number of columns\t:'))
arr1=np.random.randint(5,20,(r,c))
arr2=np.random.randint(5,20,(r,c))
arr_result2=np.vstack((arr1, arr2))
print('Result is\n',arr_result2)
print('Shape\t:',arr_result2.shape)
```

Output:

```
Enter the number of rows :3
Enter the number of columns :4
Result is
[[19 11 10 12]
 [16 12 9 18]
 [15 15 5 6]
 [ 8 14 10 17]
 [ 9 16 8 10]
 [ 7 15 18 11]]
Shape : (6, 4)
```

17.2.3 Concatenate

This function joins the input arrays along the specified axis. The parameters of the function are as follows:

The tuple of arrays (<array1, arry2, …>): The sequence of arrays

axis: This parameter represents the axis along which the operations need to be performed. The default value of this parameter is 0. If the value of this parameter is NONE, then the input arrays are flattened and then concatenated.

out: This is an optional parameter, which tells the location of the resultant array. Note that if specified, this argument must be in the correct shape.

The following examples demonstrate the usage of this function.

Illustration 17.5:

Generate two random arrays of integers having 10 values between 5 and 25. Concatenate them along

(i) axis=0
(ii) axis=1
(iii) axis=None

and show the result.

Solution:

(i)

```
arr1=np.random.randint(5,20,10)
arr2=np.random.randint(5,20,10)
arr_result1=np.concatenate((arr1, arr2), axis=0)
print('Result is\n',arr_result1)
print('Shape\t:',arr_result1.shape)
```

Output:

```
Result is
 [17  5 19 16  8 11  7  6  8 18  5 19  5 11 11  9 12 14 19  9]
Shape : (20,)
```

(ii)

```
arr1=np.random.randint(5,20,10)
arr2=np.random.randint(5,20,10)
arr1=arr1.reshape(1,arr1.shape[0])
arr2=arr2.reshape(1,arr2.shape[0])
print(arr1.shape, ' ',arr2.shape)
arr_result1=np.concatenate((arr1.T, arr2.T), axis=1)
print('Result is\n',arr_result1)
print('Shape\t:',arr_result1.shape)
```

Output:

```
(1, 10)  (1, 10)
```

```
Result is
 [[10 14]
 [11 19]
 [15 10]
 [ 7  8]
 [12  5]
 [13 12]
 [ 6  9]
 [ 5  7]
 [17  9]
 [18  6]]
Shape : (10, 2)
```

(iii)

```
arr1=np.random.randint(5,20,10)
arr2=np.random.randint(5,20,10)
arr_result3=np.concatenate((arr1, arr2), axis=None)
print('Result is\n',arr_result3)
print('Shape\t:',arr_result3.shape)
```

Output:

```
Result is
 [15  7  5 13 13 15 11 17 13 16 11  8 17 16 11 16  9  5  8 14]
Shape : (20,)
```

Illustration 17.6:

Generate two random arrays of integers having r rows and c columns. The elements of the array must be between 5 and 25. Concatenate them along

(i) axis=0
(ii) axis=1
(iii) axis=None

and show the result.

Solution:

(i)

```
r=int(input('Enter the number of rows\t:'))
c=int(input('Enter the number of columns\t:'))
```

```
arr1=np.random.randint(5,20,(r,c))
arr2=np.random.randint(5,20,(r,c))
arr_result1=np.concatenate((arr1, arr2), axis=0)#Same as vstack
print('Result is\n',arr_result1)
print('Shape\t:',arr_result1.shape)
```

Output:

```
Enter the number of rows :3
Enter the number of columns :4
Result is
[[ 5  6  8  8]
 [ 8 10 17 19]
 [12 17 18  5]
 [14 19  7 17]
 [12  8 16  5]
 [16 13 18  9]]
Shape : (6, 4)
```

(ii)

```
r=int(input('Enter the number of rows\t:'))
c=int(input('Enter the number of columns\t:'))
arr1=np.random.randint(5,20,(r,c))
arr2=np.random.randint(5,20,(r,c))
arr_result1=np.concatenate((arr1, arr2), axis=1)#Same as hstack
print('Result is\n',arr_result1)
print('Shape\t:',arr_result1.shape)
```

Output:

```
Enter the number of rows    :3
Enter the number of columns :4
Result is
[[19  5 11  5 12 13 15 17]
 [13 12  6 18  7 17 16 15]
 [ 7  9 14 16 11 19  8  6]]
Shape : (3, 8)
```

(iii)

```
arr1=np.random.randint(5,20,10)
arr2=np.random.randint(5,20,10)
```

```
arr_result1=np.concatenate((arr1, arr2), axis=0)
print('Result is\n',arr_result1)
print('Shape\t:',arr_result1.shape)
```

Output:

```
Result is
 [17  5 19 16  8 11  7  6  8 18  5 19  5 11 11  9 12 14 19  9]
Shape : (20,)
```

17.3 SPLITTING ARRAYS

Having seen the methods used for joining the arrays, let us now move to splitting the arrays. The following subsections discuss **hsplit, vsplit, split**, and **extract** for finding subsets of the given array.

17.3.1 hsplit

The **numpy.hsplit** function divides the array horizontally in equal parts if the second argument is a number. In case the second argument is an array, the function works as shown in Figure 17.2.

Figure 17.1 shows the splitting of a given array into two equal parts and four equal parts. Note that if the number of columns is not divisible by the second argument, it results in an error.

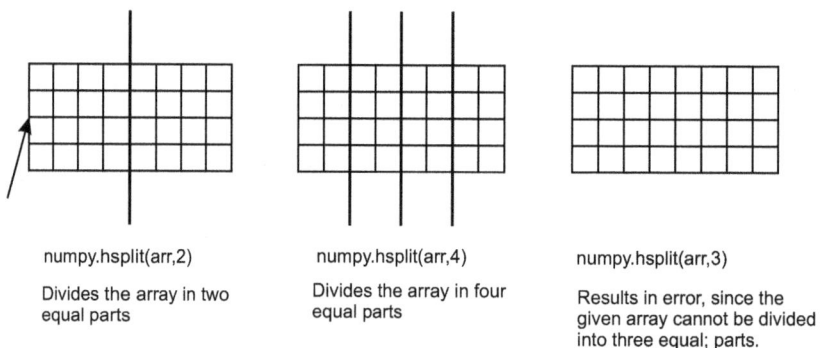

numpy.hsplit(arr,2)

Divides the array in two equal parts

numpy.hsplit(arr,4)

Divides the array in four equal parts

numpy.hsplit(arr,3)

Results in error, since the given array cannot be divided into three equal; parts.

FIGURE 17.1 hsplit(arr, <integer>).

Figure 17.2 shows how a given array is split if the second argument is an array. The number of elements in the second argument is one less than the number of splits. The elements of the second argument are the indices at which the given array is split.

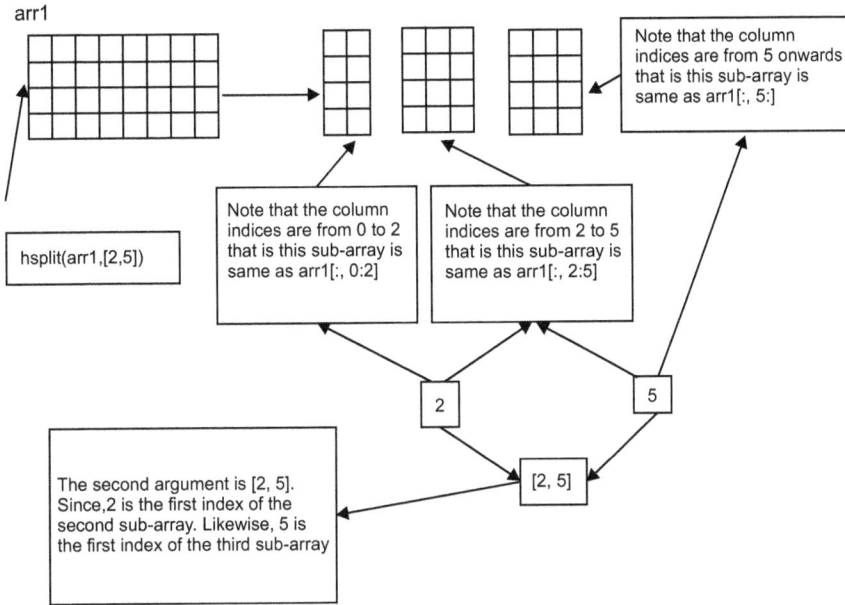

FIGURE 17.2 hsplit(arr, <array>).

17.3.2 vsplit

The **numpy.vsplit** function divides the array, in equal parts, vertically if the second argument is a number. In case the second argument is an array, the function works as shown in Figure 17.4.

Figure 17.3 shows the splitting of a given array into two equal parts and four equal parts. Note that if the number of rows is not divisible by the second argument, it results in an error.

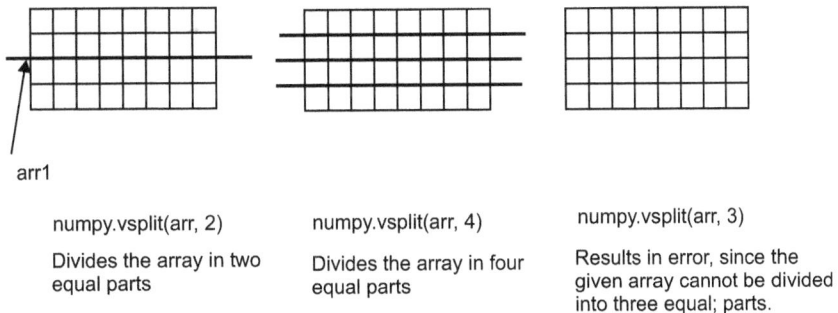

numpy.vsplit(arr, 2)

Divides the array in two equal parts

numpy.vsplit(arr, 4)

Divides the array in four equal parts

numpy.vsplit(arr, 3)

Results in error, since the given array cannot be divided into three equal; parts.

FIGURE 17.3 vsplit(arr, <integer>).

Figure 17.4 shows how a given array is split if the second argument, in **vsplit**, is an array. The number of elements in the second array is one less than the number of splits. The elements of the second argument are the indices at which the given array is split.

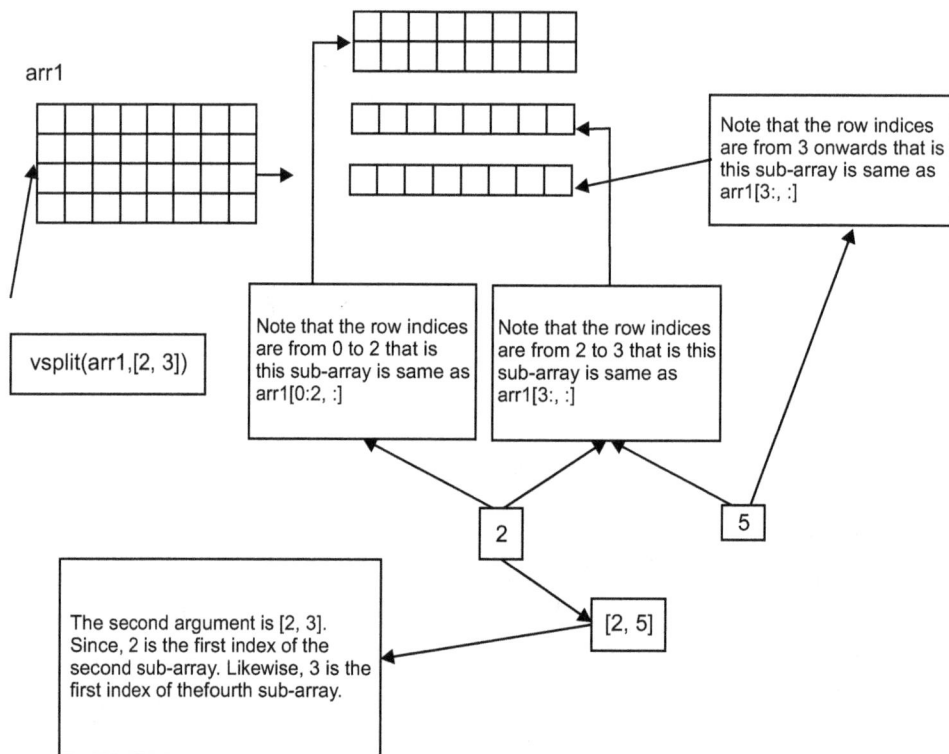

FIGURE 17.4 vsplit(arr, <array>).

The following examples demonstrate the use of the above functions for carrying out some assorted tasks.

Illustration 17.7:

Generate an array of random integers between 5 and 75 having a shape (4, 8). Generate two arrays by splitting the given array horizontally into two parts.

Solution:

The process has already been explaine(d) The code follows.

Code:

```
import numpy as np
arr1=np.random.randint(5,75,(4,8))
print(arr1)
sub1, sub2=np.hsplit(arr1,2)
print('First subarray:\n',sub1)
print('Second subarray:\n',sub2)
```

Output:

```
[[39 58 49 69 64 50 61 11]
 [74 72 22 23 11 7 8 47]
 [16 21 68 6 64 36 44 11]
 [31 5 27 61 51 8 11 38]]
First subarray:
[[39 58 49 69]
 [74 72 22 23]
 [16 21 68 6]
 [31 5 27 61]]
Second subarray:
[[64 50 61 11]
 [11 7 8 47]
 [64 36 44 11]
 [51 8 11 38]]
```

Illustration 17.8:

Generate an array of random integers between 5 and 75 having shape (4, 8). Generate two arrays by splitting the given array horizontally into four parts.

Solution:

The process has already been explained. The code follows.

Code:

```
arr1=np.random.randint(5,75,(4,8))
print(arr1)
sub1, sub2, sub3, sub4=np.hsplit(arr1,4)
print('First subarray:\n',sub1)
```

```
print('Second subarray:\n',sub2)
print('Third subarray:\n',sub3)
print('Fourth subarray:\n',sub4)
```

Output:

```
[[32 11 23 61 20 22 11 30]
 [68 62 11 68 14 69 58 50]
 [38 40 70 67 61 25 73 48]
 [54 51 54 45 49 19 68 9]]
First subarray:
[[32 11]
 [68 62]
 [38 40]
 [54 51]]
Second subarray:
[[23 61]
 [11 68]
 [70 67]
 [54 45]]
Third subarray:
[[20 22]
 [14 69]
 [61 25]
 [49 19]]
Fourth subarray:
[[11 30]
 [58 50]
 [73 48]
 [68 9]]
```

Illustration 17.9:

Generate an array of random integers between 5 and 75 having shape (4, 8). Report what happens when we try to generate arrays by splitting the given array horizontally into three parts.

Solution:

Note that this will result in an error as, the given array cannot be split into three parts horizontally. The code follows.

Code:

```
arr1=np.random.randint(5,75,(4,8))
#print(arr1)
sub1, sub2, sub3=np.hsplit(arr1,3)
print('First subarray:\n',sub1)
print('Second subarray:\n',sub2)
print('Third subarray:\n',sub3)
```

Output:

```
ValueError: array split does not result in an equal division
```

Illustration 17.10:

*Generate an array of random integers between 5 and 75 having shape (4, 8). Generate arrays by splitting the given array horizontally into subarrays using [2, 5] as the second argument in the **hsplit** function.*

Solution:

The process has already been explained. The code follows.

Code:

```
arr1=np.random.randint(5,75,(4,8))
print(arr1)
sub1, sub2, sub3=np.hsplit(arr1,[2,5])
print('First subarray:\n',sub1)
print('Second subarray:\n',sub2)
print('Third subarray:\n',sub3)
```

Output:

```
[[ 6 35 24 43 14 68 62 59]
 [23 50 58 11 12 24 69 19]
 [53 25 10 10 41 19 23 60]
 [14 7 74 36 74 39 54 34]]
```

```
First subarray:
[[ 6 35]
 [23 50]
 [53 25]
 [14  7]]
Second sub-array:
[[24 43 14]
 [58 11 12]
 [10 10 41]
 [74 36 74]]
Third subarray:
[[68 62 59]
 [24 69 19]
 [19 23 60]
 [39 54 34]]
```

Illustration 17.11:

Generate an array of random integers between 5 and 75 having shape (4, 8). Generate two arrays by splitting the given array vertically into two parts.

Solution:

The process has already been explaine(d) The code follows.

```
arr1=np.random.randint(5,75,(4,8))
print(arr1)
sub1, sub2=np.vsplit(arr1,2)
print('First subarray:\n',sub1)
print('Second subarray:\n',sub2)
```

Output:

```
[[10 40 18 64 66 11 52 53]
 [71 55 40 10 44 40 14 49]
 [24 53 57 68 57 13 5 22]
 [38 9 61 17 34 62 61 38]]
First subarray:
[[10 40 18 64 66 11 52 53]
 [71 55 40 10 44 40 14 49]]
```

```
Second subarray:
[[24 53 57 68 57 13 5 22]
 [38 9 61 17 34 62 61 38]]
```

Illustration 17.12:

Generate an array of random integers between 5 and 75 having shape (4, 8). Generate two arrays by splitting the given array vertically into four parts.

Solution:

The process has already been explained. The code follows.

```
arr1=np.random.randint(5,75,(4,8))
print(arr1)
sub1, sub2, sub3, sub4=np.vsplit(arr1,4)
print('First subarray:\n',sub1)
print('Second subarray:\n',sub2)
print('Third subarray:\n',sub3)
print('Fourth subarray:\n',sub4)
```

Output:

```
[[42 16 61 58 40 71 35  8]
 [13  7 73 32 38 52 43  6]
 [73 72 16 68  7 21 26 16]
 [43 32  8 69 23 15 71 55]]
First subarray:
[[42 16 61 58 40 71 35  8]]
Second subarray:
[[13  7 73 32 38 52 43  6]]
Third subarray:
[[73 72 16 68  7 21 26 16]]
Fourth subarray:
[[43 32  8 69 23 15 71 55]]
```

Illustration 17.13:

Generate an array of random integers between 5 and 75 having shape (4, 8). Report what happens when we try to generate arrays by splitting the given array vertically into three parts.

Solution:

Note that this will result in an error as, the given array cannot be split into three parts horizontally. The code follows.

Code:

```
arr1=np.random.randint(5,75,(4,8))
#print(arr1)
sub1, sub2, sub3=np.vsplit(arr1,3)
print('First subarray:\n',sub1)
print('Second subarray:\n',sub2)
print('Third subarray:\n',sub3)
```

Output:

```
valueError: array split does not result in an equal division
```

Illustration 17.14:

*Generate an array of random integers between 5 and 75 having shape (4, 8). Generate arrays by splitting the given array horizontally into subarrays using [2, 5] as the second argument in the **vsplit** function.*

Solution:

The process has already been explained. The code follows.

Code:

```
arr1=np.random.randint(5,75,(4,8))
print(arr1)
sub1, sub2, sub3=np.vsplit(arr1,[2,3])
print('First subarray:\n',sub1)
print('Second subarray:\n',sub2)
print('Third subarray:\n',sub3)
```

Output:

```
[[34 12 8 55 55 12 69 44]
 [ 8 21 9 30 70 9 37 25]
 [ 8 60 63 72 66 36 59 54]
 [30 45 20 13 22 36 21 45]]
```

```
First subarray:
[[34 12 8 55 55 12 69 44]
 [ 8 21 9 30 70 9 37 25]]
Second subarray:
[[ 8 60 63 72 66 36 59 54]]
Third subarray: [[30 45 20 13 22 36 21 45]]
```

17.3.3 Split

The above tasks can also be accomplished using the **split** function. This function takes the array, the number of divisions (or the array indicating divisions), and the **axis** as the arguments. The function works in almost the same manner as the above two functions. The following illustrations demonstrate the use of this function.

Illustration 17.15:

*Refer to Illustration 17.7 Accomplish the task using the **split** function.*

Solution:

```
arr1=np.random.randint(5,75,(4,8))
print(arr1)
sub1, sub2=np.split(arr1,2, axis=1)
print('First subarray:\n',sub1)
print('Second subarray:\n',sub2)
```

Illustration 17.16:

*Refer to Illustration 17.8 Accomplish the task using the **split** function.*

Solution:

```
arr1=np.random.randint(5,75,(4,8))
print(arr1)
sub1, sub2, sub3, sub4=np.split(arr1,4, axis=1)
print('First subarray:\n',sub1)
print('Second subarray:\n',sub2)
print('Third subarray:\n',sub3)
print('Fourth subarray:\n',sub4)
```

Illustration 17.17:

*Refer to Illustration 17.10. Accomplish the task using the **split** function.*

Solution:

```
arr1=np.random.randint(5,75,(4,8))
print(arr1)
sub1, sub2, sub3=np.split(arr1,[2,5], axis=1)
print('First subarray:\n',sub1)
print('Second subarray:\n',sub2)
print('Third subarray:\n',sub3)
```

Illustration 17.18:

*Refer to Illustration 17.11 Accomplish the task using the **split** function.*

Solution:

```
arr1=np.random.randint(5,75,(4,8))
print(arr1)
sub1, sub2=np.split(arr1,2, axis=0)
print('First subarray:\n',sub1)
print('Second subarray:\n',sub2)
```

Illustration 17.19:

*Refer to Illustration 17.12 Accomplish the task using the **split** function.*

Solution:

```
arr1=np.random.randint(5,75,(4,8))
print(arr1)
sub1, sub2, sub3, sub4=np.split(arr1,4, axis=0)
print('First subarray:\n',sub1)
print('Second subarray:\n',sub2)
print('Third subarray:\n',sub3)
print('Fourth subarray:\n',sub4)
```

Illustration 17.20:

*Refer to Illustration 17.14. Accomplish the task using the **split** function.*

Solution:

```
arr1=np.random.randint(5,75,(4,8))
print(arr1)
sub1, sub2, sub3=np.split(arr1,[2,3], axis=0)
print('First subarray:\n',sub1)
print('Second subarray:\n',sub2)
print('Third subarray:\n',sub3)
```

17.3.4 Extract

This function extracts the elements of the given array based on some condition. It takes two arguments:

- the condition and
- the array.

The signature of the function is as follows:

numpy.extract(condition, arr)

The condition can be a Boolean array. In this case, the locations at which **True** exist in this array (condition) would be used to extract elements from the array passed as the second argument. In the other case, the nonzero elements from this array are used for extracting elements from the second array. The following illustrations give an insight into the working of this function.

Illustration 17.21:

Generate a 2-D array of random integers having 7 rows and 9 columns. The integers in the array should be in the range of 0 to 255.

Solution:

Code:

```
import numpy as np
arr1=np.random.randint(0,255,(7,9))
print(arr1)
```

Output:

```
[[116 218 124  57 202 210 138 102   5]
 [ 21 102  18  29 241  81  10 190 177]
 [114 205 215 124 141 163  19 141   3]
 [ 38 205  22 234 194  91 238 101 127]
 [ 75  41 108   7 211   4 203 251 231]
 [ 28  50  55 253 122  84  97   5  46]
 [157 171 221 184 227 241 167 167 154]]
```

Illustration 17.22:

Now extract the elements which are divisible by 7, from the array generated in the above illustration.

Solution:

Code:

```
condition1=np.mod(arr1,7)==0
arr2=np.extract(condition1, arr1)
print(arr2)
```

Output:

```
[210  21  91 238   7 203 231  28  84 154]
```

Note that **condition1** is nothing more than an array containing **True** (at the locations where elements are divisible by 7) and **False** (at the locations where elements are not divisible by 7).

print(condition1)

```
[[False False False False False  True False False False]
 [ True False False False False False False False False]
 [False False False False False False False False False]
 [False False False False False  True  True False False]
 [False False False  True False False  True False  True]
 [ True False False False False  True False False False]
 [False False False False False False False False  True]]
```

TRY: *Execute arr1[arr2] and report the result*

The **compress** function can be used along with the **extract** function for applying a condition to the rows or columns of a 2-D array. The following illustration gives an overview of the mechanism to extract relevant subset from a given array. The illustration aims to select every third row and every third column from a given 2-D array. To accomplish the task, an array containing numbers from 0 to the number of rows is generated. This array is subjected to condition **<array> mod 3 ==0 (r_c)**. Likewise, an array containing numbers from 0 to the number of columns are generated. This array is subjected to condition **<array> mod 3 ==0 (c_c)**. The compress function is then used to select the required rows, followed by the required columns.

Illustration 17.23:

Select every third row and every third column from a given 2-D array.

Solution:

The process has already been explained, the code is as follows.

```
r=np.arange(arr1.shape[0])
c=np.arange(arr1.shape[1])
r_c=np.mod(r,3)==0
c_c=np.mod(c,3)==0
print(r_c)
print(c_c)
arr3=np.compress(r_c,arr1, axis=0)
print(arr3)
arr4=np.compress(c_c, arr1, axis=1)
print(arr4)
```

Output:

```
[ True False False  True False False  True]
[ True False False  True False False  True False False]
[[116 218 124  57 202 210 138 102   5]
 [ 38 205  22 234 194  91 238 101 127]
 [157 171 221 184 227 241 167 167 154]]
[[116  57 138]
 [ 21  29  10]
 [114 124  19]
```

```
[ 38 234 238]
[ 75   7 203]
[ 28 253  97]
[157 184 167]]
```

Illustration 17.24:

*Accomplish the task of Illustration 17.23 without using the **compress** function.*

Solution:

```
r=np.arange(arr1.shape[0])
c=np.arange(arr1.shape[1])
r_cond=np.mod(r,3)==0
c_cond=np.mod(c,3)==0
r_c1=np.arange(arr1.shape[0])
r_c=r_c1[r_cond]
arr3=arr1[r_c, :]
c_c1=np.arange(arr1.shape[1])
c_c=c_c1[c_cond]
arr3=arr3[:, c_c]
print(arr3)
```

Output:

```
[[116  57 138]
 [ 38 234 238]
 [157 184 167]]
```

17.4 VARIANCE

The meaning of the word "variance," as per Oxford Dictionary, is "The fact or quality of being different, divergent, or inconsistent." In statistics, variance is a method to calculate the spread of a set of numbers. So, if a set has low variance, it is concentrated. If the given set of numbers has a large variance, it is scattered to a larger extend.

Mathematically, variance denotes the difference of a random variable from its expected value. It can be found by taking the average of the squares of the differences between the elements of a set and the mean value. The standard deviation is the square root of the variance.

For a set $X = \{x_i, x_2, \ldots, x_n\}$

$$\text{Standard Deviation} = \sqrt{Variance} = \sqrt{\frac{\sum_{i=1}^{n}(x_i - \bar{x})^2}{n}}$$

Where,

$$\bar{x} = \frac{x_i + x_2 + \ldots, + x_n}{n}$$

The variance can be calculated as follows. First of all, the mean of the given set of numbers is found using **numpy.mean**. This is followed by finding the square of deviations of the elements from the mean and then dividing the result by the number of elements. Finally, the square root of the result so obtained is taken. The code follows.

```
x_mean=np.mean(X)
sq_deviations=(X-x_mean)**2
print(sq_deviations)
sum_sq_dev=np.sum(sq_deviations)
print(sum_sq_dev)
av1=sum_sq_dev/len(X)
var=np.sqrt(av1)
print(av1)
```

The variance of a given set can also be found using the ***numpy.var*** function. The following code finds the variance of X using this function.

Code:

```
var1=np.var(X)
print(var1)
```

17.5 COVARIANCE

The covariance of two sets is a tool that helps us to compare two arrays. The covariance gives an idea about how close the datasets are. If the value of covariance is a large positive number, it indicates that the two variables are highly correlated and the increase in one results in an increase in others. In case of a high negative correlation, the two variables are highly correlated and the increase in one results in a decrease in others. Zero correlation indicates the variables are not related. It can be calculated as follows.

For a set $X = \{x_i, x_2,...,x_n\}$ and $Y = \{y_i, y_2,...,y_n\}$

$$Cov(X,Y) = \frac{\Sigma_{i=1}^{n}(x_i - \overline{x})(y_i - \overline{y})}{n-1}$$

Where,

$$\overline{x} = \frac{x_i + x_2 +...,+x_n}{n} \text{ and } \overline{y} = \frac{y_i + y_2 +...,+y_n}{n}$$

Covariance can be found by using the following code.

Code:

```
X=np.random.randint(5,90,20)
Y=np.random.randint(5,100,20)
dev_x=X-np.mean(X)
dev_y=Y-np.mean(Y)
cov1=(np.dot(dev_x, dev_y))/20
print(cov1)
```

The **cov** function of the **numpy** package takes arguments and generates an array of covariance. That is,

$$Cov(X_1, X_2) = \frac{cov(X_1, X_1) \quad cov(X_1, X_2)}{cov(X_2, X_1) \quad cov(X_1, X_2)}$$

Where, $cov(X_1, X_1)$ is the covariance between X_1 and X_1 and $cov(X_1, X_2)$ is the covariance between X_1 and X_1. For example, for the following code

Code:

```
X=np.random.randint(5,90,20)
Y=np.random.randint(5,100,20)
print(X)
print(Y)
cov2=np.cov(X,Y)
print(cov2)
```

Output:

```
[[562.40789474 -66.98684211]
 [-66.98684211 628.02894737]] .
```

17.6 CORRELATION

Correlation is another measure of dependency on two variables. This measure gives the direction of the relationship between the variables. As per the Oxford dictionary, it is "a mutual relationship or connection between two or more things." The covariance between two sets X and Y is given by,

$$Correlation\ (X,Y) = \frac{Covariance(X,Y)}{\sigma(X) \times \sigma(X)}$$

where $Ps(X)$ is the standard deviation of X and $s(Y)$ is the standard deviation of Y.

The following code calculates the correlation between X and Y.

Code:

```
X=np.random.randint(5,90,20)
Y=np.random.randint(5,100,20)
dev_x=X-np.mean(X)
dev_y=Y-np.mean(Y)
corr=(np.dot(dev_x, dev_y))/(20*np.std(X)*np.std(Y))
print(corr)
```

The correlation coefficient can also be calculated using the **numpy.corrcoef(X, Y)**. This function calculates the coefficient of correlation between X and Y. The following code demonstrates the use of **numpy.corrcoef(X, Y)**.

Code:

```
corr2=np.corrcoef(X,Y)
print(corr2)
```

Output:

```
[[1. 0.03415175]
 [0.03415175 1. ]]
```

17.7 CONCLUSION

This chapter discussed some of the most important topics in **NumPy**. The joining and splitting of arrays would be used if you are using Python in any domain. Moreover, the concepts of variance, covariance, correlation, and regression would greatly help you in analyzing data and even in Data Science. You are advised to attempt the exercises given at the end of this chapter to get a better hold of the **numpy** package.

EXERCISES

Multiple Choice Questions

1. Which of the following is used for splitting the array horizontally?

   (a) hsplit **(b)** horzsplit

   (c) vsplit **(d)** None of the above

2. Which of the following are valid arguments in the hsplit function?

   (a) The array

   (b) The number of splits or the array containing the position of the splits

   (c) Both of the above

   (d) None of the above

3. What happens if the number of columns is not divisible by the second argument in the hsplit function

   (a) Rounding **(b)** Ceiling

   (c) Error is generated **(d)** None of the above

4. Which of the following is used for splitting the array vertically?

   (a) vsplit **(b)** vertsplit

   (c) vsplit **(d)** None of the above

5. Which of the following are valid arguments in the vsplit function?

   (a) The array

   (b) The number of splits or the array containing the position of the splits

   (c) Both of the above

   (d) None of the above

6. What happens if the number of columns is not divisible by the second argument in the vsplit function

    (a) Rounding   **(b)** Ceiling

    (c) Error is generated   **(d)** None of the above

7. Which function can do both vertical and horizontal splitting?

    (a) split   **(b)** join

    (c) Both   **(d)** None of the above

8. Which of the function stacks two arrays horizontally?

    (a) hstack   **(b)** vstack

    (c) Both   **(d)** None of the above

9. Which of the function stacks two arrays vertically?

    (a) hstack   **(b)** vstack

    (c) Both   **(d)** None of the above

10. Which of the following functions can perform both horizontal and vertical stacking?

    (a) concatenate   **(b)** stack

    (c) Both   **(d)** None of the above

11. Which of the following is represented by variance?

    (a) spread   **(b)** central tendency

    (c) Both   **(d)** None of the above

12. Which of the following functions are used for finding the variance of a given array?

    (a) var   **(b)** variance

    (c) Both   **(d)** None of the above

13. Which of the following is a measure of the relation between two variables?

    (a) Correlation   **(b)** Covariance

    (c) Both   **(d)** None of the above

14. Which of the following functions finds the coefficient of correlation between two variables?

    (a) corrcoeff   (b) corrcoef

    (c) Both   (d) None of the above

15. What should be the value of covariance if there is no relation between two variables?

    (a) 0   (b) Positive

    (c) Negative   (d) None of the above

Theory

1. Explain the purpose of following methods and discuss their main arguments

    (a) split   (b) vsplit

    (c) hsplit   (d) concatenate

    (e) hstack   (f) vstack

    (g) var   (h) covar

    (i) corrcoef

2. Explain various methods to join arrays?

3. Explain various methods to split arrays?

4. What is meant by variance? Explain the formula for finding variance.

5. What is covariance? Explain the formula for finding covariance.

6. What is the importance of correlation? Explain the formula for finding correlation.

18

Data Visualization-I

Objectives

After reading this chapter, the reader should be able to

- Understand the importance of **Visualization**
- Appreciate the features of **Matplotlib**
- Understand the parameters of the **plot** function
- Plot lines and scatter diagrams
- Plot bar charts

18.1 INTRODUCTION

The advent of **IoT** (Internet of Things) and related technologies have facilitated the collection of huge amounts of data. However, this data can help us to make informed decisions only if we can analyze it and extract information from it. This analysis can be based on the statistics and trends that originate from this data. Another way of analyzing the given data is by using **Visualization** that is by drawing plots, pi-charts, histograms, etc.

> **Visualization**
>
> It refers to a way of conveying intangible and tangible ideas using visual imagery. It may refer to the graphical representation of information using tools like scatter plots, line plots, bars, histograms, frequency graphs, etc.

Visualization helps us to understand the given data. It is important as it provides an insight into the results and may help uncover the underlying patterns. Moreover, the growing importance of data analytics and Machine

Learning has exponentially enhanced the value of **Visualization**. Figure 18.1 shows the advantages of Visualization.

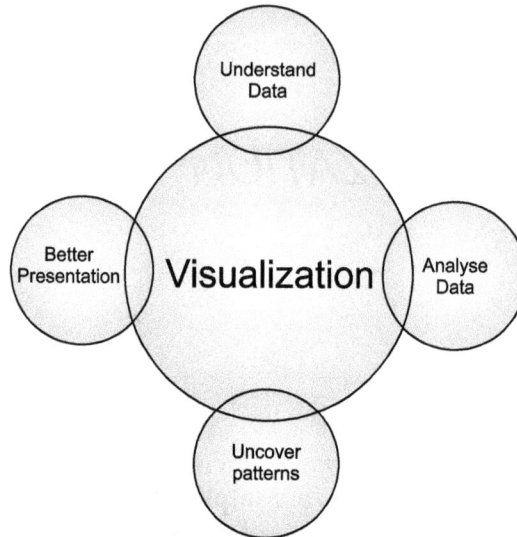

FIGURE 18.1 Visualization.

This chapter discusses the **pyplot** module of the **matplotlib** package for plotting lines, scatter diagrams, and bar charts.

Matplotlib is a package that helps us to plot various types of graphs and visualize the data. It is a plotting library. Though, initially, it was meant for 2D plotting; now, it also provides support for 3D plotting. Using **Matplotlib** one can primarily generate:

- Plots
- Histograms
- Bar graphs
- Scatter Plots
- Frequency Plots etc.

The **Seaborn** package was built on the top of Matplotlib and is used to create more attractive and informative statistical graphics to visualize univariate and bivariate data. It uses beautiful themes for decorating Matplotlib graphics as well as it is more comfortable in handling the Pandas data frames.

Though it is also used for many other kinds of plots we will focus on the stated types, in this chapter and the next one. The reader is required to take note of the methods and procedures to deal with the data discussed in the **NumPy** module.

*The **pyplot** module of **matplotlib** helps one to create graphs with various fonts, axes properties, etc. It provides us with a set of functions that help programmers to perform various tasks associated with plotting.*

Here, it may be stated that when you install the **Anaconda Navigator, matplotlib** is shipped with it. To check if it has been installed, you can go to the **Environment** and check the installed packages. In case you have not installed Anaconda, refer to *https://matplotlib.org/3.1.1/users/installing.html*

As per the official site: "Matplotlib and its dependencies are available as wheel packages for macOS, Windows and Linux distributions:"

Run the following commands in the command shell to complete the installation:

- python –m pip installs –U pip
- python –m pip install –U matplotlib.

This chapter presents some interesting illustrations. The chapter has been organized as follows. Section 18.2 introduces the **plot** function, Section 18.3 discusses the plotting of lines and curves, and Section 18.4 presents a brief discussion on some additional arguments of the **plot** function. Section 18.5 presents an overview of the bar charts, and the last section concludes.

18.2 THE PLOT FUNCTION

We start our discussion by plotting the values of a list. A plot can be instantiated using the **plot** function of the **pyplot** package. To generate a basic plot, a list **L**, having **n** values (index: 0 to (n-1)), can be passed as an argument to the **plot** function. In the plot so generated, the **X**-axis would have values from 0 to (n-1) and **Y**-axis would have the values from the given list. For example, if

$$L = [3, 5, 9, 6, 8]$$

is passed to the **plot** function. Then, **X**-axis will have values from 0 to 4 (note that there are five elements in the list, having indices from 0 to 4). The corresponding values in the **Y**-axis will be 3, 5, 9, 6, and 8. That is, (0, 3), (1, 5), (2, 9), (3, 6), and (4,8) are shown in the graph (Figure 18.2).

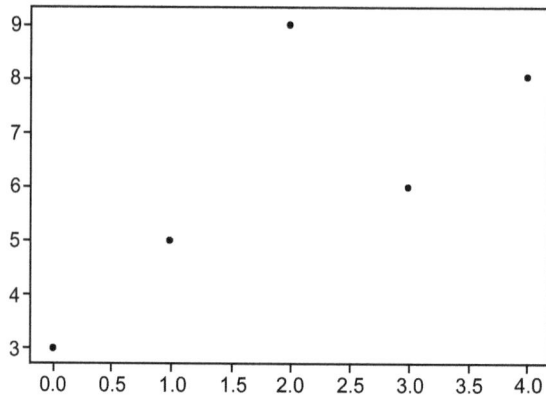

FIGURE 18.2 Plotting a list.

Now, let us move to the parameters of the **plot** function. The discussion that follows introduces parameters namely **xlabel**, **ylabel**, **axis**, **xlim**, **ylim**, **xticks**, and **yticks**.

18.2.1 xlabel

The **xlabel** attribute associates a label with the **X**-axis. The value of **xlabel** is a string.

18.2.2 ylabel

The ylable attribute associates a label with the **Y**-axis. The value of **ylabel** is a string.

18.2.3 axis

The limits of the **X** and **Y** axis can be changed using the **axis** function, which takes a list as an argument. The list passed as an argument has the following arguments

- x_{min}
- y_{min}
- x_{max}
- y_{max}

indicating the minimum value of the X-axis, that of the Y-axis, the maximum value of the X-axis, and the maximum value of the Y-axis.

18.2.4 xlim, ylim

The **pyplot** also provides the **xlim** and **ylim** arguments for setting the limits of the **X** and the **Y**-axis respectively.

18.2.5 xticks, yticks

The ticks on the **X** and the **Y**-axis can be set by using the **xticks** and **yticks** functions. These functions take a list containing the values to be displayed on the axes.

Having seen the parameters of the **plot** function, let us now move to some of the useful methods associated with the **plot** function. The following discussion introduces the **show** and the **savefig** methods associated with the **plot** module.

18.2.6 show

The **show** method displays the figure.

18.2.7 savefig

One can save the figure formed by the **plot** function using the **savefig** function. The **savefig** function takes two arguments: the path to the figure which is being saved and the optional **dpi**.

The next section uses these objects to plot various types of lines and curves.

18.3 PLOTTING LINES AND CURVES

This section presents some of the ways to plot lines, curves, and scatter diagrams. Note that, in a line plot, the points, passed as an argument to the **plot** functions, are joined by straight lines. In the case of a scatter diagram, these points are plotted but are not joined by lines. The **plot** function can be used to plot both line plots and scatter plots. By default, the function displays a line plot. This section discusses the **plot(X)**, **plot(X, Y)**, **plot(<2D array>)** and **scatter diagrams**.

18.3.1 Plot(X)

The **plot** function may take a list (or an array) as its argument and plot the values in the list. To understand this, consider the following example, in which a list **L=[1, 4, 8, 10]** is passed to the **plot** function. The **xlabel** is set to the string **X-Axis** and the **ylabel** is set to **Y-Axis**. The plotted figure would be saved as **line.png**. Figure 18.3 shows the output of the program.

Code:

```
import matplotlib.pyplot as plt
plt.plot([1,4,8,10])
plt.xlabel("X Axis")
plt.ylabel("Y Axis")
plt.show()
plt.savefig("line.png",dpi=80)
```

Output:

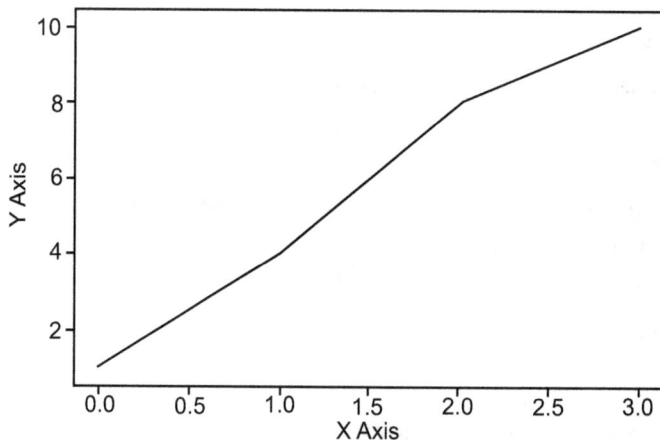

FIGURE 18.3 If a list is passed to the **plot** function, the values are plotted against the **indices** of the list. Note that the values of **xlabel** and **ylabel** are set to **X-Axis** and **Y-Axis,** respectively.

18.3.2 Plot(X, Y)

In the above example, the **plot** function takes one argument. However, it can also take two arguments indicating the values of both **X** and **Y** coordinates. In such cases, the tuple formed by taking an element from **X**, and the corresponding element from **Y** is plotted. The following example plots $y = 2x^2 - 3$ (Figure 18.4) by taking two input arguments. Note that **X** has integers from -5 to 5 and for each value of x in **X**, an element of **Y** is calculated as per the given formula.

Code:

```
X=[-5,-4,-3,-2,-1,0,1,2,3,4,5]
Y= [2*x*x-3 for x in X]
plt.plot(X,Y)
plt.xlabel("X Axis")
plt.ylabel("Y Axis")
plt.show()
plt.savefig("line.png",dpi=80)
```

Output:

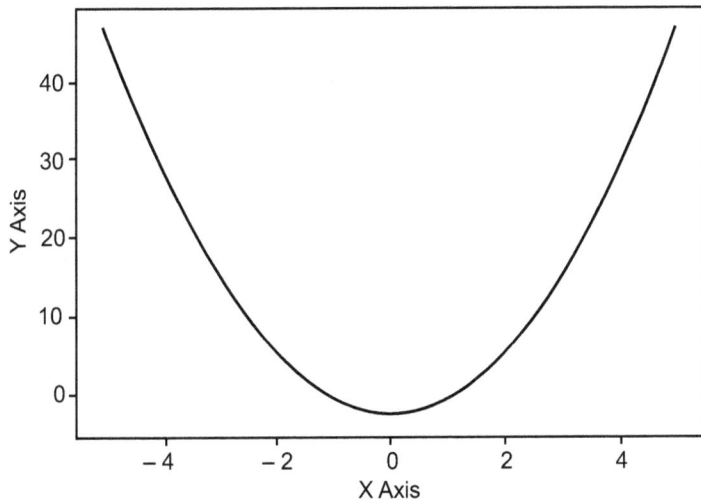

FIGURE 18.4 The **plot** function can also take two arguments; the second argument's values can be generated using generators or comprehensions.

18.3.3 Plot(<2D Array>)

One can even pass a two-dimensional array (or list of lists) in the plot, in which case the first element of each row (or list) would be plotted as a separate plot, the second element as a separate plot, and so on.

In the following program, the Y coordinates of the first line are [2, 4, 6, 8, 9, 10, 11]; the Y coordinates of the second line are [3, 6, 9, 10, 11, 18, 23] and that of the third line is [1, 3, 7, 5, 7, 12, 14]. Figure 18.5 shows the output of the program that follows.

Code:

```
X=[[2,3,1],[4,6,3],[6,9,7],[8,10,5],[9,11,7],[10,18,12],[11,23,14]]
plt.plot(X)
plt.show()
```

Output:

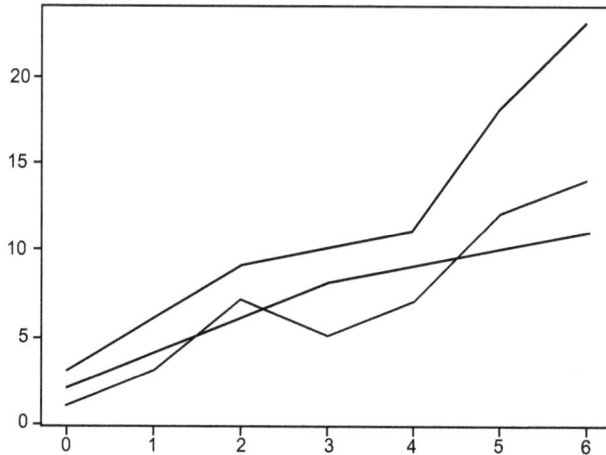

FIGURE 18.5 The **plot** function can take a two-dimensional array as its argument for plotting multiple lines.

18.3.4 Axis Function

The **axis** function has already been discussed in Section 18.2. As stated earlier, it takes four arguments: x_{min}, x_{max}, y_{min}, and y_{max}. Let us try to understand the function using an example. In the following example, the **X**-axis would span from 0 to 6 and **Y**-axis would span from 0 to 15, owing to the arguments of the **axis** function. The output of the following code would be the same as that of the first one except for the color of the plot and the axis (Figure 18.6).

Code:

```
plt.plot([1,4,8,10], color='red')
plt.xlabel("X Axis")
plt.ylabel("Y Axis")
plt.axis([0,6,0,15])
plt.show()
plt.savefig("line.png",dpi=80)
```

Output:

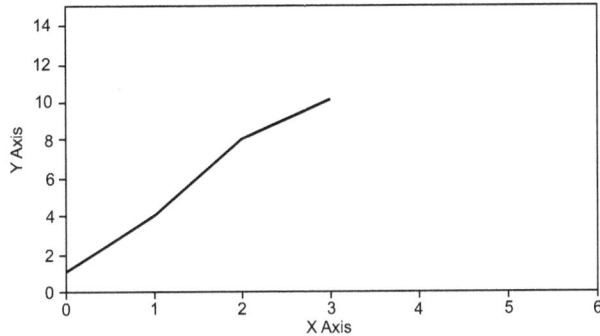

FIGURE 18.6 The **plot** function can also have an argument to set the color of the plot. The X-axis spans from 0 to 6 and the Y-axis spans from 0 to 15.

18.3.5 Plotting Points: Scatter Diagram

If one wants to plot only the points (shown by markers in the shape of circles) and not the lines, then an additional argument **o** can be passed to the **plot** function. Likewise, the plots indicated by a square and a triangle can be plotted by giving **s** and "**^**." The following codes and the respective output have been shown in Figures 18.7 and 18.8. The next subsection discusses various **markers** in detail.

Code:

```
# Plot circles
plt.plot([1,3,4],[7,8,3],'o')
plt.show()
```

Output:

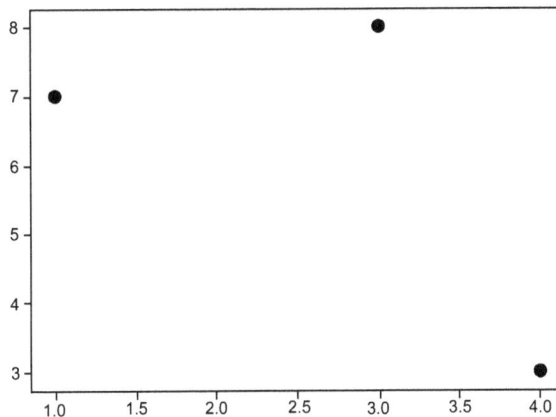

FIGURE 18.7 The **plot** function can also plot circles using an additional "o" argument.

The following code plots three pairs of lists using three distinct markers. The first pair would be shown by a circle, the second by a square, and the third by a triangle.

Code:

```
plt.plot([1,3,4],[7,8,3],'o')
plt.plot([1,2,3,4],[2,1,3,5],'s')
plt.plot([1,5,6],[9,10,11],'^')
plt.show()
```

Output:

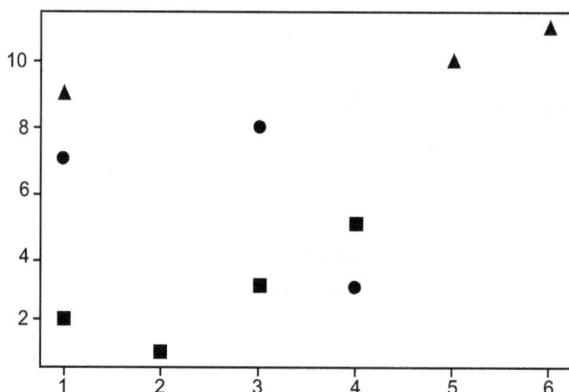

FIGURE 18.8 The **plot** function can also plot squares and triangles using additional "^" or "s" as an argument.

18.3.6 Sine and Cosine Curves

The following example shows the procedure for plotting the **Sine** and the **Cosine** function using **Matplotlib**. The **plot**, **show**, and **savefig** functions have already been explained, in the above discussion. In the following code, the **X**-axis is divided into 256 parts (from –22/7 to 22/7). The **linspace** function helps accomplish this task (refer to the chapter on **Numpy** for details regarding the **linspace** function). The **sine** of the X values can be calculated using the **sin** function of **NumPy**. Likewise, the cosine can be calculated using the **cos** function of **NumPy**. Both plots are plotted in the same area. The output is shown in Figure 18.9.

Code:

```
from matplotlib import pyplot as plt
import numpy as np
```

```
# Plotting sin and cos on the same graph
plt.figure(figsize=(8, 6), dpi=80)
plt.subplot(1, 1, 1)
X = np.linspace(-np.pi, np.pi, 256, endpoint=True)
S, C = np.cos(X), np.sin(X)
plt.plot(X, C, color="blue", linestyle="-")
plt.plot(X, S, color="red", linestyle="-")
plt.xlim(-4.0, 4.0)
plt.xticks(np.linspace(-4, 4, 9, endpoint=True))
plt.ylim(-1.0, 1.0)
plt.yticks(np.linspace(-1, 1, 5, endpoint=True))
plt.savefig("SinCos.png", dpi=180)
plt.show()
```

Output:

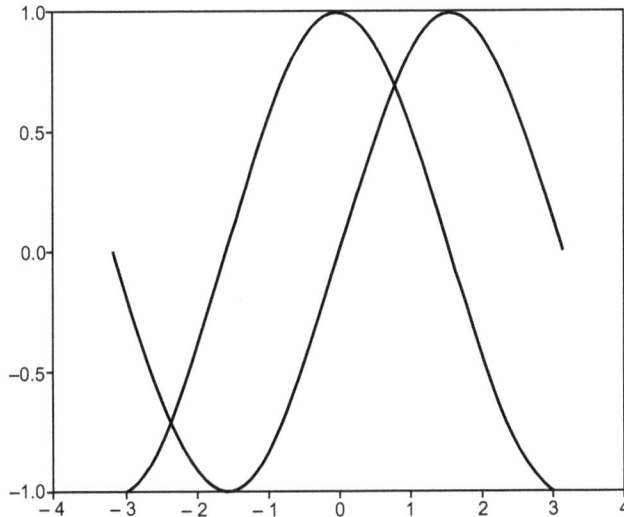

FIGURE 18.9 The sine and cosine function in the same plot.

18.3.7 Comparing Functions

The plots as described above are an excellent way of comparing the functions (say x^2, x^3, and x^4) using the power function of **numpy**. The **plot** function has a **label** argument which can be set to the type of curve. The following code illustrates plotting. The limits of the **X**-axis are set to from 1 to 20 and that of the **Y**-axis is from 0 to 800. The output of the code is shown in Figure 18.10.

Code:

```
x = np.linspace (0, 10, 50)
y1 = np.power(x, 2)
y2=np.power(x,3)
y3 = np.power(x, 4)
plt.plot(x, y1, label='$x^2$')
plt.plot(x, y2, label='$x^3$')
plt.plot(x, y3, label='$x^4$')
plt.xlim((0 , 20))
plt.ylim((0 , 800))
plt.xlabel('X Axis')
plt.ylabel('Y : Powers')
plt.title('First :$x^2$ Second:$x^3$ Third:$x^4$')
#plt.legend()
plt.savefig("powers.png",dpi=80)
plt.show()
```

Output:

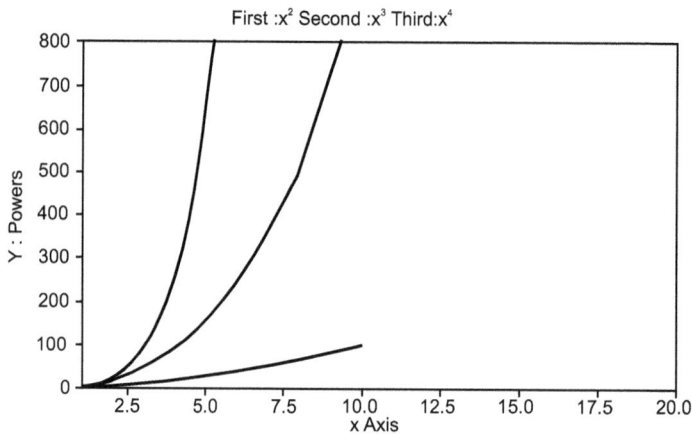

FIGURE 18.10 The plots of x^2, x^3, and x^4.

18.3.8 Plotting Multiple Lines

In case two lists are passed as arguments in the **plot** function, the second list's values are plotted considering the firsts as the indices. You can also give the

color, **linestyle**, **width**, etc. of the line as in the case of a single list containing the values.

18.4 ADDITIONAL ARGUMENTS

At times, we need to see the points on the graphs and not the complete lines. In such cases, **markers** come to our rescue. This section discusses the **markers** and explains their importance.

18.4.1 Markers

The **plot** method can have an additional argument called **marker**. For example, to print a circle or a triangle, **markers** "o," "^," etc., can be used as the second argument. The following table (Table 18.1) shows the various markers.

TABLE 18.1 Markers.

Symbol	Description
"."	point
","	pixel
"o"	circle
"v"	triangle_down
"^"	triangle_up
"<"	triangle_left
">"	triangle_right
"1"	tri_down
"2"	tri_up
"3"	tri_left
"4"	tri_right
"8"	octagon
"s"	square
"p"	pentagon
"P"	plus (filled)
"*"	star
"h"	hexagon1
"H"	hexagon2
"+"	plus
"x"	x
"X"	x (filled)
"D"	diamond

Symbol	Description
"d"	thin_diamond
"|"	vline
"_"	hline

Source: *https://matplotlib.org/api/markers_api.html#module-matplotlib.markers*

18.4.2 Color

The **plot** method can have an additional argument called **color** which helps in setting the color of the plot to the requisite value. The default color is **blue** and it can be changed easily. The **color** argument of the **plot** function can be set to a particular value, say **red** (color = **r**) to generate a plot of red color. If you want to show a specific color the alphabet indicating the color can be written as one of the arguments of the **plot** method. The following table (Table 18.2) shows the various colors supported.

TABLE 18.2 Color.

Symbol	Description
'b'	blue
'g'	green
'r'	red
'c'	cyan
'm'	magenta
'y'	yellow
'k'	black
'w'	white

18.4.3 Linestyle

The style of a line can be set using the **linestyle** argument. One can also mention the value of the optional **linestyle** argument. The following table (Table 18.3) shows the various values of the **linestyle argument**.

TABLE 18.3 Linestyle.

Line style	Description
'-' or 'solid'	solid line
'--' or 'dashed'	dashed line
'-.' or 'dashdot'	dash-dotted line
':' or 'dotted'	dotted line

18.4.4 Linewidth

The **linewidth** argument takes a float as an argument. It sets the **width** of the line.

Having studied these arguments, let us now move to some illustrations, to understand the usage of the above arguments.

Illustration 18.1:

*Ask the user to enter a list and plot it using the **plot** function.*

Solution:

The procedure to accomplish the task has already been discussed. The output is shown in Figure 18.11.

Code:

```
from matplotlib import pyplot as plt
import numpy as np
L=[]
n=int(input('Enter the number of elements\t:'))
for i in range(n):
    item=int(input('Enter an element\t:'))
    L.append(item)
print('List\t:',L)
plt.plot(L)
plt.show()
```

Output:

```
Enter the number of elements :5
Enter an element :2
Enter an element :7
Enter an element :1
Enter an element :4
Enter an element :9
List : [2, 7, 1, 4, 9]
```

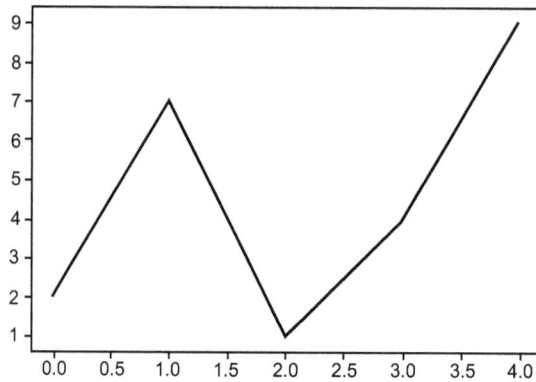

FIGURE 18.11 Illustration 18.1's plot.

Illustration 18.2:

In the above illustration what changes must be made to print a square at the points indicated by the list passed as an argument?

Solution:

To accomplish the given task, an additional argument "s" needs to be passed in the **plot** function.

In the code of Illustration 18.1, change the second last line to

plt.plot(L,'s'). The output is shown in Figure 18.12.

Output:

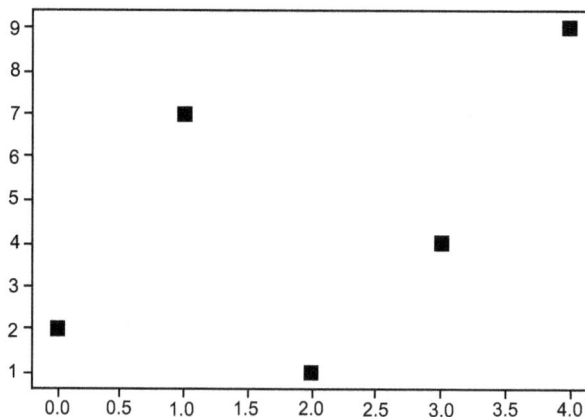

FIGURE 18.12 Illustration 18.2's plot.

Illustration 18.3:

Plot sine wave between – 2π and 2π. Also use color, linewidth, and linestyle functions to plot a purple, dashed curve.

Solution:

The **sin** of a given list can be calculated by using the **numpy.sin** function. The output is shown in Figure 18.13.

Code:

```
from matplotlib import pyplot as plt
import numpy as np
x=np.linspace(-2*np.pi,2*np.pi,256,endpoint=True)
s=np.sin(x)
plt.plot(x,s,color='purple',linewidth=5,linestyle='dashed')
#linestyle or ls=solid,dashed,dashdot,dotted
plt.show()
```

Output:

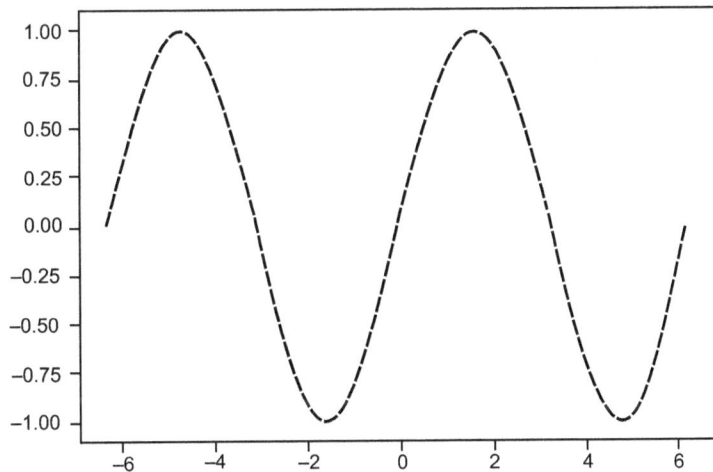

FIGURE 18.13 The **Sine** curve.

Illustration 18.4:

Plot cosine wave between –2π and 2π. Also use color, linewidth, and linestyle functions to plot a yellow, dashdot type curve.

Solution:

The **cosine** of a given list can be calculated by using the **numpy.cos** function. The rest of the functions have already been explained in the above discussion. The output is shown in Figure 18.14.

Code:

```
from matplotlib import pyplot as plt
import numpy as np
x=np.linspace(-2*np.pi,2*np.pi,256,endpoint=True)
c=np.cos(x)
plt.plot(x,c,color='y',linewidth=4,linestyle='dashdot')
plt.show()
```

Output:

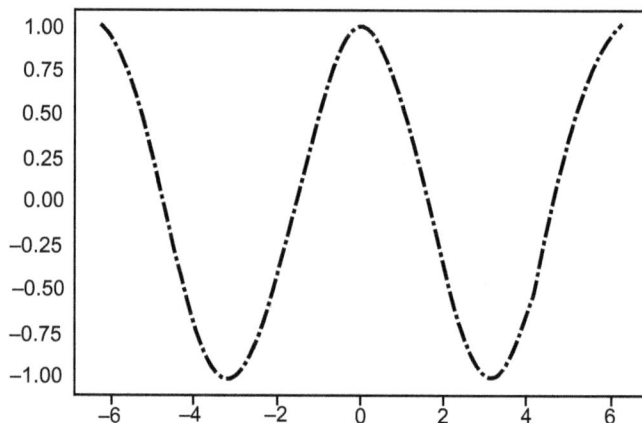

FIGURE 18.14 The cosine curve.

Illustration 18.5:

Enter a list and mark the list points with customization.

Solution:

The **marker** attribute and **markeredgecolor** can be used to show the points in the desired form and set the color of the edges.

The procedure to accomplish the task has already been discussed. The output is shown in Figure 18.15.

Code:

```
from matplotlib import pyplot as plt
import numpy as np
a=[2,4,6,3,6]
plt.plot(a,'c*',markersize=15,markeredgecolor='orange')
plt.xlabel('X-Axis')
plt.ylabel("Y-Axis")
plt.show()
```

Output:

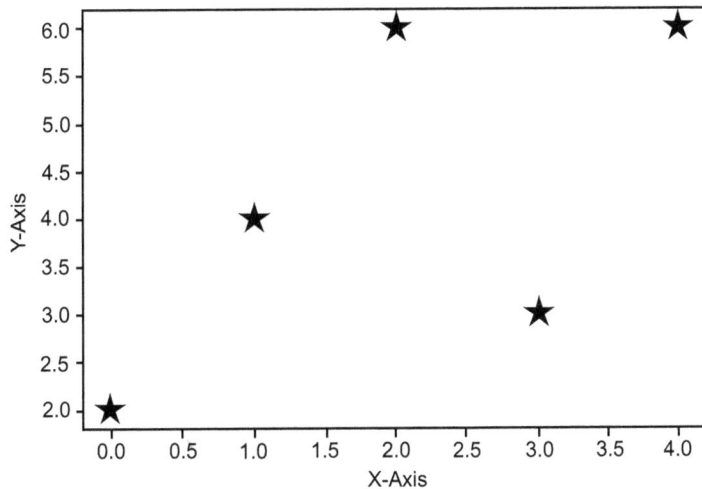

FIGURE 18.15 Plotting data using markers and setting the color of the edges.

Illustration 18.6:

Enter the student marks in n tests of a single subject and plot the graph.

Solution:

The marks of the student can be saved in a list and the list can be printed as explained in the previous examples. The output is shown in Figure 18.16.

Code:

```
from matplotlib import pyplot as plt
import numpy as np
```

```
L=[]
n=int(input('Enter number of tests\t:'))
i=0
while(i<n):
      str1='Enter mark['+str(i)+']:'
      num=int(input(str1))
      L.append(num)
      i+=1
print(L)
plt.plot(L,marker='s',color='red')
plt.xlabel('X-Axis')
plt.ylabel('Y-Axis')
plt.show()
```

Output:

```
Enter number of tests  :4
Enter mark[0]:34
Enter mark[1]:56
Enter mark[2]:78
Enter mark[3]:67
[34, 56, 78, 67]
```

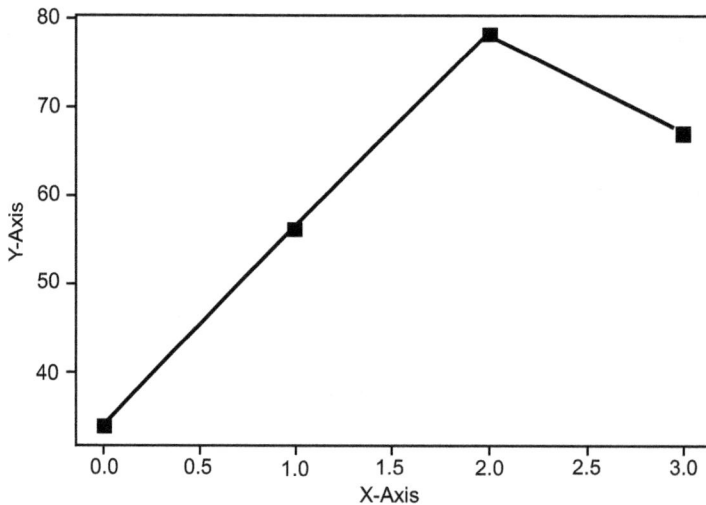

FIGURE 18.16 Marks of a student in n tests.

Illustration 18.7:

Enter the student marks in n tests of a single subject and plot it in by using the marker indicating a square.

Solution:

The marks of the student can be saved in a list and the list can be printed as explained in the subsection on markers. The output is shown in Figure 18.17.

Code:

```
L=[]
n=int(input('Enter number of tests\t:'))
i=0
while(i<n):
    str1='Enter mark['+str(i)+']:'
    num=int(input(str1))
    L.append(num)
    i+=1
    print(L)
plt.plot(L,'s',markersize=15,color='red')
plt.xlabel('X-Axis')
plt.ylabel('Y-Axis')
plt.show()
```

Output:

```
Enter number of tests  :4
Enter mark[0]:34
Enter mark[1]:56
Enter mark[2]:78
Enter mark[3]:67
[34, 56, 78, 67]
```

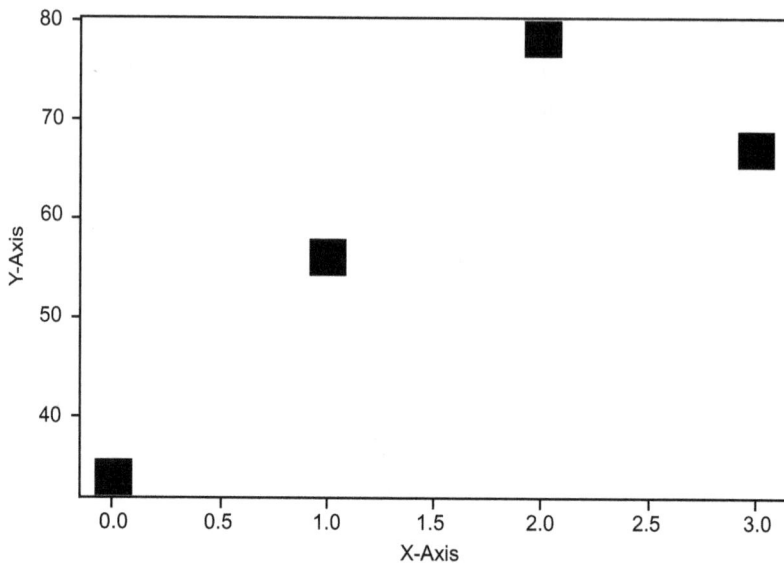

FIGURE 18.17 Plot of Illustration 18.7.

Illustration 18.8:

Enter the student marks of three subjects and plot them in the same graph.

Solution:

The list containing the marks of a particular student will have three values. The list of lists would represent the marks of n students in 3 subjects. The passing of a 2D array (or a list of lists) has already been discussed. The output is shown in Figure 18.18.

Code:

```
X=[[20,30,40],[40,50,70],[80,90,20],[70,80,100],[30,60,80],[20,4
    0,56]]
#points in line=2,4,8,73,2 and 3,5,9,8,6,4 and 4,7,2,10,8,11
plt.plot(X)
plt.xlabel('X-Axis')
plt.ylabel('Y-Axis')
plt.show()
```

Output:

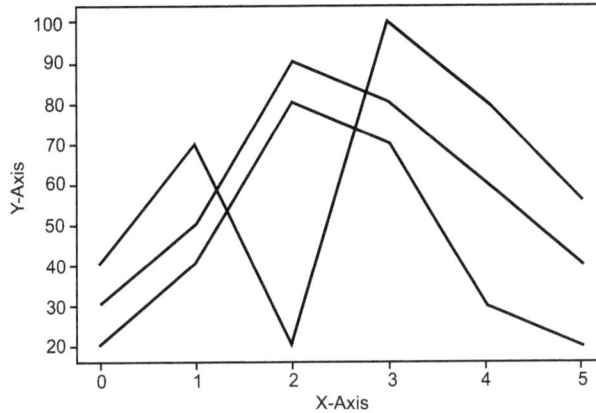

FIGURE 18.18 Plot of Illustration 18.8.

18.5 THE BAR CHART

In a **Bar Chart**, the categorical values are represented using vertical or horizontal bars. The lengths of these bars are proportional to the represented values. The **matplotlib.pyplot.bar** package is used to plot a bar plot.

The **bar** so plotted starts from the bottom and the dimensions are given by the **width** and the **height** parameters of the **bar** method. The parameters of the **bar** method are as follows:

- **x:** This parameter represents the coordinates of bars.
- **height:** This parameter represents the heights of the bar(s).
- **width:** This parameter represents the width of the bar(s).
- **bottom:** This parameter represents the vertical string position(s).

Each of the above can be a scalar if the value is to be applied to each bar, else they can be sequences. In addition to the above, the bars can have the following parameters:

- **color:** This argument sets the color of the bar. It is an optional argument. It can be a single number or a sequence of n elements.
- **edgecolor:** This argument sets the color of the edges.
- **linewidth:** This argument sets the width of the edges.
- **orientation:** This argument can have the following values:
  - "horizontal"
  - "vertical"

To understand the usage of the above parameters, consider the following examples.

Illustration 18.9:

Ask the user to enter the marks of five students and their names. Display the data using a bar chart showing marks along with the names on the x-axis.

Solution:

In this case, the **bar** function will take two lists as arguments: The first for representing the labels in the **X-axis** and the second for representing the height of the bars. The output is shown in Figure 18.19.

Code:

```
from matplotlib import pyplot as plt
import numpy as np
a=['Yash','Suraj','Sahil','Mahendra']
b=[24,45,67,29]
plt.bar(a,b)
plt.xlabel('Names---->')
plt.ylabel('Marks---->')
plt.show()
```

Output:

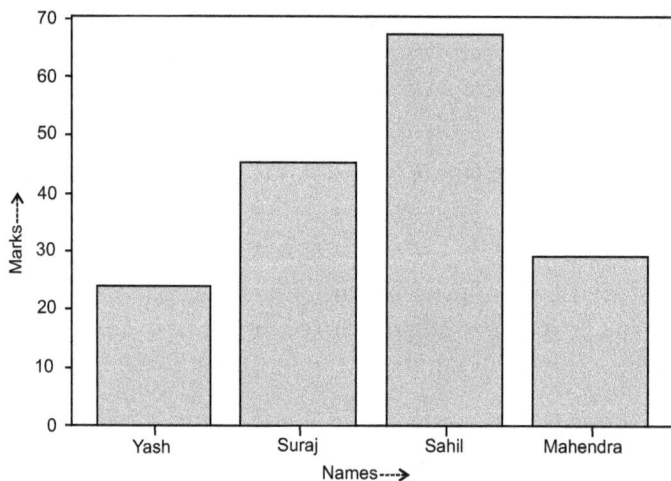

FIGURE 18.19 Figure of Illustration 18.9.

Illustration 18.10:

Generate 20 random numbers between 100 to 200 and twenty numbers between 0 to 50. Plot first list on Y-axis and second on X-axis.

Solution:

In this case, two lists would be passed as arguments in the **plot** function. The requisite functions have already been explained in Section 18.4. The output is shown in Figure 18.20.

Code:

```
A=np.random.randint(low=100, high=200, size=20)
B=np.random.randint(50, size=20)
print(A)
print(B)
plt.bar(B,A, width=0.8)
plt.xlabel('X-axis--->')
plt.ylabel('Y-Axis--->')
plt.show()
```

Output:

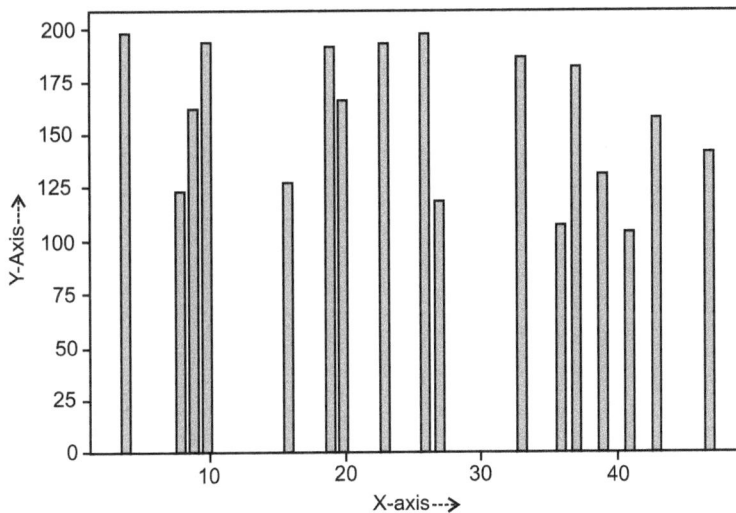

FIGURE 18.20 Figure of Illustration 18.10.

Illustration 18.11:

Generate two lists, A and B. Draw a bar chart where B represents the height of the bar and A represents the x-axis. Also, represent all different bars with different colors.

Solution:

To accomplish the desired task, the **color** attribute can be set to the list containing the values of the colors of each bar. The output is shown in Figure 18.21.

Code:

```
a=[]
n=int(input('Enter number of names\t:'))
i=0
while(i<n):
    str1='Enter name['+str(i)+']:'
    num=str(input(str1))
    a.append(num)
    i+=1
print(a)
b=[]
n=int(input('Enter number of marks\t:'))
i=0
while(i<n):
    str1='Enter mark['+str(i)+']:'
    num=int(input(str1))
    b.append(num)
    i+=1
print(b)
plt.bar(a,b,width=0.6,color=['green','red','blue'])
plt.xlabel('Names---->')
plt.ylabel('Marks---->')
plt.show()
```

Output:

```
Enter number of names  :3
Enter name[0]:pulin
```

```
Enter name[1]:rakesh
Enter name[2]:suraj
['pulin', 'rakesh', 'suraj']
Enter number of marks  :3
Enter mark[0]:97
Enter mark[1]:56
Enter mark[2]:45
[97, 56, 45]
```

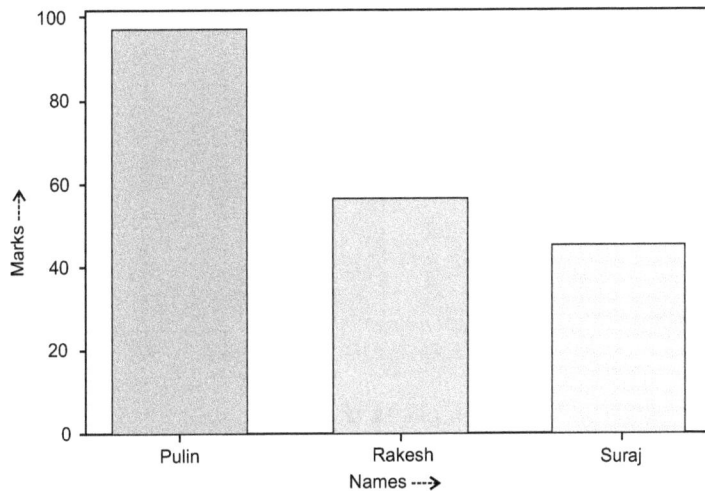

FIGURE 18.21 Plot for Illustration 18.11.

Illustration 18.12:

Plot two lists in the same bar, showing the value of the given index side by side.

Solution:

Refer to the explanation of the bar function. Note that the values of the X-axis have been displaced 0.35 to the left for the first list and 0.35 to the right for the second. The output is shown in Figure 18.22.

Code:

```
A=[2,4,6,8]
B=[2.4,6.5,3.9,8.5]
X=np.arange(len(A))
```

```
plt.bar(X,A,color='red',width=0.35)
plt.bar(X+0.35,B,color='blue',width=0.35)
plt.xlabel('Names Code---->')
plt.ylabel('Marks---->')
plt.show()
```

Output:

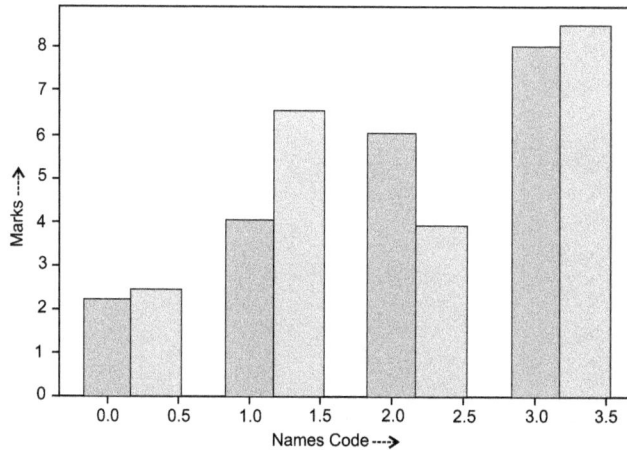

FIGURE 18.22 Figure of Illustration 18.12.

Illustration 18.13:

In Figure 18.22, what changes would generate a horizontal bar?

Solution:

The **barh** function can be used to accomplish the given task. The output is shown in Figure 18.23.

Code:

```
A=[2,4,6,8]
B=[2.4,6.5,3.9,8.5]
X=np.arange(len(A))       #i.e 4
plt.barh(X,A,color='red',height=0.35)
plt.barh(X+0.35,B,color='blue',height=0.35)
plt.xlabel('Marks---->')
plt.ylabel('Names Code---->')
plt.show()
```

Output:

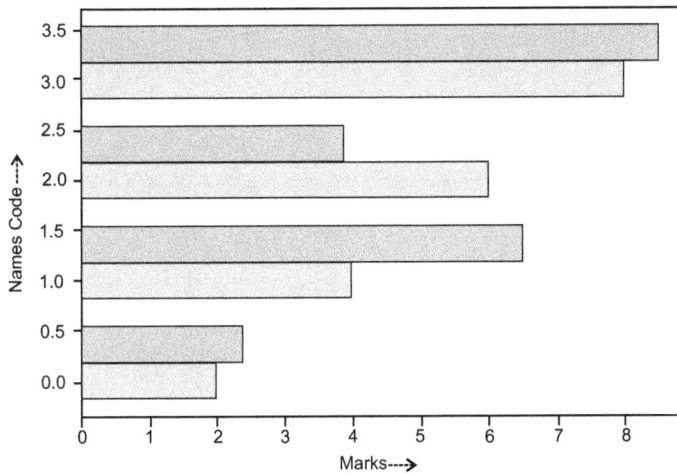

FIGURE 18.23 Output of Illustration 18.13.

18.6 CONCLUSION

This chapter starts with the importance of visualization and introduced **matplotlib**, which is an immensely important collection, used for visualization. Visualization helps us to understand the input data, the experimental results, and for comparisons. **Matplotlib** is used to generate production level figures, in a very simple manner. This chapter explained the basics of plotting and explained the **bar** plot in detail. A significant number of examples are included in the chapter to make the things comprehendible. It may also be stated that the next chapter takes this discussion forward and introduces frequency plots, pie charts, and box plots. The reader is advised to go attempt the exercises to get a better insight into the topics presented in this chapter.

EXERCISES

Multiple Choice Questions

1. Using **matplotlib** one can primarily generate

   (a) Plots   **(b)** Histograms

   (c) Bar graphs   **(d)** Scatter plots

   (e) All of the above

2. The **axis** function can take which of the following arguments.

   (a) x_{min} **(b)** y_{min}

   (c) x_{max} **(d)** y_{max}

   (e) all of the above

3. The marker "o" stands for

   (a) circle   **(b)** Zero

   (c) The letter "O"   **(d)** None of the above

4. Which function is used to plot a histogram?

   (a) plot   **(b)** hist

   (c) pie   **(d)** all of the above

5. What argument plots a cumulative graph?

   (a) commutative   **(b)** sum

   (c) total   **(d)** None of the above

6. You are required to show the Cartesian coordinates in a plot? How will you do it?

   (a) By passing two lists   **(b)** By passing a single list

   (c) By passing pie   **(d)** None of the above

7. Which function is used to plot a bar plot in **matplotlib**?

   (a) bar   **(b)** pie

   (c) both   **(d)** None of the above

8. Which function is used to plot a bar?

   (a) plt.bar   **(b)** plt.pie

   (c) plt.hist   **(d)** None of the above

9. Which argument is used to set the widths of bars in the bar method?

   (a) width   **(b)** w

   (c) d   **(d)** None of the above

10. Is width argument a scalar or a sequence?

    (a) Scalar   **(b)** Sequence

    (c) It can be both   **(d)** None of the above

11. Which argument is used to set the bottoms of bars in the bar method?

    (a) bottom   **(b)** b

    (c) h   **(d)** None of the above

Theory

1. Explain the importance of visualization.

2. Explain the features of **matplotlib**.

3. How do you plot a single line using the plot function?

4. How do you plot multiple lines using the plot function of **matplotlib**?

5. How do you plot the coordinates using markers in a plot?

6. Which function is used to plot a bar in **matplotlib**? Explain its arguments.

7. Which function is used to plot a horizontal bar in **matplotlib**? Explain its arguments.

8. Suggest a method to plot two bars side by side.

9. What is the purpose of **xticks**?

10. Explain the procedure to plot multiple lines in the same graph, using a two-dimensional array.

DATA VISUALIZATION–II

After reading this chapter, the reader should be able to

- Understand the importance of **Frequency Plots**
- Appreciate the features of a **Boxplot**
- Plot a **Histogram**
- Plot **Pie Charts**

19.1 INTRODUCTION

The importance of visualization has already been discussed in the last chapter. This chapter takes the discussion forward and introduces some more methods of visualization.

This chapter has been organized as follows. The second section presents an overview of the box plots, the third section discusses histograms and frequency plots. The fourth section presents a discussion on the pie charts. The last section concludes.

19.2 BOX PLOT

The **Box Plot** of data shows the minimum value of the input, the first quartile, the median, the third quartile, and the maximum value. There is a box spanning from the first quartile to the third quartile of the data.

The points at the beginning and the ends show the minimum and the maximum value of the input. The beginning of the box indicates the position of the first quartile and the end of the box indicates the position of the third quartile. The line in the middle indicates the median of the data. Figure 19.1 shows the structure of the box plot.

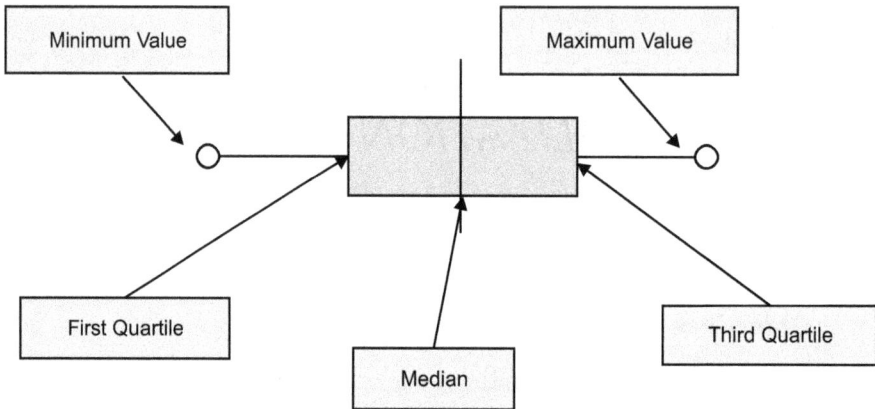

FIGURE 19.1 The structure of a Box Plot.

The **boxplot** method of the **matplotlib.pyplot** package helps us to plot a **Box Plot**. This method takes the following parameters (Table 19.1).

TABLE 19.1 Arguments of the boxplot method.

Argument	Data Type	Explanation
x	Array or a Sequence of vectors	This parameter presents the input data.
notch	bool	This is an optional parameter. The default value of this parameter is **False**. If the value of this parameter is **True**, a notched box plot is produced. It represents the Confidence Interval around the median.
vert	bool	If this parameter is **True**, the plots are drawn vertically; if **False**, the plots are horizontal.

This method returns a **dictionary**. The **dictionary** has the following keys (As per the official documentation at *https://matplotlib.org/3.1.1/api/_as_gen/matplotlib.pyplot.boxplot.html*):

* **boxes:** The quartile's and median's confidence intervals.
* **medians:** Lines at the medians of each column of the data.
* **whiskers:** The lines extending to the most extreme, nonoutlier data points.
* **caps:** The lines at the ends of the whiskers.

- **fliers:** Points representing data that extend beyond the whiskers.
- **means:** Points or lines representing the means.

The following example shows the **boxplot** of the **Iris** dataset. The dataset has 150 samples (rows) and four features (columns). The four features are:

- **sepal length** in cm
- **sepal width** in cm
- **petal length** in cm
- **petal width** in cm

The following code uses the first 100 samples of the data to plot a **boxplot**. Note the **boxplots** of each of the four features in Figure 19.2.

Code:

```
from sklearn.datasets import load_iris
Data=load_iris()
Data=Data.data[:100,:]
plt.boxplot(Data)
```

Output:

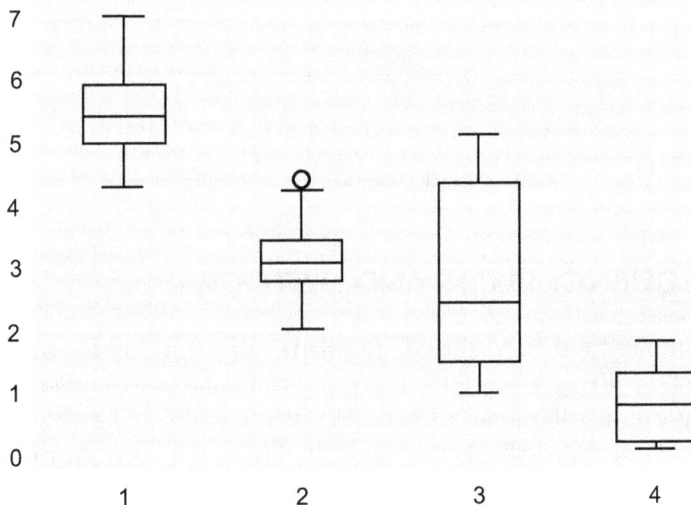

FIGURE 19.2 The **boxplot** of **Iris** data.

Having seen the **boxplot** of the **Iris** dataset, let us have a look at what happens if the value of **showmeans** parameter is passed as **True**.

EXAMPLE 19.1:

*The following code uses the **showmeans = True**. The output is shown in Figure 19.3. Observe the indicators at the means of each of the **boxplot** (of each feature).*

Code:

```
plt.boxplot(Data, showmeans=True)
```

Output:

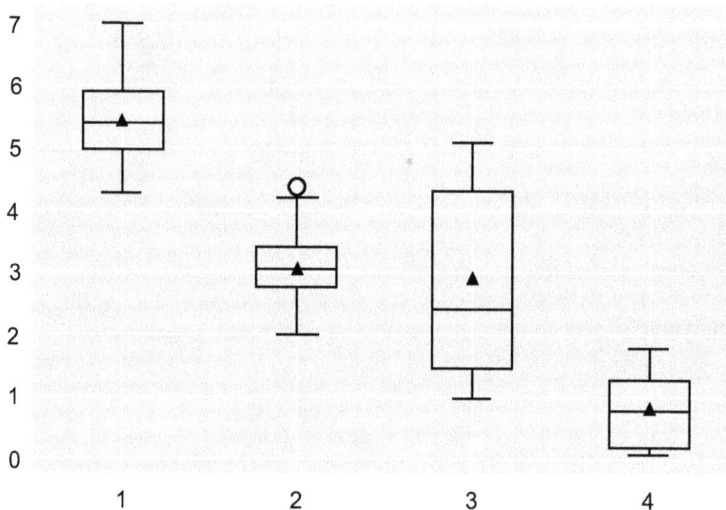

FIGURE 19.3 The **boxplot** of Iris data with **showmeans = true.**

19.3 FREQUENCY PLOTS AND HISTOGRAM

According to the **National Institute of Standards and Technology (NIST)**, a **Frequency Plot** is a graphical data analysis method for summarizing the distributional information of a variable. The response variable is divided into equal-sized intervals, called **bins**. The **number of occurrences** of the response variable is then calculated for each bin.

A **Frequency Plot** consists of:

- **Vertical Axis:** Frequencies
- **Horizontal axis:** The midpoint of each interval

As per **NIST**, there are four types of **Frequency Plots** (Figure 19.4)

(a) Basic Frequency Plot, which shows the absolute counts

(b) Relative Frequency Plot, which converts counts to proportions

(c) Cumulative Frequency Plot and

(d) Cumulative Relative Frequency Plot

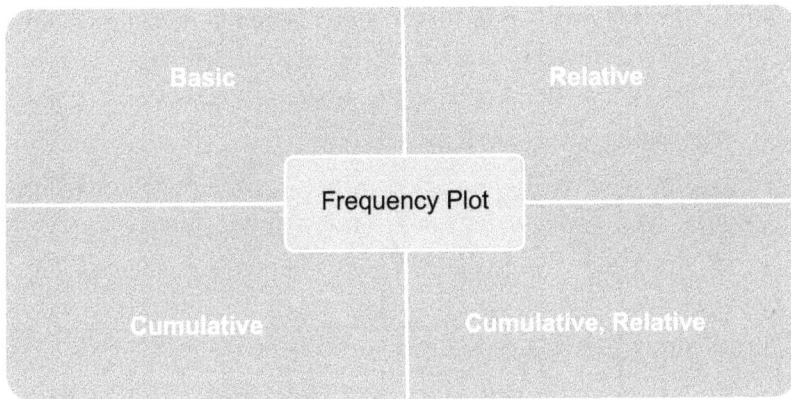

FIGURE 19.4 Types of Frequency Plots.

How to draw a Frequency Plot

To draw the frequency plot,

▪ Draw the histogram.
▪ Join the center of the bars of the histogram.

EXAMPLE 2:

Consider the following array:

[13, 15, 19, 26, 30, 12, 27, 8, 8, 24, 31, 30, 8, 6, 28, 20, 5, 20, 8, 22, 31, 27, 23, 24, 6, 16, 8, 26, 13, 29]

1. Let the number of bins be **10.** Find the number of elements in each bin using the tenth, twentieth, … percentiles of the data.

- The tenth percentile is 7.8 and there are three numbers less than 7.8 in the array, therefore the first bin will have 3.

- The twentieth percentile is 8; therefore, the number of elements in the second bin will be 5.

- The thirteenth percentile is 12.7; therefore, the number of elements in the third bin will be 1.

- The fortieth percentile is 15.4; therefore, the number of elements in the fourth bin will be 1.

- The fiftieth percentile is 18.; therefore, the number of elements in the fifth bin will be 3. Likewise, find the number of elements in the rest of the bins.

2. Draw histogram

The histogram would be:

(array([3., 5., 1., 3., 1., 3., 2., 2., 5., 5.]), array([5. , 7.6, 10.2, 12.8, 15.4, 18., 20.6, 23.2, 25.8, 28.4, 31.]), <a list of 10 Patch objects>)

3. Now, join the midpoints of the bars of the histogram (Figure 19.5).

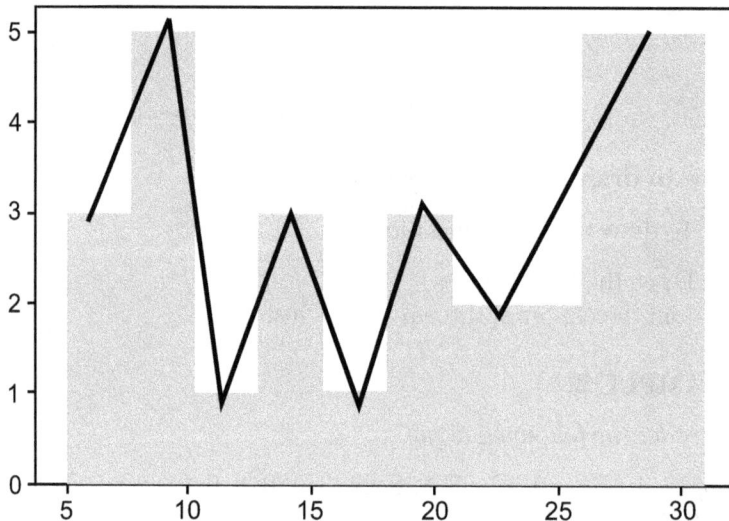

FIGURE 19.5 The Frequency Plot of the above example.

As per Karl Pearson "A **histogram** is an approximate representation of the distribution of numerical or categorical data."

The **hist** method of the **pyplot** module of **matplotlib**, allows us to plot a histogram. The arguments of the method are shown in Table 19.2 and its attributes are shown in Table 19.3.

TABLE 19.2 The arguments of the hist method.

Argument	Explanation
x	**Data:** Array or sequence of arrays
bins	It is an integer, which represents the number of equal-width bins. However, it can also be a sequence, in which case this represents the width of different bins.
histtype	This argument can have one of the following values.
	It is an optional parameter, and its default value is "bar."
	▪ **bar:** This creates a bar-type histogram. If any arguments are passed, bars are created side by side.
	▪ **step:** This argument creates a lineplot. By default, such histograms are not filled.
	▪ **stepfield:** This argument creates a lineplot, which is filled.
	▪ **barstacked:** This argument helps us to create histograms stacked on one another.
align	This argument controls how the histogram is placed. The possible values of this argument are:
	▪ **left:** This argument centers the bars to the left,
	▪ **mid:** This argument centers the bars between the edges and
	▪ **right:** This argument centers the bars to the right of the edges.
orientation	It is an optional argument, and the default value is "mid."
	This argument orients the bars along with the vertical or horizontal orientation. This argument is optional, and the default value of this argument is vertical.
color	This argument represents the color or array-like of colors or None. It is an optional argument. And the default value of this argument is None.
Stacked	It is a Boolean argument. If the value of this argument is True, then multiple bars are stacked on top of each other. If the value of this argument is False, then multiple bars are arranged side by side if **histtype** is "bar" or on top of each other if histtype is "step"

This method returns the following attributes (Table 19.3):

TABLE 19.3 The Attributes of the hist method.

Attribute	Explanation
narray or list of arrays	This array (or list of arrays) shows the values of the histogram bins.
binsarray	As per the official documentation this attribute represents. "The edges of the bins. Length **nbins + 1** (nbins left edges and right edge of the last bin)."
patcheslist or list of lists	As per the official documentation this attribute represents a "Silent list of individual patches used to create the histogram or lists of such list if multiple input datasets."

The following illustrations will help the reader understand the use of the above arguments in plotting a histogram.

Illustration 19.1:

Create an array of 30 numbers between 5 and 35. Draw a histogram having 10 bins. Set the **color** *of the histogram =* **"gray."**

Solution:

The **hist** method has already been explained. The code follows and the output is shown in Figure 19.6.

Code:

```
# Creating an array of 30 numbers between 5 and 30
array_sample = np.random.randint(5,35,30)
# Plotting a histogram
plt.hist(array_sample, 10,color='gray')
plt.show()
```

Output:

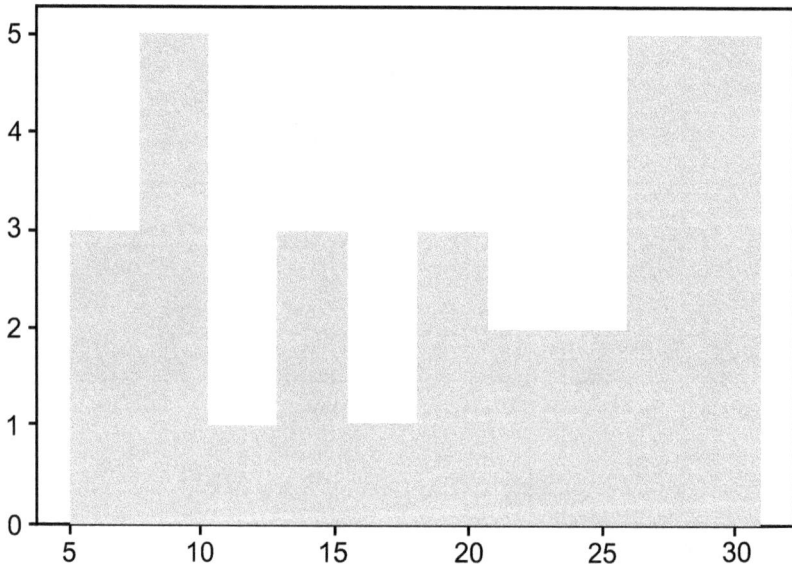

FIGURE 19.6 Output of Illustration 19.1.

Illustration 19.2:

In the illustration above, draw a histogram having 5 bins.

Solution:

The **hist** method has already been explained. The code follows and the output is shown in Figure 19.7.

Code:

```
# Creating an array of 30 numbers between 5 and 30
array_sample = np.random.randint(5,35,30)
# Plotting a histogram
plt.hist(array_sample, 5)
plt.show()
```

Output:

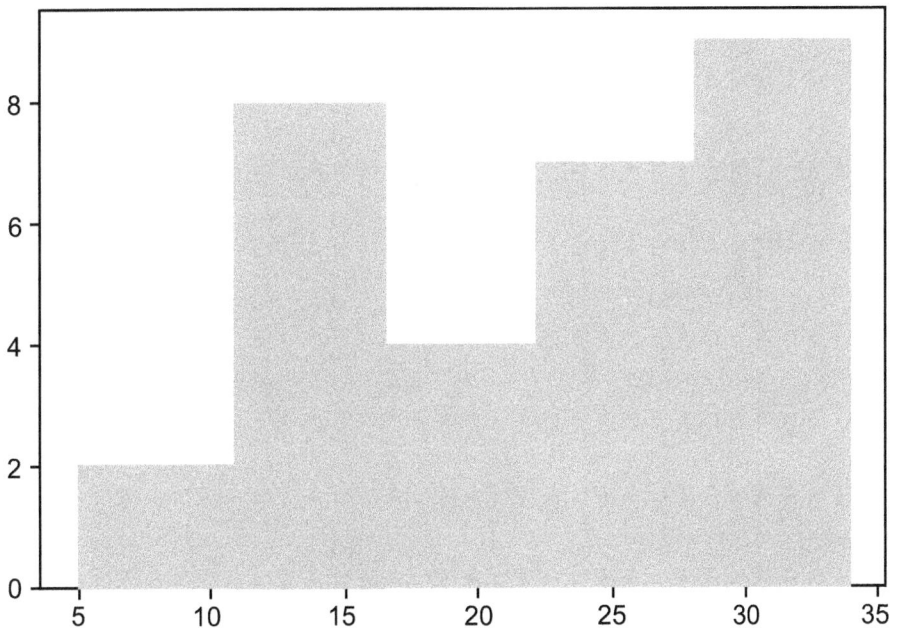

FIGURE 19.7 Output of Illustration 19.2.

Illustration 19.3:

Create an array having 30 elements, between 5 and 35. Draw a cumulative histogram having 10 bins.

Solution:

The **hist** method has already been explained. Set the **cumulative** argument of the **hist** method equal to **True**. The code follows and the output is shown in Figure 19.8.

Code:

```
# Creating an array of 30 numbers between 5 and 30
array_sample = np.random.randint(5,35,30)
# Plotting a histogram
plt.hist(array_sample, 10, cumulative=True)
plt.show()
```

Output:

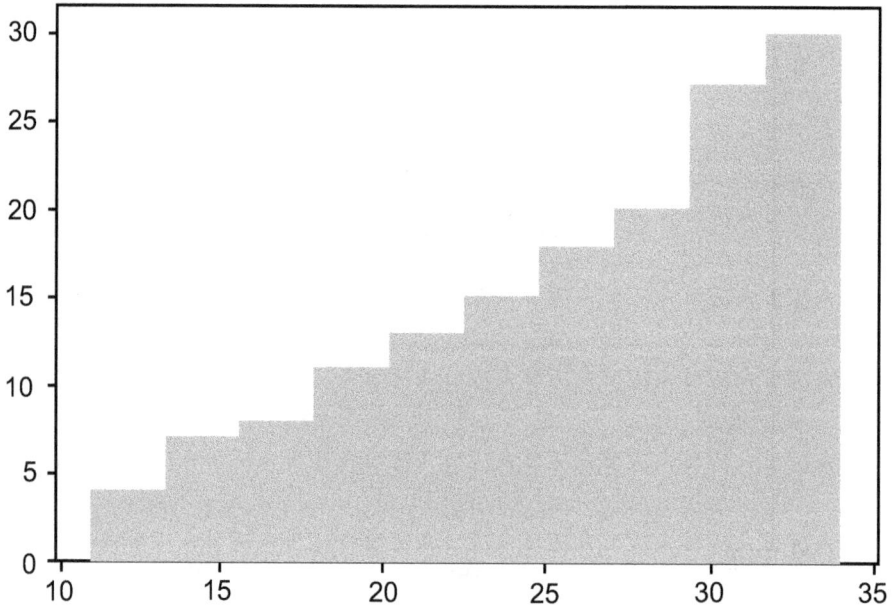

FIGURE 19.8 Output of Illustration 19.3.

Illustration 19.4:

*Create an array having 30 elements, between 5 and 35. Draw a histogram having 20 bins, with **histtype** = "step."*

Solution:

The **hist** method has already been explained. Set the **histtype** argument equal to **step**. The code follows and the output is shown in Figure 19.9.

Code:

```
# Creating an array of 30 numbers between 5 and 30
array_sample = np.random.randint(5,35,30)
# Plotting a histogram
plt.hist(array_sample, bins=20, histtype='step')
plt.show()
```

Output:

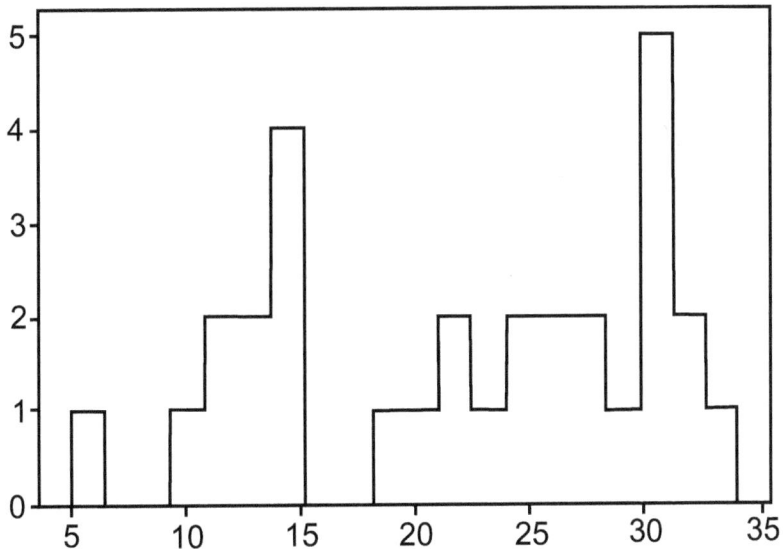

FIGURE 19.9 Output of Illustration 19.4.

Illustration 19.5:

Create an array having 30 elements, between 1 and 50. Create another array having 30 elements between 5 and 35. Now, draw a histogram showing the histograms of both the arrays, in the same plot.

Solution:

The **hist** method has already been explained. Pass both the arrays in the **hist** method (separated by a comma, in square brackets). The code follows and the output is shown in Figure 19.10.

Code:

```
# Creating an array of 30 numbers between 1 and 50
array_sample1 = np.random.randint(1,50,30)
print(array_sample1)
# Creating an array of 30 numbers between 5 and 35
array_sample2 = np.random.randint(5,35,30)
print(array_sample2)
# Plot the histogram
plt.hist([array_sample1, array_sample2])
plt.show()
```

Output:

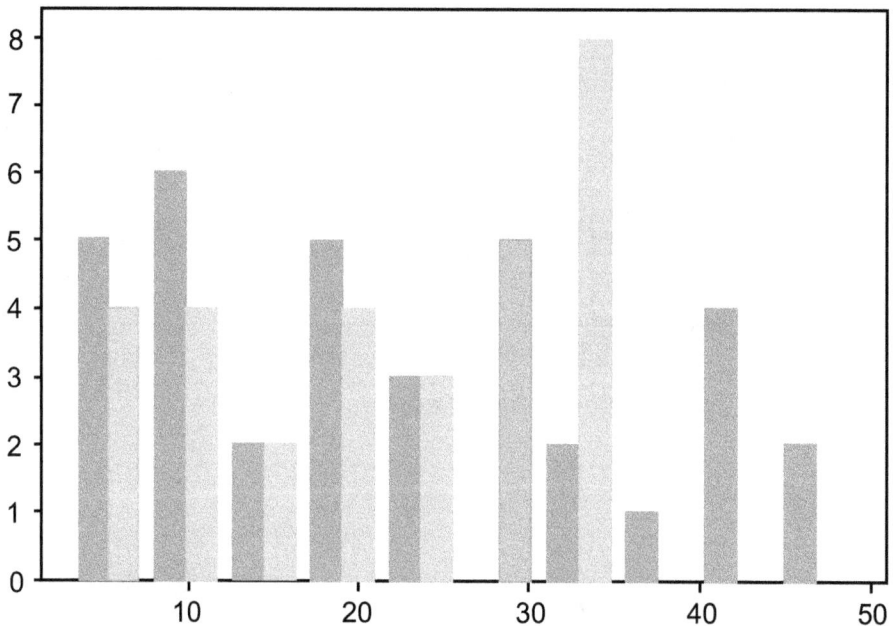

FIGURE 19.10 Output of Illustration 19.5.

Illustration 19.6:

In the above illustration, what changes must be made to stack the histograms on each other.

Solution:

Set the **histtype** argument of the **hist** method to **barstacked**. The code follows and the output is shown in Figure 19.11.

Code:

```
plt.hist([array_sample1, array_sample2], histtype='barstacked')
plt.show()
```

Output:

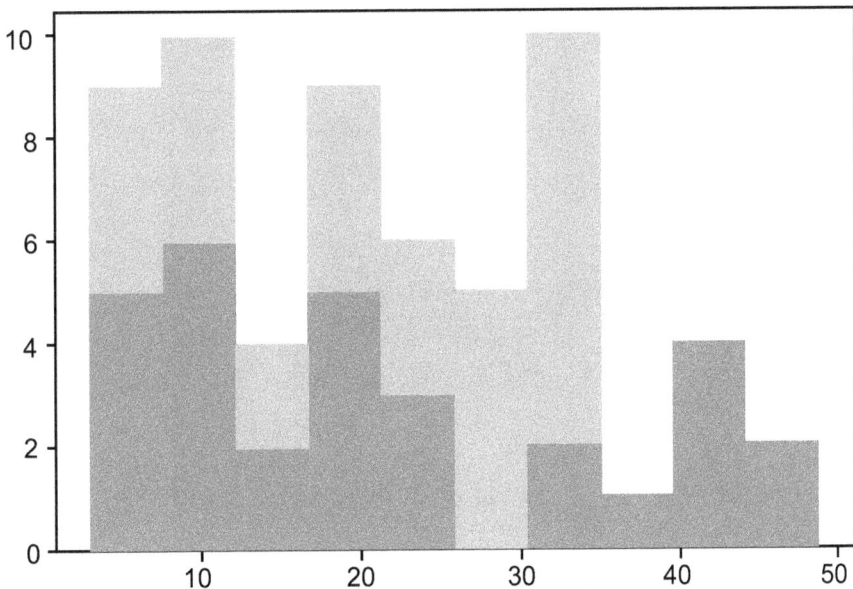

FIGURE 19.11 Output of Illustration 19.6.

Illustration 19.7:

In Illustration 19.5, what changes must be done to stack the cumulative histograms on each other.

Solution:

Set the **histtype** argument of the **hist** method to **barstacked** and the **cumulative** argument to **True**. The code follows and the output is shown in Figure 19.12.

Code:

```
plt.hist([array_sample1, array_sample2], histtype='barstacked',
    cumulative= True)
plt.show()
```

Output:

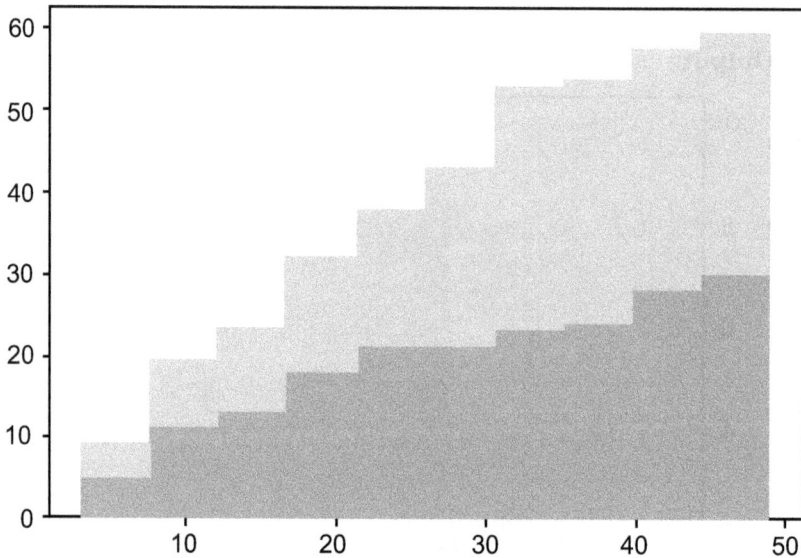

FIGURE 19.12 Output of Illustration 19.7.

Illustration 19.8:

In Illustration 19.5, what changes must be done to stack the histograms on each other and set the orientation to horizontal.

Solution:

Set the **histtype** argument of the **hist** method to **barstacked** and the **orientation** argument to **horizontal**. The code follows and the output is shown in Figure 19.13.

Code:

```
plt.hist([array_sample1, array_sample2],
histtype='barstacked',orientation='horizontal')
plt.show()
```

Output:

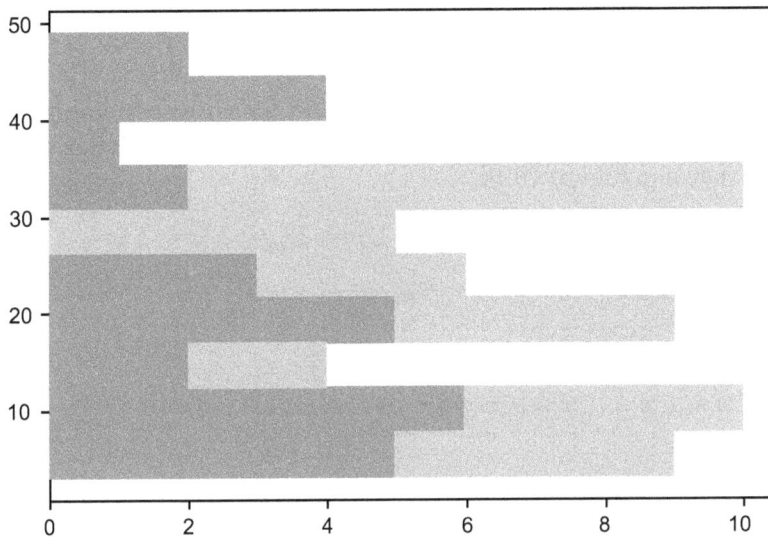

FIGURE 19.13 Output of Illustration 19.8.

19.4 THE PIE CHART

The **matplotlib.pyplot.pie** plots a pie chart using the array x, given as an input. Each wedge, in a Pie chart, has an area proportionate to the value. In the chart, the wedges are plotted counterclockwise. The arguments of the **pie** function are as follows.

- **x:** This argument represents the sequence of the values.
- **explode:** This argument gives the offset of each wedge. It is an optional argument.
- **labels:** This argument represents a sequence of strings depicting the labels for each wedge. It is an optional argument.
- **colors:** This argument sets the colors of each wedge.
- **autopct:** It is an optional argument. This argument sets the label showing the percentage will be placed inside the wedge.

The following illustrations would help us to understand the plotting of a pie chart, using the above arguments. The reader is expected to analyze the effect of each argument by observing the output.

Illustration 19.9:

Ask the user to enter its top five expenses and draw its pie chart.

Solution:

The expenses can be stored in a list and the corresponding pie chart can be drawn using the **pie** function of the **pyplot**. The code follows and the output is shown in Figure 19.14.

Code:

```
from matplotlib import pyplot as plt
import numpy as np
exp=[]
n=int(input('Enter number of expenses\t:'))
i=0
while(i<n):
    str1='Enter expense['+str(i)+']:'
    num=int(input(str1))
    exp.append(num)
    i+=1
print(exp)
plt.pie(exp)
plt.show()
```

Output:

```
Enter number of expenses  :4
Enter expense[0]:45
Enter expense[1]:56
Enter expense[2]:34
Enter expense[3]:78
[45, 56, 34, 78]
```

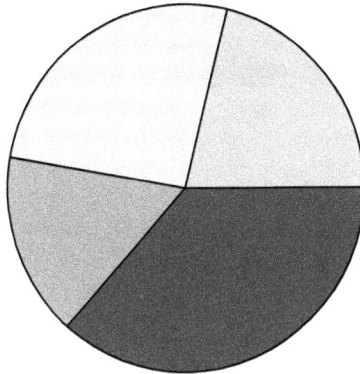

FIGURE 19.14 Plot of Illustration 19.9.

Illustration 19.10:

In the above illustration, put labels in the pie chart.

Solution:

The labels can be set using the **labels** attribute of the **pie** method. The code follows and the output is shown in Figure 19.15.

Code:

```
expenses=[45,56,34,78]
names=['zen','pong','anny','master']
plt.pie(expenses,labels=names)
plt.show()
```

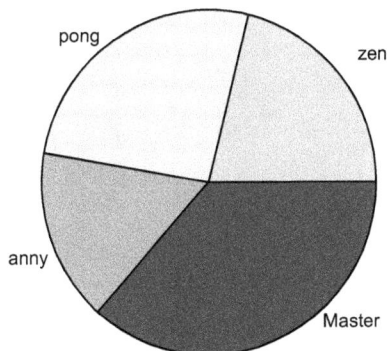

FIGURE 19.15 Output of Illustration 19.10.

Illustration 19.11:

In the above pie chart, display the percentage with each slice, up to two deci-mal places.

Solution:

The **output** attribute can be used to show the percentage of the data, as shown in the following code. The output is shown in Figure 19.16.

Code:

```
expenses=[45,56,34,78]
names=['zen','pong','anny','master']
plt.pie(expenses,labels=names,autopct="%1.2f%%")
plt.show()
```

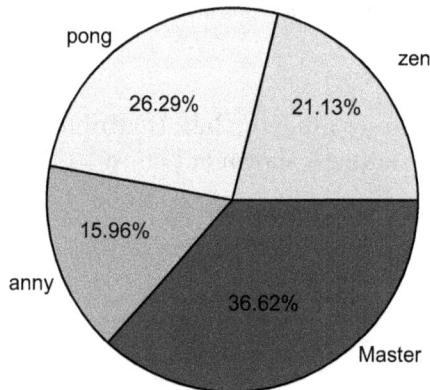

FIGURE 19.16 Output of Illustration 19.11.

Illustration 19.12:

Change the color of slices of the above pie chart as per specification.

Solution:

The **colors** attribute can be set to the requisite list, to accomplish the desired task. The code follows and the output is shown in Figure 19.17.

Code:

```
expenses=[45,56,34,78]
names=['zen','pong','anny','master']
colr=['r','g','m','y']
plt.pie(expenses,labels=names,colors=colr,autopct="%1.2f%%")
plt.show()
```

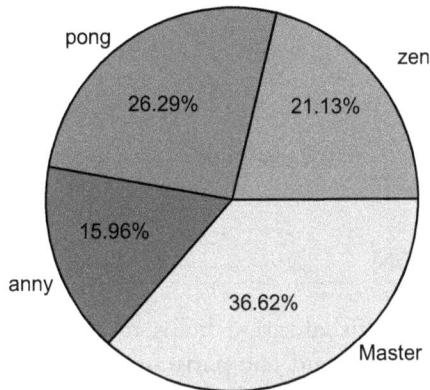

FIGURE 19.17 Output of Illustration 19.12.

Illustration 19.13:

Explode a slice of the above pie chart in Illustration 19.10, by a given distance.

Solution:

The desired task can be accomplished using the **explode** attribute, as shown. The output is shown in Figure 19.18.

Program:

```
expenses=[45,56,34,78]
names=['zen','pong','anny','master']
colr=['r','g','m','y']
expl=[0,0,0.3,0]
plt.pie(expenses,labels=names,colors=colr,autopct="%1.2f%%",
    explode=expl)
plt.show()
```

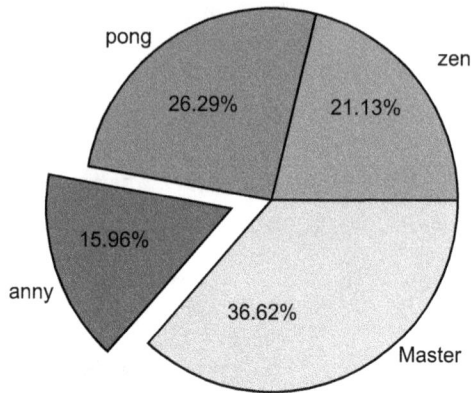

FIGURE 19.18 Figure of Illustration 19.13.

19.5 CONCLUSION

As stated earlier, Visualization helps us to understand the input data, the experimental results, and comparisons. **matplotlib** is an immensely important collection, used for visualization. It is used to generate production level figures, in a very simple manner. Some of the methods of **matplotlib** were introduced in the previous chapter. This chapter takes the discussion forward and introduces the **Frequency Plots, Box Plots, Histograms,** and **Pie Charts**. A sufficient number of examples are included in the chapter to make the things comprehendible. The reader is advised to go attempt the exercises to get a better insight into the topics presented in this chapter.

EXERCISES

Multiple Choice Questions

1. Which of the following are types of **Frequency Plots**?

    (a) Basic Frequency Plot

    (b) Relative Frequency Plot

    (c) Cumulative Frequency Plot

    (d) Cumulative Relative Frequency Plot

    (e) All of the above

2. How do you draw a **Frequency Plot**?

    (a) Join the midpoints of bars of a histogram

    (b) Find frequency of each unique item and plot

    (c) Find the frequency of numbers in each bin

    (d) None of the above

3. Which of the following is shown in a **Box Plot**?

    (a) Median   **(b)** Minimum and Maximum value

    (c) Q1 and Q3   **(d)** All of the above

4. Can a box plot be horizontal?

    (a) Yes   **(b)** No

    (c) Depends   **(d)** None of the above

5. Which of the following are shown in a **Box Plot**?

    (a) boxes   **(b)** medians

    (c) whiskers   **(d)** caps

    (e) fliers   **(f)** means

    (g) All of the above

Theory

1. Explain the importance of the **Frequency Plots**.

2. Explain the steps to create a **Frequency Plot**.

3. What is the difference between a **Frequency plot** and a **Box Plot**?

4. Explain the importance of the **Box Plots**.

5. What are the components of a **Box Plot**?

6. Explain the importance of **Pie Charts**.

7. What is a histogram?

8. Explain the importance of a Pie Chart.

PANDAS–I

Objectives

After reading this chapter, the reader should be able to

- Understand the difference between a **Pandas Series** and a **Data Frame**.
- Understand various methods to create a **Pandas Series**.
- Manipulate a **Pandas Series**.
- Understand slicing and indexing in a **Pandas Series**.
- Understand various methods to create a **Data Frame**.
- Understand operations on rows and columns of a **Data Frame**.
- Iterate through a **Data Frame**.

20.1 INTRODUCTION

Pandas is a Software Library, primarily used for Data manipulation and Analysis. It was developed by Wes McKinney and was released on January 11, 2008. It is a free library released under the three-clause BSD license. As per Wikipedia, this library derives its name from **Panel Data**. The term **Panel Data** is used in Statistics and Economics and it involves measurement over time.

The two most important data structures in **Pandas** are:

- Series and
- DataFrame

The former represents one-dimensional indexed data and the latter represents two-dimensional indexed data. Formally, these data structures are defined as follows.

Series: A **Pandas Series** represents a one-dimensional array of indexed data. It can be created by an array of actual data and an associated array of indices. The **Series** function helps us to create a **Series** data type.

Data Frame: A **DataFrame** is a two-dimensional labeled array that stores an ordered collection of columns.

Pandas provides us with

- Label-based data access
- Data alignment
- Ways to handle missing data
- Hierarchical indexing
- Pivoting and reshaping
- Aggregating and grouping
- Spreadsheet-style pivot tables
- Performance and use for large datasets

and many more features, which make this library distinctive and useful. The library is efficient, primarily because a major portion of the code is written in C.

This chapter provides an insight into **Pandas** and discusses some of its most important features. The chapter has been organized as follows (Figure 20.1):

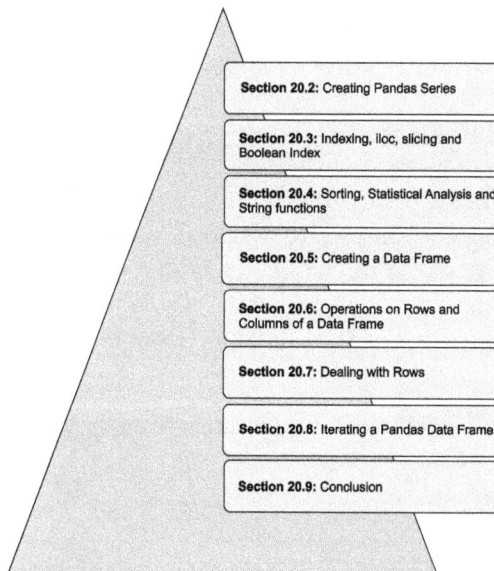

Section 20.2: Creating Pandas Series

Section 20.3: Indexing, iloc, slicing and Boolean Index

Section 20.4: Sorting, Statistical Analysis and String functions

Section 20.5: Creating a Data Frame

Section 20.6: Operations on Rows and Columns of a Data Frame

Section 20.7: Dealing with Rows

Section 20.8: Iterating a Pandas Data Frame

Section 20.9: Conclusion

FIGURE 20.1 Organization of the chapter.

Let's now dive into **Pandas** and move toward **Data Science**.

20.2 CREATING PANDAS SERIES

A **Pandas Series** can be created by using a **List**, a **Numpy array**, or a **dictionary** (Figure 20.2). This section explains the three ways of creating a **Series** and presents examples of each method.

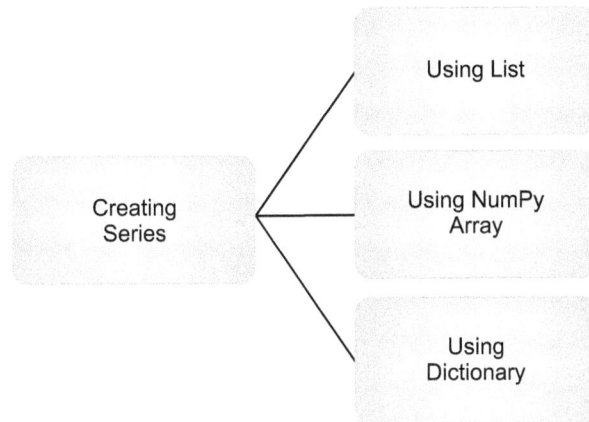

FIGURE 20.2 Creating Pandas Series.

20.2.1 Using List

A Python list can be converted into a **Pandas Series** by passing the list in the **pandas.Series** method. For example, in the following code, a list called **L1** is passed in the **Series** function to create a **Pandas Series**.

The list passed in the **Series** function can also contain **NaN** values, like in the case of **L2**. The **Series S2** is created by passing **L2** in the **Series** function of **Pandas**. Note that the output contains the index along with the values.

Code:

```
import pandas as pd
L1=[90, 81, 72, 63, 54, 45, 36]
S1=pd.Series(L1)
print(S1)
L2=[3, 6, 7, 1, np.nan, 8]
S2=pd.Series(L2)
print(S2)
```

Output:

```
0    90
1    81
2    72
3    63
4    54
5    45
6    36
dtype: int64
0    3.0
1    6.0
2    7.0
3    1.0
4    NaN
5    8.0
dtype: float64
```

20.2.2 Using NumPy Arrays

A **NumPy** array can be converted into a **Pandas Series** by passing the array in the **pandas.Series** function. For example, in the following code, a **NumPy** array called **arr_np** is passed in the **Series** function to create a **Pandas Series**.

The array, passed in the **Series** function, can also contain **NaN** values, like in the case of **arr_np2**. The Series **S3** is created by passing **arr_np2** in the **Series** function of **Pandas**.

Code:

```
arr_np=np.random.randint(2,89,10)
S3=pd.Series(arr_np)
print(S3)
arr_np2=np.array([2, np.nan, 9])
S3=pd.Series(arr_np2)
print(S3)
```

Output:

```
0    23
1    20
2    50
```

```
3    70
4    35
5    58
6    46
7    65
8    68
9    44
dtype: int32
0    2.0
1    NaN
2    9.0
dtype: float64
```

20.2.3 Using Dictionary

A **Dictionary** can be converted into a **Pandas Series** by passing the dictionary in the **pandas.Series** function. For example, in the following code, a dictionary called **dict_1** is passed in the **Series** function to create a **Pandas Series**. In such cases, the key to the dictionary becomes the index of the **Series**.

Code:

```
dict_1={'Harsh':97,'Naks':90,'Sahil':91}
S4=pd.Series(dict_1)
print(S4)
```

Output:

```
Harsh    97
Naks     90
Sahil    91
dtype: int64
```

20.3 INDEXING, ILOC, SLICING, AND BOOLEAN INDEX

This section presents the ways to access the elements of a **pandas Series** and getting a segment of a **Series** out of a given **Series**. This section also presents an overview of the **loc[]** and **iloc[]** objects and discusses slicing in **Series**. The first five elements and the last five elements of a **Series** can be displayed using the **head** and the **tail** methods respectively. The next two subsections cover the **head** and the **tail** methods of the **Pandas Series**. Finally, the **index** and **describe** the methods of the **Pandas Series** have been discussed. It may be stated that the **Boolean indexing, explained in the last subsection**, can also be used in **Series**.

20.3.1 Indexing: loc

An element of a **Series** can be accessed by passing the value of the index in square brackets. For example, **S1[0]** is used to access the element at index 0, of the series S1. Likewise, **S1[2]** is used to access the third element. Note that **S1** used in the following code is the same as that created in Section 20.2.1 and **S4** is the same as that created in Section 20.2.3.

Code:

```
print(S1[2])
print(S1[0])
print(S4['Harsh'])
```

Output:

```
72
90
97
```

20.3.2 Indexing Continued: iloc

Accessing elements using 0 based indices: The **iloc** object helps to access elements in the same manner as a **NumPy** array. Note that, **iloc** can also give you the last element using **iloc[-1]**, like in the case of a **NumPy** array. This object also helps in slicing a given **Series**, in the same manner as a **NumPy** array. Note that **S3** in the following code is the same as that created in Section 20.2.2 and **S4** is the same as that created in Section 20.2.3.

Code:

```
#To access the last and the second element
S4.iloc[-1]
S4.iloc[1]
```

Output:

```
91
90
```

Code:

```
#To retrieve elements from the second index to the fourth index
    (element at index 5 will not be #included).
S3.iloc[2:5]
#before S3 is changed
```

Output:

```
2    50
3    70
4    35
dtype: int32
```

20.3.3 Slicing

Slicing is used to create a new **Series** from a given **Series** by specifying the first and the last required index, separated by a ":". For example, in the following code **S5** is created by taking the elements of **S2** from index 3 to 7 and **S6** is created from **S4**, containing values from index "Harsh" to Index "Sahil." Note that **S2** used in the following code is the same as that created in Section 20.2.1 and **S4** is the same as that created in Section 20.2.3.

Code:

```
S5=S2[3:7]
print(S5)
print(S4['Harsh'])
S6=S4['Harsh':'Sahil']
print(S6)
```

Output:

```
3    1.0
4    NaN
5    8.0
dtype: float64
97
Harsh    97
Naks     90
Sahil   ·91
dtype: int64
```

20.3.4 Functions: Head, Tail, Describe, and index

20.3.4.1 *head()*

The **head()** function displays the top 5 values of the **Series**. For example, **S3.head()** displays the first five values of **S3**. Note that **S3** used in the following code is the same as that created in Section 20.2.2.

Code:

```
S3.head()
```

Output:

```
0   23
1   20
2   50
3   70
4   35
dtype: int32
```

20.3.4.2 *tail()*

The **tail()** function displays the last 5 values of the **Series**. For example, **S3.tail()** displays the last 5 values of **S3**. Note that **S3** used in the following code is the same as that created in Section 20.2.2.

Code:

```
S3.tail()
```

Output:

```
5   58
6   46
7   65
8   68
9   44
dtype: int32
```

20.3.4.3 *index*

This method displays the **index**(s) of the given **Series**. For example, **S4.index** shows the index of the series **S4**. Note that **S4** used in the following code is the same as that created in Section 20.2.3.

Code:

```
S4.index
```

Output:

```
Index(['Harsh', 'Naks', 'Sahil'], dtype='object')
```

20.3.4.4 describe()

The **describe** function of the **Series** object displays the following information about a **Series**

- **count:** This gives the number of the items in the given Series
- **mean:** This gives the mean of the items in the given Series
- **min:** This gives the minimum of the items in the given Series
- **max:** This gives the maximum of the items in the given Series
- **25%, 50%, 75%:** These three denote the 25th, 50th, and 75th percentile.
- **std:** This gives the standard deviation of the items in the given Series

An example of the describe function is as follows. Note that **S3** used in the following code is the same as that created in Section 20.2.2.

Code:

```
S3.describe()
```

Output:

```
count 10.000000
mean 47.900000
std 17.872698
min 20.000000
25% 37.250000
50% 48.000000
75% 63.250000
max 70.000000
dtype: float64
```

20.3.5 Boolean Index

In a **Series**, the required condition can be specified inside the square brackets, to get the elements that satisfy the given condition. For example, to get elements greater than 30 from **S3**, the following code can be written. Note that **S3** used in the following code is the same as that created in Section 20.2.2.

Code:

```
S3[S3>30]
```

Output:

```
2    50
3    70
4    35
5    58
6    46
7    65
8    68
9    44
dtype: int32
```

Likewise, you can state any Boolean statement involving the series, in the square brackets, alongside the Series.

20.4 SORTING, STATISTICAL ANALYSIS, AND STRING FUNCTIONS

This section discusses the various methods of the **Series** data structure of **Pandas**: to sort the values of a **Series**, find the minimum value, maximum value, the mean, median, and standard deviation of the values in a **Series**.

20.4.1 sort_values()

This method sorts the items of the given Series. This also displays the indexes of the sorted arrays. For example, **S3.sort_values()** sorts the elements of the Series **S3**. Note that **S3** used in the following code is the same as that created in Section 20.2.2.

Code:

```
S3.sort_values()
```

Output:

```
1    20
0    23
4    35
```

```
9   44
6   46
2   50
5   58
7   65
8   68
3   70
dtype: int32
```

20.4.2 Statistical Functions

Except for the above, the following methods (Table 20.1) help us to get the required statistics from a series. Note that in the following table **S1** and **S2** are the same as those created in Section 20.2.1.

TABLE 20.1 Functions to find the maximum, minimum, sum, median, standard deviation, and value count of a Series.

Name of the function	Explanation	Example	Output
max	This function finds the maximum value from a given series.	max_S2=S2.max()	8.0
min	This function finds the minimum value from a given series.	min_S2=S2.min()	1.0
sum	This function finds the sum of values from a given series.	sum_S2=S2.sum()	25.0
median	This function finds the median value of the given series.	median_S2=S2.median()	6.0
value_counts	This function counts the frequencies of values in a given Series.	S6=pd.Series(np.random.randint (2,10,20)) print('Value Counts') c=S6.value_counts() print(c)	Value Counts 8 6 9 3 5 3 7 2 6 2 3 2 4 1 2 1

20.4.3 String Functions

The functions applicable to strings can also be applied to a **Pandas Series**. For example, the following code converts all the strings in a given **Series** to upper case and then in the lower case. The reader is expected to experiment with other string functions with the **Pandas Series**.

Code:

```
S7 = pd.Series(['Harsh', 'Bharsh', 'Carsh', 'Arsh',np.nan,
    'Darsh', 'ABC'])
S8=S7.str.lower()
print(S8)
S9=S7.str.upper()
print(S9)
```

Output:

```
0 harsh
1 bharsh
2 carsh
3 arsh
4 NaN
5 darsh
6 abc
dtype: object
0 HARSH
1 BHARSH
2 CARSH
3 ARSH
4 NaN
5 DARSH
6 ABC
dtype: object
```

20.5 CREATING A DATA FRAME

The definition of a Pandas DataFrame has already been discussed in the first section of this chapter. This section revisits the topic and sheds light on the creation and manipulation of a Data Frame.

The following points are worth noting regarding a Data Frame.

■ A Data Frame has a row axis (axis=0). The index of a row is known as the **index**. These indices can be Strings or Integers.

▪ A Data Frame has a column axis (axis=1). The index of a column is known as the **column name**.
▪ The values of cells in a Data Frame can be changed.

A Data Frame can be created using any of the following methods (Figure 20.3):

▪ By passing a dictionary in the **DataFrame** method of **Pandas**
▪ By passing a two-dimensional **NumPy** array in the **DataFrame** method of **Pandas**
▪ By passing some Series in the **DataFrame** method of **Pandas**
▪ By passing a **Dataframe** object in the **DataFrame** method of **Pandas**

This section discusses the above methods in detail and presents examples of each of the above.

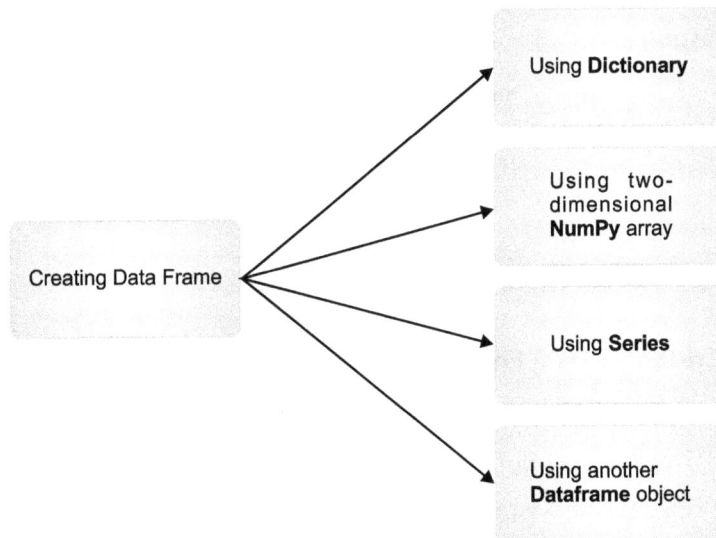

FIGURE 20.3 Creating a Data Frame.

20.5.1 Creating a Data Frame Using a Dictionary

The following steps will create a **Data Frame** using a dictionary:

1. Create a dictionary in which each index is associated with a list of values. Make sure that each list contains the same number of items.

2. Pass the dictionary in the **DataFrame** method.

The following code creates a **DataFrame** called **df1** using a dictionary called **book_details**.

Code:

```
import pandas as pd
import numpy as np
#Dictionary
book_details={'Author' : ['A', 'B', 'C', 'D', 'E'], 'Books':
    ['5', '6', '4', '6', '3'], 'Country': ['INDIA', 'ITALY',
    'INDIA', 'CHINA', 'USA']}
#Data Frame From Dictionary
df1=pd.DataFrame(book_details)
print(df1)
```

Output:

–	Author	Books	Country
0	A	5	INDIA
1	B	6	ITALY
2	C	4	INDIA
3	D	6	CHINA
4	E	3	USA

20.5.2 Creating a Data Frame Using a Two-Dimensional Array

The following steps will create a **Data Frame** using a two-dimensional array:

1. Create a two-dimensional **NumPy** array.

2. Pass the above array in the **DataFrame** method.

The following code creates a **DataFrame** called **df2** using a **NumPy** array called **array1**. This array, created using the **np.random.randint** method contains three rows and four columns. The elements of this array are between 7 and 89.

Code:

```
#Create a two-dimensional array
array1=np.random.randint(7,89,(3,4))
#Data Frame from array
df2=pd.DataFrame(array1)
print(df2)
```

Output:

–	0	1	2	3
0	61	53	51	53
1	52	80	29	54
2	69	40	17	9

20.5.3 Creating the Data Frame Using a Series

The following steps will create a **Data Frame** using Pandas Series:

1. Create **Series** data structures. The number of **Series** will be the same as the number of columns in the required Data Frame. Each **Series** here represents a column of the resultant Data Frame. The number of elements in each **Series** should be the same.

2. Pass a dictionary (in which each **Series** is associated with the required index) in the **DataFrame** method.

The following code creates a **DataFrame** called **df3** using three **Series** namely **name**, **salary**, and **post**. These **Series** are then associated with the required index in a dictionary, which is passed to the **DataFrame** method.

Code:

```
#Creating Series
name=['Duck', 'Rat', 'Cat', 'Snake']
name=pd.Series(name)
salary=[1000, 3421, 1127, 2379]
salary=pd.Series(salary)
post=['HR', 'CEO', 'ME', 'CFO']
post=pd.Series(post)
#DataFrame from Series
df3=pd.DataFrame({'Emp_Name': name, 'EMP_Salary':salary,
    'Emp_Post':post})
print(df3)
```

Output:

_	Emp_Name	EMP_Salary	Emp_Post
0	Duck	1000	HR
1	Rat	3421	CEO
2	Cat	1127	ME
3	Snake	2379	CFO

Students_df Data Frame

The following examples use the **Students_df Data Frame**. The student is expected to create the **Data Frame**, using the code that follows.

Code:

```
import pandas as pd
Students={'Names':['Amit', 'Ajay', 'Atul', 'Abhay',' Amay' ],
    'Age':[18,19,17,18,19],'Perc_Prev':[78,90.2,87.2,79.3,92] }
Students_df=pd.DataFrame(Students)
Students_df
```

Output:

_	Names	Age	Perc_Prev
0	Amit	18	78.0
1	Ajay	19	90.2
2	Atul	17	87.2
3	Abhay	18	79.3
4	Amay	19	92.0

20.6 OPERATIONS ON ROWS AND COLUMNS OF A DATA FRAME

Having seen the creation of a **Data Frame**, let us move on to the operations that can be applied to rows and columns of a **Data Frame**. This section presents some of the most important methods and procedures to access the rows and columns of a **Data Frame**.

20.6.1 Adding a Column in a Data Frame

In a **Data Frame**, we can add a column in many ways, one of which is using a list. To add a column in any given **Data Frame**, perform the following steps:

- Create a List, say **L**
- Assign the list to the name of the Data Frame followed by the name of the column in square brackets. That is,
  <name of the Data Frame>['<Column Name'>] = L

For example, the following code adds a column named **Perc** in a **Data Frame** called **Students_df**.

Code:

```
#Addition of Columns
Students_df['Perc']=[90,87.2,72,69,89]
Students_df
```

Output:

	Names	Age	Perc_Prev	Perc
0	Amit	18	78.0	90.0
1	Ajay	19	90.2	87.2
2	Atul	17	87.2	72.0
3	Abhay	18	79.3	69.0
4	Amay	19	92.0	89.0

20.6.2 Deleting Column from the Data Frame

The **drop** function is used to drop rows/columns from a **Data Frame**. This function takes the following parameters:

- A list containing the names of the columns to be removed.
- **axis**, which in this case is assigned the value 1 (as we wish to delete a column/columns).
- **inplace**, which must be **True** if the changes are to be made permanent.

For example, the following code deletes the **Perc** column from a **Data Frame** called **Students_df**.

Code:

```
#Deleting column
Students_df.drop(['Perc_Prev'], axis=1)
```

Output:

_	Names	Age	Perc	Perc
0	Amit	18	90.0	90.0
1	Ajay	19	87.2	87.2
2	Atul	17	72.0	72.0
3	Abhay	18	69.0	69.0
4	Amay	19	89.0	89.0

Note that since we have not used the **inplace** parameter. The changes are not permanent. If **Students_df** is displayed, the column still appears.

Code:

```
Students_df
```

Output:

_	Names	Age	Perc_Prev	Perc
0	Amit	18	78.0	90.0
1	Ajay	19	90.2	87.2
2	Atul	17	87.2	72.0
3	Abhay	18	79.3	69.0
4	Amay	19	92.0	89.0

However, if **inplace=True** is passed as an argument to the **drop** method, the changes are done for good.

Code:

```
#Deleting column
Students_df.drop(['Perc_Prev'], axis=1, inplace=True)
Students_df
```

_	Names	Age	Perc
0	Amit	18	90.0
1	Ajay	19	87.2
2	Atul	17	72.0
3	Abhay	18	69.0
4	Amay	19	89.0

20.6.3 Adding a Row in a Data Frame

The **concat** function helps to add a row to a **Data Frame**. A **Data Frame** containing the values in the row to be added can be concatenated with the given **Data Frame**, as shown in the following code.

Code:

```
#Create a Data Frame
df1=pd.DataFrame([['Kim',17,92.8]], columns=['Names','Age',
    'Perc'])
print(df1)
#Concatenate with the existing Data Frame
Students_df = pd.concat([Students_df,df1])
```

Output:

_	Names	Age	Perc
	Amit	18	90.0
1	Ajay	19	87.2
2	Atul	17	72.0
3	Abhay	18	69.0
4	Amay	19	89.0
0	Kim	17	92.8

20.6.4 Deleting Row from the Data Frame

The **drop** function is used to drop rows/columns from a **Data Frame**. This function takes the following parameters:

- A list containing the names of the rows
- **axis**, which in this case is assigned the value 0 (as we wish to delete a row)
- **inplace**, which must be True if the changes are to be made permanent

For example, the following code deletes the fifth row from a **Data Frame** called **Students_df**.

Code:

```
Students_df.drop([4], axis=0, inplace=True)
```

Output:

_	Names	Age	Perc
0	Amit	18	90.0
1	Ajay	19	87.2
2	Atul	17	72.0
3	Abhay	18	69.0
0	Kim	17	92.8

20.7 DEALING WITH ROWS

Having seen the creation of a **DataFrame** and basic operations on a **DataFrame**, let's have a brief look at the various operations on rows of a DataFrame.

20.7.1 loc[] and iloc[]

The **loc[]** object can be used to access the rows of a **Data Frame**. This takes the name of the row as its input and helps us to access the desired row. The **iloc[1]** object, on the other hand, helps in locating a row using its index. As stated earlier, **iloc** comes to the rescue when only the zero-based indexes can be used for accessing the data.

For example, the following code accesses the fourth row (having index=3) using **loc[]** and **iloc[]**.

Code:

```
Students_df.loc[3]
```

Output:

```
Names Abhay
18 Perc
69 Name: 3, dtype: object
```

Code:

```
Students_df.iloc[2]
```

Output:

```
Name Atul
Age 17
Perc 72
Name: 2, dtype: object
```

20.7.2 rename

The **rename** method helps us to rename the rows or columns of a given **Data Frame**. This function takes the following parameters:

- Name of the Column/ row followed by the new name, separated by a colon.
- **inplace** = True, if the changes are intended to be permanent

For example, the following code renames the **Perc** column of the **Students_df Data Frame** to **Percentage**.

```
Students_df.rename(columns={"Perc":"Percentage"})
```

_	Names	Age	Percentage
0	Amit	18	90.0
1	Ajay	19	87.2
2	Atul	17	72.0
3	Abhay	18	69.0

Using this method, many columns can also be renamed. Also, if the value of **inplace = True**, then changes made are permanent.

TIPS

- <name of the Data Frame>.*columns gives the list of columns of the Data Frame.*

EXAMPLE:

*To see the names of the columns of the **Students_df Data Frame**, issue the following command.*

```
Students_df.columns
```

Output:

```
Index(['Names', 'Age', 'Percentage'], dtype='object')
```

- The **rename** function renames the rows or columns of a **Data Frame**.
- The **unique** function finds unique values in a column of a **Data Frame**.

EXAMPLE:

*To see the unique values of the **age** column of the **Students_df** Data Frame issue the following command.*

```
Students_df.Age.unique()
```

Output:

```
array([18, 19, 17], dtype=int64)
```

- The **nunique** function finds the number of unique values in a **Data Frame** column.

EXAMPLE:

Students_df.Age.nunique()

Output:

3

- String functions can be used to apply string operations to the names of Columns

20.8 ITERATING A PANDAS DATA FRAME

This section presents various methods to iterate through a **Pandas Data Frame**. The first subsection deals with the methods for rows and the second deals with the method for columns. This section uses the **Students_df1 Data Frame**. The reader is requested to create this **Data Frame** before proceeding any further.

Code:

```
import pandas as pd
Students={'Name':['Amit',  'Ajay',  'Atul',  'Abhay',  'Amay'],
'Age':[18,  19,  17,  18,  19],'Marks':[78,  90.2,  87.2,  79.3,  92],
'City':['Faridabad', 'New Delhi', 'Faridabad', 'Delhi', 'New Delhi']}
```

```
Students_df1=pd.DataFrame(Students)
Students_df1
```

Output:

	Name	Age	Marks	City
0	Amit	18	78.0	Faridabad
1	Ajay	19	90.2	New Delhi
2	Atul	17	87.2	Faridabad
3	Abhay	18	79.3	Delhi
4	Amay	19	92.0	New Delhi

20.8.1 Iterating Pandas Data Frame Rows

There are many ways to iterate through **Pandas Data Frame rows**. This subsection discusses some of the most important methods to do so.

20.8.1.1 *iterrows()*

The **pandas.DataFrame.iterrows** method helps us to iterate through the rows of a **Data Frame**. This method does not take any argument and yields the **index** of a row and a **Series** representing the row.

For example, the following code iterates through the **Students_df1 Data Frame** using the **iterrows** method.

Code:

```
for index, row in Students_df1.iterrows():
    print(index,' : ', row)
```

Output:

```
0 : Name Amit
Age 18
Marks 78
City Faridabad
Name: 0, dtype: object
```

```
1 : Name Ajay
Age 19
Marks 90.2
City New Delhi
Name: 1, dtype: object
2 : Name Atul
Age 17
Marks 87.2
City Faridabad
Name: 2, dtype: object
3 : Name Abhay
Age 18
Marks 79.3
City Delhi
Name: 3, dtype: object
4 : Name Amay
Age 19   .
Marks 92
City New Delhi
Name: 4, dtype: object
```

20.8.1.2 index

The **pandas.DataFrame.index** attribute may also be used to iterate over a given **Data Frame** rows. This attribute gives the **index** of the rows and is particularly useful if some attributes of the **Data Frame** are to be seen/ manipulated iteratively. For example, the following code prints *"<Name of the student> lives in <city>"* for each row of the **Students_df1 Data Frame**.

Here, *<name of the student>* is the value of the attribute name in each row, and <city> is the value of the attribute city in each row.

Code:

```
for ind in Students_df1.index:
    print(Students_df1['Name'][ind],' lives in ', Students_
    df1['City'][ind])
```

Output:

```
Amit lives in Faridabad
Ajay lives in New Delhi
```

```
Atul lives in Faridabad
Abhay lives in Delhi
Amay lives in New Delhi
```

20.8.1.3 itertuples()

The **DataFrame.itertuples**() method can also be used to iterate through a **Pandas Data Frame**. This method gives the rows as tuples that map the attribute and the value of each item in a row.

For example, the following code iterates through the **Students_df1 Data Frame** using the **itertuples** method.

Code:

```
for t in Students_df1.itertuples():
    print(t)
```

Output:

```
Pandas(Index=0, Name='Amit', Age=18, Marks=78.0, City='Faridabad')
Pandas(Index=1, Name='Ajay', Age=19, Marks=90.2, City='New Delhi')
Pandas(Index=2, Name='Atul', Age=17, Marks=87.2, City='Faridabad')
Pandas(Index=3, Name='Abhay', Age=18, Marks=79.3, City='Delhi')
Pandas(Index=4, Name='Amay', Age=19, Marks=92.0, City='New Delhi')
```

TIP
- *The **loc** and **iloc** can also be used to iterate through a **Data Frame**.*
- *The **apply** method can also be used to iterate through a **Data Frame**. However, it requires the know-how of **lambda functions**, which is beyond the scope of this book.*

20.8.2 Iterating Over Columns

There are many ways to iterate through **Pandas Data Frame** columns. This section discusses some of the most important methods to do so.

20.8.2.1 iteritems()

The **pandas.DataFrame.iteritems** can be used to iterate over **Pandas Data Frame** columns. The method does not take any argument and yields the following.

- **label:** Name of the column
- **content:** A **Series** representing the column.

For example, the following code iterates through the **Students_df1 Data Frame** using the **iteritems** method.

Code:

```
for label, col in Students_df1.iteritems():
    print(label,' : ', col)
```

Output:

```
Name : 0 Amit
1 Ajay
2 Atul
3 Abhay
4 Amay
Name: Name, dtype: object
Age : 0 18
1 19
2 17
3 18
4 19
Name: Age, dtype: int64
Marks : 0 78.0
1 90.2
2 87.2
3 79.3
4 92.0
Name: Marks, dtype: float64
City : 0 Faridabad
1 New Delhi
2 Faridabad
3 Delhi
4 New Delhi
Name: City, dtype: object
```

20.8.2.2 list

The **list** method extracts the names of the columns of a **DataFrame**. This list can be used to access the required data using the **for** loop.

For example, the following code iterates through the **Students_df1 Data Frame** using the **list** method.

Code:

```
for col_name in list(Students_df1):
    print(Students_df1[col_name])
```

Output:

```
0 Amit
1 Ajay
2 Atul
3 Abhay
4 Amay
Name: Name, dtype: object
0 18
1 19
2 17
3 18
4 19
Name: Age, dtype: int64
0 78.0
1 90.2
2 87.2
3 79.3
4 92.0
Name: Marks, dtype: float64
0 Faridabad
1 New Delhi
2 Faridabad
3 Delhi
4 New Delhi
Name: City, dtype: object
```

TIP *The **head** method shows the first five rows of a Data Frame and the **tail** method shows the last five rows of the Data Frame.*

20.9 CONCLUSION

This chapter discussed some of the common ways to create **Pandas Series** and **Data Frames**. The topics like indexing, slicing, and Boolean Index have also been discussed. This chapter also introduced functions for sorting the values, finding maximum, minimum, median, standard deviation, mean, and count of values along with head, tail, and description. The operations on Rows and Columns of a **Data Frame** have also been dealt with in detail.

The next chapter takes the discussion forward and introduces topics like aggregation, pivoting, joining, Merging, and Concatenation. The chapter also discusses the methods to import and export Data between CSV files and **Data Frames**.

The reader is expected to attempt exercises given at the end of this chapter and those in the Workbook to develop a better understanding of the topic.

EXERCISES

Multiple Choice Questions

1. Which of the following represents an array of actual data and an associated array of indices?

    (a) Series   **(b)** Data Frame

    (c) Both   **(d)** None of the Above

2. Which of the following represents two-dimensional data and an associated array of indices?

    (a) Series   **(b)** Data Frame

    (c) Both   **(d)** None of the Above

3. Which of the following can be used to create a **Series**?

    (a) List   **(b)** NumPy array

    (c) Dictionary   **(d)** All of the above

4. Which of the following can be used to create a Data Frame?

    (a) List of Lists

    (b) 2-D NumPy array

(c) Dictionary, in which elements are lists

(d) All of the above

5. Which of the following can be used to access an element from a **Series**?

(a) loc   (b) iloc

(c) both   (d) None of the above

6. Which of the following can take -1 as an argument?

(a) loc   (b) iloc

(c) Both   (d) None of the above

7. Which of the following functions display the first five rows of a Data Frame?

(a) head   (b) tail

(c) both   (d) None of the above

8. Boolean Indexing can be applied to which of the following?

(a) Series   (b) Data Frame

(c) Both   (d) None of the above

9. axis=0, in a Data Frame, represents

(a) rows   (b) columns

(c) Both   (d) None of the above

10. Which method finds unique values in a column of a Data Frame?

(a) unique   (b) nunique

(c) Both   (d) None of the above

11. Which method finds the number of unique values in a Data Frame column?

(a) unique   (b) nunique

(c) Both   (d) None of the above

12. Which function is used to find the means of a **Series**?

(a) mean   (b) average

(c) Both   (d) None of the above

Theory

1. What is a **Pandas Series**?

2. What is a **Pandas DataFrame**?

3. Discuss various methods to create a **Pandas Series**. Give examples of each.

4. Discuss various methods to create a **Pandas DataFrame**. Give examples of each.

5. Differentiate between loc and iloc.

6. Explain slicing in **Pandas Series**.

7. Explain the describe function of the **Pandas Series**.

8. Explain the procedure to add a column in a **DataFrame**.

9. Explain the procedure to delete a column from a **DataFrame**. What is the importance of **inplace**?

10. Explain the procedure to add a row in a **DataFrame**.

11. Explain the procedure to delete a row from a **DataFrame**. What is the importance of **inplace**?

12. How can you iterate through a **DataFrame**?

PANDAS–II

Objectives

After reading this chapter, the reader should be able to

- Understand the importance of **head()**, **tail()**, and **describe()**
- Understand **Boolean Indexing**.
- Use methods for showing descriptive statistics of a **Data Frame**.
- Read from a **CSV** file
- Handle missing values.

21.1 INTRODUCTION

The last chapter introduced **Pandas** and discussed the two most important data structures in **Pandas: Series** and the **Data Frame**. The chapter provided an insight into **Pandas** and discussed some of its most important features.

This chapter takes the discussion forward and explains some of the most important methods of **Pandas Data Frame**. This chapter also discusses reading data from a **csv**. The procedures to deal with missing values have also been discussed in this chapter. This chapter has been organized as follows (Figure 21.1).

Let's now dive into advanced topics in **Pandas** and move toward **Data Science**.

21.2 DATA FRAME METHODS: HEAD, TAIL, AND DESCRIBE

This section gives an overview of the methods that help us to see a **Data Frame**, to describe it, and to apply mathematical functions to the columns of a **Data Frame**. This section also revisits **Boolean indexing**.

The Section that follows makes use of the following **Data Frame**

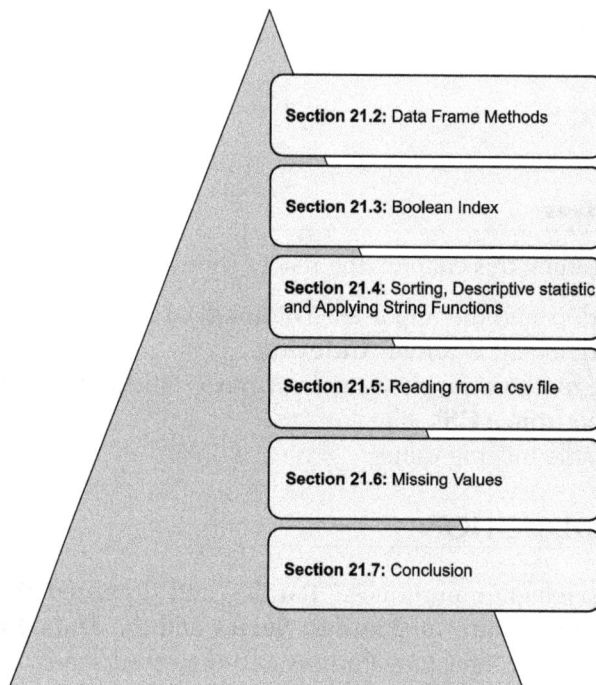

Section 21.2: Data Frame Methods

Section 21.3: Boolean Index

Section 21.4: Sorting, Descriptive statistic and Applying String Functions

Section 21.5: Reading from a csv file

Section 21.6: Missing Values

Section 21.7: Conclusion

FIGURE 21.1 Organization of the chapter.

Code:

```
Students={'Names':['Amit', 'Ajay', 'Atul', 'Abhay', 'Amay', 'Biyoy',
'Bimal', 'Binay', 'Darvesh', 'Durgesh', 'Ela'], 'Age':[18, 19,
17, 18, 19, 20, 21, 20, 19, 17, 20], 'Perc_Prev':[78, 90.2, 87.2, 79.3,
92, 91, 90, 89, 92, 80, 67] }

Students_df=pd.DataFrame(Students)

Students_df
```

Output:

—	Names	Age	Prev_Perc
0	Amit	18	78.0
1	Ajay	19	90.2
2	Atul	17	87.2
3	Abhay	18	79.3
4	Amay	19	92.0
5	Biyoy	20	91.0
6	Bimal	21	90.0
7	Binay	20	89.0
8	Darvesh	19	92.0
9	Durgesh	17	80.0
10	Ela	20	67.0

21.2.1 Functions: Head, Tail, and Describe

21.2.1.1 *head()*

The **head**() function displays the first five rows of a **Data Frame**. For example, **Students_df.head**() displays the first five rows of **Students_df**.

21.2.2 tail()

The **tail**() function displays the last five rows of a **Data Frame**. For example, **Students_df.tail**() displays the last five values of **Studentes_df**.

21.2.3 columns

This displays the **column**(s) of the given **Data Frame**. For example, **Students_df.columns** show the columns of the **Data Frame Students_df**.

Code:

```
Students_df.columns
```

Output:

```
Index(['Names', 'Age', 'Perc_Prev'], dtype='object')
```

21.2.4 describe()

The **describe** method of the **Data Frame** object displays the following information about the numeric columns of a **Data Frame**

- **count:** This gives the number of the items in the Column.
- **mean:** This gives the average of the items in the Column.
- **min:** This gives the minimum of the items in the Column.
- **max:** This gives the maximum of the items in the Column.
- **25%, 50%, 75%:** These three denote the 25%, 50% and the 75% values.
- **std:** This gives the standard deviation of the items in the Column.

An example of the describe function is as follows.

Code:

```
Students_df.describe()
```

Output:

	Age	Perc_Prev
count	11.000000	11.000000
mean	18.909091	85.063636
std	1.300350	7.983267
min	17.000000	67.000000
25%	18.000000	79.650000
50%	19.000000	89.000000
75%	20.000000	90.600000
max	21.000000	92.000000

21.3 BOOLEAN INDEX

In a **Data Frame**, the required condition can be specified inside the square brackets, to get the rows that satisfy the given condition. For example, to display the records in which the age of a student is greater than 19 from the **Students_df**, the following code can be written.

Code:

```
Students_df[Students_df['Age']>18]
```

Output:

Names	Age	Perc_Prev	Perc_Prev
1	Ajay	19	90.2
4	Amay	19	92.0
5	Biyoy	20	91.0
6	Bimal	21	90.0
7	Binay	20	89.0
8	Darvesh	19	92.0
10	Ela	20	67.0

Likewise, you can state any Boolean statement involving a column, in the square brackets.

21.4 SORTING, DESCRIPTIVE STATISTICS, AND APPLYING STRING FUNCTIONS

This section discusses the various methods of the **Data Frame** data structure of **Pandas**: to sort the values of a **Column**, find the minimum value, maximum value, the mean, median, mode, and standard deviation of the values in a column.

21.4.1 sort_values()

This method sorts the items of the given **Column** of a **Data Frame**. This also displays the indexes of the sorted arrays. For example, **Students_df['Age']. sort_values()** sorts the elements of the columns "**Age**."

Code:

```
Students_df['Age'].sort_values()
```

Output:

2 17
9 17
0 18
3 18
1 19

```
4 19
8 19
5 20
7 20
10 20
6 21
Name: Age, dtype: int64
```

21.4.2 Finding Maximum, Minimum, Median, Standard Deviation, Mean, and Count of Values

The following methods (Table 21.1) help us to get the required statistics from a column of a given **Data Frame**.

TABLE 21.1 Functions to find the maximum, minimum, sum, median, standard deviation, variance, mode, quantile, and value count of a column of a Data Frame.

Name of the function	Explanation	Example	Output
max	This function finds the maximum value from the specified column of a **Data Frame**.	Students_df['Age'].max()	21
min	This function finds the minimum value from the specified column of a **Data Frame**.	Students_df['Age'].min()	17
sum	This function finds the sum of values from the specified column of a **Data Frame**.	Students_df['Age'].sum()	208
median	This function finds the median of the value from the specified column of a **Data Frame**.	Students_df['Age'].median()	19.0
std	This function finds the standard deviation of values from a given column of a **Data Frame**.	Students_df['Age'].std()	1.3003
var	This function finds the variance of values from the specified column of a **Data Frame**.	Students_df['Age'].var()	1.6909
mode	This function finds the mode of values from the specified column of a **Data Frame**.	Students_df['Age'].mode()	0 19 1 20 dtype: int64

Name of the function	Explanation	Example	Output
quantile	This function finds the qauntile of values from the specified column of a **Data Frame**. For example, quantile(0.25) finds the 25th quantile and quantile(0.50) finds the 50th quantile of a given column of a Data Frame.	Students_df['Age']. quantile(0.25) Students_df['Age']. quantile(0.75)	18.0 19.0
value_counts	This function counts the frequencies of values in the specified column of a **Data Frame**.	c=S Students_df['Age']. value_counts() print(c)	Value Counts 20 3 19 3 18 2 17 2 21 1 Name: Age, dtype: int64

21.4.3 String Functions

The functions applicable to strings can also be applied to the columns of a **Pandas Data Frame**. For example, the following code converts all the strings in the **Names** column of the **Students_df Data Frames** to the upper case and then in the lower case respectively. The reader is expected to try other string functions with the **str** type **Columns** of a **Data Frame**.

Code:

```
Students_df['Names'].str.upper()
```

Output:

```
0 AMIT
1 AJAY
2 ATUL
3 ABHAY
4 AMAY
5 BIYOY
6 BIMAL
7 BINAY
8 DARVESH
```

```
9 DURGESH
10 ELA
Name: Names, dtype: object
```

Code:

```
Students_df['Names'].str.lower()
```

Output:

```
0  amit
1  ajay
2  atul
3  abhay
4  amay
5  biyoy
6  bimal
7  binay
8  darvesh
9  durgesh
10 ela
Name: Names, dtype: object
```

21.5 READING FROM A CSV FILE: PANDAS.READ_CSV

The **read_csv** method of **Pandas** helps us to read a **Comma Separated File** or a **csv** file. However, it may be stated that the separator in a **csv** file is not always a comma. Table 21.2 presents the parameters of the method and the examples that follow illustrate the usage of this method.

TABLE 21.2 The arguments of read_csv method.

Argument	Type	Explanation
filepath_or_buffer	str	This argument represents the path of the file. This path may include any of the following: (*i*) http (*ii*) ftp or (*iii*) file
sep	str	The default value of this argument is ",". This argument represents the delimiter to use.

Argument	Type	Explanation
header	int, list of int	This argument represents the row number(s) to use as the start of the data. The default behavior is to infer the column names.
names	array-like	This is an optional argument. It represents the list of column names to use.
squeeze	bool	The default value of this argument is **False**. If the parsed data only contains one column then return a Series.
mangle_dupe_cols	bool	The default value of this argument is **True**. This helps us to retain only the first one of the duplicate columns if they exist.

Alert: *The .csv files used in the following illustrations are given in the Appendix of this Book.*

Illustration 21.1:

*Read the **Data_Pandas1.csv** file.*

Solution:

The **read_csv** method is used to read a **csv** file. It takes the name of the file as an argument. The following code reads the **"Data_Pandas1.csv."**

Code:

```
Data=pd.read_csv('Data_Pandas1.csv')
Data
```

Output:

```
0    Tanu\t30\t1\tFalse
1    Manu\t28\t2\tTrue
2    Kim\t32\t1\tTrue
3    Lakshay\t24\t2\tFalse
4    Krishna\t18\t0\tTrue
```

Illustration 21.2:

*Note that the original file contained "**\t**" as a separator; it is being shown in the **Data Frame**. Read the file correctly.*

Solution:

To read the file correctly, the **sep** argument is set to **"\t."** The following code reads the **"Data_Pandas1.csv,"** setting **sep=**"\t."

Code:

```
Data=pd.read_csv('Data_Pandas1.csv', sep = '\t')
Data
```

Output:

	Harsh	36	10	True
0	Tanu	30	1	False
1	Manu	28	2	True
2	Kim	32	1	True
3	Lakshay	24	2	False
4	Krishna	18	0	True

Illustration 21.3:

In the above two illustrations, the first row is wrongly read as a header, by the function. Rectify this.

Solution:

To rectify this, the **header** argument is set to **None**.

Code:

```
Data=pd.read_csv('Data_Pandas1.csv', sep='\t', header=None)
Data
```

Output:

	0	1	2	3
0	Harsh	36	10	True
1	Tanu	30	1	False
2	Manu	28	2	True
3	Kim	32	1	True
4	Lakshay	24	2	False
5	Krishna	18	0	True

Illustration 21.4:

Suppose in the above illustration, we want to read the data from a csv file from the fourth row onwards (after row number = 2). What changes must be made in the code to accomplish this task?

Solution:

To accomplish the given task, the **header** argument is set to **2**.

Code:

```
Data=pd.read_csv('Data_Pandas2.csv', header=2)
Data
```

Output:

	Manu	28	2	True
0	Kim	32	1	True
1	Lakshay	24	2	False
2	Krishna	18	0	True

21.6 MISSING VALUES

The data imported from a **csv** file, or for that matter any other source may contain missing values. **None** and **NaN** are two standard missing value representations in **Pandas**. The **None** is a python object, which is recognized by **Pandas**. The **NaN** means "Not a Number." It is recognized by all the systems. The following discussion will help us to recognize and replace missing values in a **Pandas Data Frame**.

21.6.1 To Check Null Values

The following methods will help us to check null values.

- **isnull():** This function returns a Boolean **Data Frame** containing **True** at the positions having null values and **False** at the rest of the positions.
- **notnull():** This function returns a Boolean **Data Frame** containing **False** at the positions having null values and **True** at the rest of the positions.

The examples that follow use the following **Data Frame**, which is created using the **Data_Pandas6**.csv file.

Code:

```
import pandas as pd
DataFrame1=pd.read_csv('Data_Pandas6.csv', header=None)
DataFrame1
```

Output:

	0	1	2	3
0	Harsh	36.0	10	True
1	Tanu	30.0	1	False
2	Manu	28.0	2	True
3	Kim	32.0	1	True
4	Lakshay	24.0	2	False
5	NaN	NaN	NaN	NaN
6	Krishna	18.0	NaN	True

Illustration 21.5:

Apply **isnull()** method to a **Data Frame** and analyze the result.

Solution:

The following code applies the **isnull()** method to the **Data Frame** called **DataFrame1** and returns a **Data Frame**, which contains **True** at the positions having **NaN** and **False** at the other positions.

Code:

```
DataFrame1.isnull()
```

Output:

	0	1	2	3
0	False	False	False	False
1	False	False	False	False
2	False	False	False	False
3	False	False	False	False
4	False	False	False	False
5	True	True	True	True
6	False	False	False	False

Illustration 21.6:

*Apply the **notnull()** method to a **DataFrame** and analyze the result.*

Solution:

The following code applies **notnull**() method to the **DataFrame** called **DataFrame1** and returns a **DataFrame**, which contains **False** at the positions having **NaN** and **True** at the other positions.

Code:

```
DataFrame1.notnull()
```

Output:

	0	1	2	3
0	True	True	True	True
1	True	True	True	True
2	True	True	True	True
3	True	True	True	True
4	True	True	True	True
5	False	False	False	False
6	True	True	True	True

21.6.2 dropna()

The data imported from a **csv** file (or other sources for that matter) may contain missing values. The **Data Frame** contains **NaN** at these positions. The **dropna**() method helps the programmer to deal with these values. This method takes the following parameters:

- **axis:** The data type of this argument is integer or string. In the case of integers, the value of the **axis** can be 0 or 1 and in the case of a string, its value can be **"index"** or **"column."**
- **how:** This argument decides, which row/column is to be dropped. If the value of this argument is **"any,"** it drops the row/column having any element as **Null**. If the value of this argument is all, it only drops the rows or columns wherein all the elements are **Null**.

- **thresh:** In case the above argument is **"any,"** it makes sense to tell the method the minimum number of elements that must be **null** to drop the row/column.
- **inplace:** if the value of **inplace** is **True**, the **Data Frame** itself is changed.

Illustration 21.7:

*Use the **dropna()** method to a **Data Frame** to remove rows containing all NaN's.*

Solution:

The following code drops the rows containing all **NaN's** from the **Data Frame** called **DataFrame1**.

Code:

```
DataFrame1.dropna()
```

Output:

–	0	1	2	3
0	Harsh	36.0	10	True
1	Tanu	30.0	1	False
2	Manu	28.0	2	True
3	Kim	32.0	1	True
4	Lakshay	24.0	2	False
6	Krishna	18.0	NAN	True

21.6.3 fillna()

The **fillna()** method is used to replace the **Null** values with some object. This method takes the following arguments.

- **axis:** The datatype of this argument is integer or string. In the case of integers, the value of the **axis** can be 0 or 1 and in the case of a string, its value can be **index** or **column**.
- **inplace:** if the value of **inplace** is **True**, the **Data Frame** itself is changed.
- **value:** This argument represents the value to be replaced for the **Null** elements.

- **method:** This argument is used if the argument value is not passed. The following methods replace the null values with those stated.
  - **bfill:** It replaces the place with value in the Previous index.
  - backfill or
  - ffill which fills the place with value in the Forward index or
    limit: The **datatype** of this argument is an integer. This specifies the maximum number of consecutive forward/backward **NaN** value fills.
- **replace():** The **replace** method is used to replace a string, regular expression, list, dictionary, etc. from a **Data Frame**. As a matter of fact, the regular expression is one of the most powerful techniques to deal with strings. However, it is beyond the scope of this book.
- **to_replace:** [str, regex, list, dict, Series, numeric, or None] This argument represents the pattern that we are trying to replace in **Data Frame**.
- value, inplace, limit, and method are the same as those explained in the previous subsection.
- **interpolate():** The **interpolate()** method is used to fill NA values in the **Data Frame** or **Series**. This method uses a different interpolation technique to fill the missing values.

The illustrations that follow exemplify the usage of this method.

Illustration 21.8:

*Replace the NaN's in a **Data Frame** with the string "**Not Known.**"*

Solution:

The following code replaces the **NaN's** from the **Data Frame** called **DataFrame1** to **Not Known**.

Code:

```
DataFrame1.fillna('Not Known')
```

Output:

	0	1	2	3
0	Harsh	36	10	True
1	Tanu	30	1	False
2	Manu	28	2	True
3	Kim	32	1	True

4	Lakshay	24	2	False
5	Not Known	Not Known	Not Known	Not Known
6	Krishna	18	NAN	True

Illustration 21.9:

*Replace the NaN's in a **Data Frame** with the string "**bfill**."*

Solution:

The following code replaces the **NaN's** from the **Data Frame** called **DataFrame1** using the "**bfill**" method.

Code:

```
DataFrame1.fillna(method='bfill')
```

Output:

_	0	1	2	3
0	Harsh	36.0	10	True
1	Tanu	30.0	1	False
2	Manu	28.0	2	True
3	Kim	32.0	1	True
4	Lakshay	24.0	2	False
5	Krishna	18.0	NAN	True
6	Krishna	18.0	NAN	True

Illustration 21.10:

*Replace the NaN's in a **Data Frame** by the string "**ffill**."*

Solution:

The following code replaces the **NaN's** from the **Data Frame** called **DataFrame1** using the "**ffill**" method.

Code:

```
DataFrame1.fillna(method='ffill')
```

Output:

	0	1	2	3
0	Harsh	36.0	10	True
1	Tanu	30.0	1	False
2	Manu	28.0	2	True
3	Kim	32.0	1	True
4	Lakshay	24.0	2	False
5	Lakshay	24.0	2	False
6	Krishna	18.0	NAN	True

21.7 CONCLUSION

This chapter discussed methods related to **Data Frames** and ways to deal with missing data. Reading of data from a **csv** file and SQL database have also been discussed in the chapter. The Appendix of this book takes the discussion forward and introduces some ideas for developing a small project using **Pandas**. The web resources also contain the files used in this chapter. The reader is encouraged to have a look at the references given at the end of the book for a detailed discussion on some of the assorted topics. Let us now explore the exercises.

EXERCISES

Multiple Choice Questions

1. Which method displays the first 5 rows of a **Data Frame**?

   (a) head   **(b)** tail

   (c) describe   **(d)** hover

2. Which method displays the last five rows of a **Data Frame**?

   (a) head   **(b)** tail

   (c) describe   **(d)** hover

3. Which method displays the most important statistics of numeric columns of a **Data Frame**?

    (a) head   **(b)** tail

    (c) describe   **(d)** hover

4. To show all the rows of the **Students_df Data Frame**, for which the age of the student is less than and equal to 17, which of the following needs to be issued?

    (a) Students_df[Students_df['Age']<18]

    (b) Students_df['Age']>18

    (c) Students_df[Students_df['Age']]>18

    (d) None of the above

5. Which of the following is used to sort the values of a given column?

    (a) sort_values   **(b)** sort

    (c) sorted   **(d)** All of the above

6. Which method can be used to find the variance of a numeric column of a **Data Frame**?

    (a) variance   **(b)** var

    (c) Both   **(d)** None of the above

7. Which method can be used to find the standard deviation of a numeric column of a **Data Frame**?

    (a) standard_d   **(b)** std

    (c) Both   **(d)** None of the above

8. Which method can be used to count the unique values in a column of a **Data Frame**?

    (a) value_count   **(b)** count

    (c) Both   **(d)** None of the above

9. Which of the following can be used to read a **csv** file in **Pandas**?

    (a) read_csv   **(b)** csv_read()

    (c) read()   **(d)** All of the above

10. Which argument in the **read_csv** method is used to specify the **delimiter**?

    (a) sep   **(b)** separator

    (c) dist   **(d)** None of the above

11. Which of the following are used to check **Null** values in a **Data Frame**?

    (a) isnull()   **(b)** notnull()

    (c) Both   **(d)** None of the above

12. Which of the following is used to replace **Null** values in a **Data Frame**?

    (a) fillna()   **(b)** nafill()

    (c) fill()   **(d)** None of the above

Theory

1. Discuss the ways to find missing values in a **Data Frame**?
2. How are the above values replaced in a **Data Frame**?
3. Explain the describe method of **Pandas Data Frame**.
4. Write the steps to read data from a **csv** file.
5. Discuss the **sep** and **header** argument of **read_csv**.
6. What are the types of **Null** values in **Pandas**?
7. Write a short note on how to deal with **Null values**?
8. Write a short note on why do we generally have **Null** values in a **CSV** file?

PROBLEMS FOR PRACTICE: PROGRAMMING QUESTIONS

SECTION I: PROCEDURAL PROGRAMMING

Conditional statements

1. Ask the user to enter a four-digit number and check whether the second digit is one more than the third digit.

2. In the above question if the condition is false, swap the digits at the third and the second place and increment the digit and the units' place by one if it is not 9. If the digit at the units' place is 9, then do not change the digit.

3. Ask a user to enter her monthly salary, her house rent (or EMI of the home loan), her car EMI, bill of the newspaper, the amount she spends on other things in a month. Now find if the amount left is sufficient enough to start an SIP. Note that an SIP can be started even with $500.

4. Ask the user to enter his total savings. In India, if the savings are above 10,000,000 rupees, then the person does not need to pay any taxes. Also, this person is entitled to get a subsidy from the government. If the savings are above 1,000,000 rupees, but below the above specified amount, he is liable to pay 30% of his savings as tax, plus a surcharge of 2% on the tax. Calculate the total tax paid by the person.

5. Ask the user to enter a three-digit number and find the largest digit of the number. Also find the sum of the digits and find if the sum of the digits is same as twice the largest digit.

6. Ask the user to enter marks obtained by a student in 5 subjects. If the person scores more than 90% in a subject, then he gets "A+." If the score is

less than 90% but greater than 85%, he gets "A." If the score is greater than 80%, he gets "A–." Likewise, if the score is greater than 75%, he gets "B+." "B" is awarded to a person scoring more than 70%. A person getting more than 65% but less than 70% gets "B–" and the one getting more than 60% but less than 65% gets "C+." A person getting more than 55% (and less than 60%) gets a "C" and the one getting more than 50% (and less than 55%) gets a "C–." Furthermore, for each grade the corresponding CGPA is as follows.

Grade	CGPA
A+	9
A	8
A–	7
B+	6
B	5
B–	4
C+	3
C	2
C–	1

Find the average CGPA of the student.

7. Find whether the year entered by the user is a multiple of 7, without using the % operator.

8. Find whether the number entered by the user is a multiple of both 5 and 7, without using the % operator.

9. Ask the user to enter a string and find the number of occurrences of vowels in the string.

Looping

10. Ask the user to enter a number and find the number obtained by reversing the order of the digits.

11. Ask the user to enter a decimal number and find its binary equivalent.

12. Ask the user to enter a decimal number and find its octal equivalent.

13. Ask the user to enter a decimal number and find its hexadecimal equivalent.

14. Ask the user to enter an n-digit number and find the digit which is maximum among them.

15. Ask the user to enter a list of numbers (he must enter 0 to quit) and find the maximum number.

16. In the above question find the minimum number.

17. Ask the user to enter in numbers and find their standard deviation and mean of the number entered.

18. In the above question, find the mean deviation.

19. Write a program to generate the pattern of rule 30, as described in the following link.

    https://en.wikipedia.org/wiki/Cellular_automaton

Functions

A data is given to you. The data has many features (columns) and the last column states the class to which it belongs (0 or 1). Each features' data can be segregated into X and Y, where X is the data that belongs to class 0 and Y is the data that belongs to class 1. The relevance of a particular feature can be calculated by numerous methods, one of which is the Fisher Discriminate Ratio.

The Fisher Discriminate Ratio of a feature (a column vector) is calculated using the following formula:

$$\text{FDR} = (\mu_X^1 - \mu_Y^1)^2 / (\sigma_X^2 - \sigma_Y^2)$$

where μ_X is the mean of the data X, μ_Y is the mean of the data Y. The standard deviation of X is σ_X and that of the Y data is σ_Y.

Ask the user to enter the elements of two lists Feature and Label.

20. Create a function **Segregate** which takes the Feature and Label as input and find the vectors X and Y.

21. The **calculate_mean** function should calculate the mean of the input vector.

22. The calculate **standard_deviation** function should calculate the standard deviation of the input vector.

23. The **FDR** function should calculate the **FDR** of a feature.

24. Finally write a program that takes 2D data as input and calculates the FDR of each feature.

    The relevance of a particular feature can also be calculated by the coefficient of correlation.

    The coefficient of correlation of a feature (a column vector) is calculated using the following formula:

$$CC = \frac{X.Y}{X|\times|Y|}, \text{where } |X| \text{ is } \sqrt{x_1^2 + x_2^2 + \ldots + x_m^2}. \text{ For } X = [x_1, x_2, \ldots x_m].$$

    Likewise, $|Y|$ is $\sqrt{y_1^2 + y_2^2 + \ldots + y_m^2}$. For $Y = [y_1, y_2, \ldots y_n]$.

25. Create a function **Segregate** which takes the Feature and Label as input and find the vectors X and Y.

26. The **calculate_mod** function should calculate the |X| for the input vector, X.

27. The **calculate_dot** function should calculate X.Y.

28. The **CORR** function should calculate the correlation coefficient of a feature.

29. Finally write a program that takes 2D data as input and calculate the coefficient of correlation of each feature.

File Handling/Strings

30. Create a file called data and insert data from a text file containing 5 news articles from a news site.

31. Now open the file and find the words beginning with vowels. Make 5 lists of works beginning with each vowel.

32. Draw a histogram of the above data.

33. Make the first letter of each word capital and write the words in 5 separate files.

34. Now, from each file find the words that end with a vowel and place the words in 5 separate files.

35. Check which of these words begin and end with a vowel?

36. From the original file find the word which is repeated maximum number of times.

37. Do the above task for all the words and plot the frequency of each word in a graph.

38. From the original file find which alphabet is used maximum number of times.

39. The reader is expected to read about Huffman code from the following link and encode the file using Huffman code.

https://users.cs.cf.ac.uk/Dave.Marshall/Multimedia/node210.html

40. From the original file, find the string having maximum length.

41. From the original file find the string having "cat" as the substring.

42. From the original file find the strings which are substrings of some other strings in the file.

43. From the original file find the strings which begin with a capital letter.

44. From the original file, find all the email IDs.

45. Find the email IDs which are on yahoo server.

46. Create a regular expression for the land line number in Germany and find all the land line numbers from the file.

47. From the above list find the phone numbers which belong to a particular area (for example, Berlin: 30).

48. Find the words which appears in all the five articles.

49. Find the words, which end with a consonant and contain a vowel.

SECTION II: OBJECT ORIENTED PROGRAMMING

Classes and Objects

You are required to develop software for a **car wash** company. The company wants software that can store the details of a **car** and generate invoices. After due deliberations, it was decided that a class called **car** with the following members can be created.

Data members

(a) Registration Number

(b) Model

(c) Make

(d) Year

(e) Name of the owner

Methods

(f) getdata()	:	Takes data from the users
(g) putdata()	:	Displays data
(h) __init__()	:	Initializes members
(i) del	:	Destructor
(j) capacity	:	Current capacity for strye.

1. Create a class called **car** to facilitate the development of the required software.

2. Make two instances of the class and display the data. The first instance should display the data entered using the **putdata()** function and the second should display the data assigned using the **__init__(self)** method.

3. Create an array of cars. Ask the user to enter the data of n cars and display the data.

4. Find the cars whose registration numbers contain "HR51."

5. Find the cars which were manufactured by "Maruti."

6. Find the cars which were manufactured before 2007.

7. Find the car whose owners name is "Harsh."

8. Find the cars, whose owners' name begin with "A" and were manufactured after 2014.

9. Find the cars which have a certain type of engine (entered by the user).

10. Find the car with maximum capacity.

Operator Overloading

11. Create a class called **vector**, which has three data members

    (a) x1: The x component of the vector

    (b) y1: The y component of the vector

    (c) z1: The z component of the vector

    The class should have a method called **getdata()**, which takes data from the user; **putdata()**, which displays the data; **__init__**, the constructor.

12. Create a class called vectors and make two instances of vector: **v1** and **v2**. Display the data of the two objects.

13. The mod of a vector can be defined as follows. If $v_1 = x_i \; \vec{i} + y_1 \; \vec{j} + z_1 \; \vec{k}$, then $|v_1| = \sqrt{x_1^2 + y_1^2 + z_1^2}$. Create an array of vectors. Ask the user to enter the data of n vectors and find the vector having maximum mod.

14. In the above vectors (Question 13) Find the vectors which have the y component 0.

15. Two vectors v_1 and v_2 can be subtracted by subtracting the corresponding components of the two vectors. That is if $v_1 = x_1 \; \vec{i} + y_1 \; \vec{j} + z_1 \; \vec{k}$ and $v_2 = x_2 \; \vec{i} + y_2 \; \vec{j} + z_2 \; \vec{k}$ then

$$v_1 - v_2 = (x_1 - x_2) \vec{i} + (y_1 - y_2) \vec{j} + (z_1 - z_2) \vec{k}.$$

Using the above concept, overload the - operator for the class.

16. Two vectors v_1 and v_2 can be added by adding the corresponding components of the two vectors. That is if $v_1 = x_1 \; \vec{i} + y_1 \; \vec{j} + z_1 \; \vec{k}$ and $v_2 = x_2 \; \vec{i} + y_2 \; \vec{j} + z_2 \; \vec{k}$ then

$$v_1 - v_2 = (x_1 + x_2) \vec{i} + (y_1 + y_2) \vec{j} + (z_1 + z_2) \vec{k}.$$

Using the above concept, overload the + operator for the class.

17. The dot product of two vectors can be obtained by adding the products obtained by multiplying the corresponding components of the two vectors. That is, if $v_1 = x_1 \; \vec{i} + y_1 \; \vec{j} + z_1 \; \vec{k}$ and $v_2 = x_2 \; \vec{i} + y_2 \; \vec{j} + z_2 \; \vec{k}$ then

$$v_1. \, v_2 = (x_1. \, x_2) + (y_1. \, y_2) + (z_1. \, z_2)$$

Using the above concept, overload the . operator for the class.

18. A hypothetical operation called increment can be defined as follows. If $v_1 = x_1 \; \vec{i} + y_1 \; \vec{j} + z_1 \; \vec{k}$ then

$$v_1 + + = (x_1 + +) \vec{i} + (y_1 + +) \vec{j} + (z_1 + +) \vec{k}$$

Using the above concept, overload the ++ operator for the class.

19. A hypothetical operation called decrement can be defined as follows. If $v_1 = x_1 \vec{i} + y_1 \vec{j} + z_1 \vec{k}$ then

$$v_1 -- = (x_1 --)\vec{i} + (y_1 --)\vec{j} + (z_1 --)\vec{k}$$

Using the above concept, overload the — operator for the class.

20. For the vector class, overload the unarray (–) operator.

Inheritance

21. Create a class called **Book** having the following members.

    (a) Name of the book: String

    (b) Authors(s): List

    (c) Year: Year of publication

    (d) ISSN: String

    (e) Publisher: Name of the publisher

    The class should have **getdata()**, **putdata()** and **__init__()** as its methods.

22. Create two subclasses: **TextBook** and **ReferenceBook** having requisite data members. Demonstrate the use of overriding in the above hierarchy.

23. Now, create three subclasses of the **TextBook** class, namely **SocialScience, Engineering, and Management**. Each class should define its version of **getdata()** and **putdata()**. Make instances of these subclasses and call the method of the derived classes.

24. Create a class called **XBook**, which is a subclass of both **TextBook** and **ReferenceBook** and demonstrate how this can lead to ambiguity.

25. Create a class called **ABC** and craft a class method and an instance method of the class.

Exception Handling

26. Create a class called **array**, which contains an array and **max** which is the maximum number of elements the array can have and methods **getdata()** and **putdata()**, which perform the requisite tasks.

27. Now create a class to raise customized exception. The exception should be raised so that the user cannot enter more elements than **max**.

28. If the user enters anything except for integer, an exception should be raised, and requisite message should be displayed.

29. Now ask the user to enter two indices and divide the numbers into those positions. If the number at the second position is 0, an exception should be raised.

30. Ask the user to enter three indices. These three indices contain the values of "a," "b," and "c" of the quadratic equation $ax^2 + bx + c = 0$. Find the discriminant and the roots of the equation. If the value of $b^2 - 4ac < 0$, an exception should be raised.

SECTION III: DATA STRUCTURES (OPTIONAL)

Sorting and Searching

1. Implement linear search and binary search. Compare the time for searching an element from a list of 500 random numbers.

2. Repeat the experiment for a list of 5000 integers and compare the time for searching an element by the two algorithms. Does increasing the number of elements 10 times increases the running time by 10 folds?

3. Implement Counting Sort. (Reference at the end of this Appendix).

4. Implement Bucket sort. (Refer to the links at the end of this Appendix).

5. Implement a version of Selection Sort which takes O(n log n) time.

6. Now take a list of 500 integers and compare the time for Selection Sort and Bucket Sort.

7. Which of the two: Bucket Sort or Counting Sort takes less time. Are they really comparable?

8. Implement Quick Sort and Merge sort using lists.

9. Take an array of 5000 random integers and compare the time of running of Quick Sort and Merge Sort.

10. Can the average case complexity of Quick Sort be case better?

Stacks and Queues

11. Implement a dynamic stack in which a single placeholder is added when overflow occurs.

12. Implement a dynamic stack in which the number of placeholders is doubled when overflow occurs.

13. Implement a dynamic stack in which the number of placeholders is randomly increased when overflow occurs.

14. Using stacks, convert an infix expression into a postfix expression.

15. Using stacks, convert an infix expression into a prefix expression.

16. Using stacks, find the nth Fibonacci term.

17. Using stacks, find the number obtained by reversing the order of digits for a given number.

18. Using queues, implement priority.

19. Using queues, implement First Come First Serve Scheduling.

20. Using queues, implement First Come First Serve with time slice.

Linked List

21. Write a program to find whether a given linked list has a cycle.

22. Write a program to join two linked lists.

23. Write a program to merge two linked lists.

24. Write a program to remove duplicate elements from a given linked list.

25. Write a program to find the second maximum element from a given linked list.

26. Write a program to find the element greater than the mean (assume that the linked list has only integers in the data part).

27. Write a program to find the common elements from two given linked lists.

28. Write a program to find the union of elements of two linked lists.

29. Write a program to arrange the elements of the linked list in descending order.

30. Write a program to partition a linked list as per the algorithm in the following reference.

Graphs and Trees

A graph can be represented using a two-dimensional array. The array would contain 0's and 1's. If the element at the ith row and the jth column has 1, it indicates the presence of an edge from vertex i to j. Ask the user to enter the number of vertices of a graph and create a two-dimensional array depicting the graph.

31. Find the number of edges in the graph. (Note that the number of 1's in the 2-D array is not same as the number of edges in the graph).

32. Find the vertex connected to the maximum number of edges.

33. Find if the graph has a cycle.

34. Ask the user to enter the initial vertex and the final vertex and find if there is a path from the initial to the final vertex.

35. In the above question find whether there are more than one paths from the initial to the final vertex and find the shortest path.

36. Now, in place of 1's asks the user to enter a finite number representing the cost of the edge from the vertex i to the vertex j. Find the shortest path from the source vertex to all other vertices.

37. Write a program to find the spanning tree of the graph.

38. Write a program to find whether the graph is a tree.

39. A tree can be represented using a two-dimensional array having n rows and two columns. In each row the first column is i and the second column is j, means that there is an edge from i to j. Ask the user to enter the requisite data and display the tree (just the list of vertices and edges associated with them).

Create a binary tree using a doubly linked list. For this tree accomplish the following tasks.

40. Write a program to implement the Postorder traversal of a binary tree.

41. Write a program to implement the Preorder traversal of a binary tree.

42. Write a program to implement the in-order traversal of the tree.

43. Check if the given tree is a Binary Search Tree.

44. In each Binary Search Tree, find the leftmost node of the right subtree of a given node.

45. In each Binary Search Tree, find the rightmost node of the left subtree of a given node.

46. Write a program to insert an element in a Binary Search Tree.

47. Write a program to delete a given node from a given Binary Search Tree.

48. Write a program to create a Heap from a given list.

49. Implement Heap Sort.

ANSWERS TO MCQS

CHAPTER 1

1. c	**2.** c	**3.** b	**4.** b	**5.** c
6. a	**7.** b	**8.** c	**9.** d	**10.** a
11. d	**12.** b, c	**13.** a	**14.** d	

CHAPTER 2

1. c	**2.** b	**3.** b	**4.** c	**5.** a
6. d	**7.** b	**8.** c	**9.** b	**10.** a
11. d	**12.** d	**13.** d	**14.** d	**15.** a

CHAPTER 3

1. d	**2.** a	**3.** a	**4.** c	**5.** d
6. a	**7.** b	**8.** c	**9.** a	**10.** b
11. b	**12.** b	**13.** b	**14.** b	**15.** a

CHAPTER 4

1. a	**2.** a	**3.** b	**4.** a	**5.** b
6. b	**7.** a	**8.** b	**9.** b	**10.** b

CHAPTER 5

1. d	**2.** c	**3.** b	**4.** d	**5.** a
6. a	**7.** a	**8.** a	**9.** a	**10.** d

CHAPTER 6

1. a	**2.** a	**3.** a	**4.** a	**5.** d
6. a, b, c	**7.** c	**8.** c	**9.** a	**10.** a

CHAPTER 7

1. c	**2.** a	**3.** d	**4.** a	**5.** c
6. b	**7.** b	**8.** a	**9.** a	**10.** b
11. d	**12.** a	**13.** b	**14.** a	**15.** b
16. a	**17.** a	**18.** b	**19.** a	**20.** a
21. b	**22.** c	**23.** b	**24.** d	**25.** d

CHAPTER 8

1. c	**2.** d	**3.** a	**4.** b	**5.** a
6. b	**7.** a	**8.** a	**9.** a	**10.** a, c
11. b	**12.** b			

CHAPTER 9

1. e	**2.** a	**3.** a	**4.** a	**5.** b
6. a	**7.** d	**8.** d	**9.** d	**10.** a

CHAPTER 10

1. a	**2.** a	**3.** c	**4.** a	**5.** b
6. a	**7.** b	**8.** a	**9.** b	**10.** b
11. b	**12.** c	**13.** a	**14.** c	**15.** a

| 16. a | 17. b | 18. b | 19. b | 20. b |
| 21. a | 22. a | 23. b | 24. b | 25. c |

CHAPTER 11

1. a	2. d	3. a	4. a	5. a
6. a	7. a	8. b	9. a	10. a
11. b	12. a	13. d	14. d	15. a
16. d	17. b	18. a	19. b	20. d

CHAPTER 12

1. a	2. d	3. a	4. c	5. a
6. c	7. b	8. a	9. a	10. a
11. a	12. a	13. a	14. c	15. b
16. a	17. a	18. a	19. a	20. b

CHAPTER 13

1. a	2. c	3. b	4. b	5. a
6. b	7. a	8. c	9. a	10. b
11. a	12. a	13. c	14. b	15. a

CHAPTER 14

1. a	2. a	3. b	4. b	5. b
6. a	7. b	8. b	9. c	10. b
11. a	12. a			

CHAPTER 15

| 1. c | 2. b | 3. c | 4. d | 5. a |
| 6. d | 7. c | 8. c | 9. b | 10. c |

CHAPTER 16

1. c	**2.** b	**3.** b	**4.** b	**5.** c
6. a	**7.** b	**8.** a	**9.** d	**10.** a
11. b	**12.** a	**13.** b	**14.** c	**15.** d
16. c	**17.** a	**18.** a	**19.** a	**20.** a

CHAPTER 17

1. a	**2.** c	**3.** c	**4.** a	**5.** c
6. c	**7.** a	**8.** a	**9.** b	**10.** a
11. a	**12.** a	**13.** c	**14.** b	**15.** a

CHAPTER 18

1. e	**2.** e	**3.** a	**4.** b	**5.** d
6. d	**7.** a	**8.** a	**9.** a	**10.** c
11. a				

CHAPTER 19

1. a	**2.** a	**3.** d	**4.** a	**5.** g

CHAPTER 20

1. c	**2.** b	**3.** d	**4.** d	**5.** c
6. b	**7.** a	**8.** c	**9.** a	**10.** a
11. b	**12.** a			

CHAPTER 21

1. a	**2.** b	**3.** c	**4.** a	**5.** a
6. b	**7.** b	**8.** a	**9.** a	**10.** a
11. c	**12.** a			

REFERENCES

1. Mark Lutz, *Learning Python, Fifth Edition*, O'Reilly, 2013.

2. Stef Maruch and Aahz Maruch, *Python for Dummies*, John Wiley & Sons, 2006, ISBN: 9780471778646.

3. David Beazley, *Python Essential Reference, Third Edition*, Sams Publishing, USA, 2006.

4. Allen Downey, *Think Python, How to Think Like a Computer Scientist, Version 2.0.16*, Green Tea Press, Needham, Massachusetts.

5. Wes McKinney, *Python for Data Analysis*, Wes McKinney. USA, 2013, ISBN: 978-1-449-31979-3.

6. Andrew Johansen, *Python, The Ultimate Beginner's Guide!*

7. Wesley J. Chun, *Core Python Programming, First Edition*, Prentice Hall PTR, 2000, ISBN: 0-13-026036-3, 8.

8. Peter Harrington, *Machine Learning in Action*, Manning Publishing Company, 2012.

9. Richard L. Halterman, *Learning to Program with Python*, Copyright © 2011 Richard L. Halterman.

10. Willi Richert, Luis Pedro Coelho, *Building Machine Learning Systems with Python, Building Machine Learning Systems with Python*, Packt Publishing, 2013.

11. Bhasin, H., *Algorithms Design and Analysis*, Oxford University Press, 2015.

12. Bhasin, H., *Programming in C#*, Oxford University Press, 2014.

13. Bhasin, H., *Python for Beginners*, New Age International, 2018.

14. Horowitz, Shini et al., *Computer Algorithms*, University Press, 2017.

15. Rao, R. N., *Core Python Programming*, Dreamtech, 2019.

WEB RESOURCES

Python

1. Official documentation:
   http://www.python.org

2. For notes:
   http://www.cheeseshop.python.org/

3. The wiki:
   http://www.wiki.python.org

4. At SciPy portal:
   https://scipy-lectures.org/intro/language/python_language.html

Data Structures

5. Lecture Notes at CMU portal:
   https://www.cs.cmu.edu/~fp/courses/15122-f15/lectures/09-stackqueue.pdf

6. Lecture notes at Washington.edu:
   https://courses.cs.washington.edu/courses/cse373/14wi/lecture1.pdf

7. Lecture notes at UTK portal:
   http://web.eecs.utk.edu/~jplank/plank/classes/cs140/Notes/Linked/

Online Compilers

If you do not wish to install anaconda, you can run your program online. Explore the following options:

- *https://repl.it/languages/python3*
- *https://www.tutorialspoint.com/execute_python3_online.php*
- *https://www.python.org/shell/*

Additional resources

The official Python documentation is available at: *https://docs.python.org/3/*. You can use stack overflow to search answers to questions related to coding and algorithms.

INDEX